I AM A
STRANGE
LOOP

I AM A STRANGE LOOP

DOUGLAS HOFSTADTER

BASIC BOOKS

A Member of the Perseus Books Group
New York

Published by Basic Books
A Member of the Perseus Books Group

Books published by Basic Books are available at special discount rates for bulk purchases within the United States by corporations, institutions, and other organizations. For further information, please contact the Special Markets Department at the Perseus Books Group, 11 Cambridge Center, Cambridge, Massachusetts 02142, or else telephone either (617) 252-5298 or (800) 255-1514, or send an email to special.markets@perseusbooks.com.

A CIP catalogue record for this book is available from the United States Library of Congress in Washington D.C.

ISBN-13: 978-0-465-03078-1
ISBN-10: 0-465-03078-5

2 4 6 8 10 9 7 5 3 1

To my sister Laura,

       who can understand,

and to our sister Molly,

       who cannot.

A note from the Publisher

Doug Hofstadter, who over the years has been a friend to Basic Books in so many ways, has kindly lent us this page to remember a late colleague. We gratefully dedicate this book

To Liz Maguire

1958–2006

who lives on in all of us.

TABLE OF CONTENTS

ತಿ ತಿ ತಿ

~ ~ ~

WORDS OF THANKS

ঌ ঌ ঌ

SINCE my teen-age years, I have been fascinated by what the mind is and does, and have pondered such riddles for many decades. Some of my conclusions have come from personal experiences and private musings, but of course I have been profoundly marked by the ideas of many other people, stretching way back to elementary school, if not earlier.

Among the well-known authors who have most influenced my thinking on the interwoven topics of minds, brains, patterns, symbols, self-reference, and consciousness are, in some vague semblance of chronological order: Ernest Nagel, James R. Newman, Kurt Gödel, Martin Gardner, Raymond Smullyan, John Pfeiffer, Wilder Penfield, Patrick Suppes, David Hamburg, Albert Hastorf, M. C. Escher, Howard DeLong, Richard C. Jeffrey, Ray Hyman, Karen Horney, Mikhail Bongard, Alan Turing, Gregory Chaitin, Stanislaw Ulam, Leslie A. Hart, Roger Sperry, Jacques Monod, Raj Reddy, Victor Lesser, Marvin Minsky, Margaret Boden, Terry Winograd, Donald Norman, Eliot Hearst, Daniel Dennett, Stanislaw Lem, Richard Dawkins, Allen Wheelis, John Holland, Robert Axelrod, Gilles Fauconnier, Paolo Bozzi, Giuseppe Longo, Valentino Braitenberg, Derek Parfit, Daniel Kahneman, Anne Treisman, Mark Turner, and Jean Aitchison. Books and articles by many of these authors are cited in the bibliography. Over the years, I have come to know quite a few of these individuals, and I count the friendships thus formed among the great joys of my life.

On a more local level, I have been influenced over a lifetime by thousands of intense conversations, phone calls, letters, and emails with family members, friends, students, and colleagues. Once again, listed in some rough semblance of chronological order, these people would include: Nancy Hofstadter, Robert Hofstadter, Laura Hofstadter, Peter Jones, Robert Boeninger, Charles Brenner, Larry Tesler, Michael Goldhaber, David Policansky, Peter S. Smith, Inga Karliner, Francisco Claro, Peter Rimbey, Paul Csonka, P. David Jennings, David Justman, J. Scott Buresh, Sydney Arkowitz, Robert Wolf, Philip Taylor, Scott Kim, Pentti Kanerva, William Gosper, Donald Byrd, J. Michael Dunn, Daniel Friedman, Marsha

Meredith, Gray Clossman, Ann Trail, Susan Wunder, David Moser, Carol Brush Hofstadter, Leonard Shar, Paul Smolensky, David Leake, Peter Suber, Greg Huber, Bernard Greenberg, Marek Lugowski, Joe Becker, Melanie Mitchell, Robert French, David Rogers, Benedetto Scimemi, Daniel Defays, William Cavnar, Michael Gasser, Robert Goldstone, David Chalmers, Gary McGraw, John Rehling, James Marshall, Wang Pei, Achille Varzi, Oliviero Stock, Harry Foundalis, Hamid Ekbia, Marilyn Stone, Kellie Gutman, James Muller, Alexandre Linhares, Christoph Weidemann, Nathaniel Shar, Jeremy Shar, Alberto Parmeggiani, Alex Passi, Francesco Bianchini, Francisco Lara-Dammer, Damien Sullivan, Abhijit Mahabal, Caroline Strobbe, Emmanuel Sander, Glen Worthey — and of course Carol's and my two children, Danny and Monica Hofstadter.

I feel deep gratitude to Indiana University for having so generously supported me personally and my group of researchers (the Fluid Analogies Research Group, affectionately known as "FARG") for such a long time. Some of the key people at IU who have kept the FARGonauts afloat over the past twenty years are Helga Keller, Mortimer Lowengrub, Thomas Ehrlich, Kenneth Gros Louis, Kumble Subbaswamy, Robert Goldstone, Richard Shiffrin, J. Michael Dunn, and Andrew Hanson. All of them have been intellectual companions and staunch supporters, some for decades, and I am lucky to be able to count them among my colleagues.

I have long felt part of the family at Basic Books, and am grateful for the support of many people there for nearly thirty years. In the past few years I have worked closely with William Frucht, and I truly appreciate his open-mindedness, his excellent advice, and his unflagging enthusiasm.

A few people have helped me enormously on this book. Ken Williford and Uriah Kriegel launched it; Kellie Gutman, Scott Buresh, Bill Frucht, David Moser, and Laura Hofstadter all read chunks of it and gave superb critical advice; and Helga Keller chased permissions far and wide. I thank them all for going "way ABCD" — way above and beyond the call of duty.

The many friends mentioned above, and some others not mentioned, form a "cloud" in which I float; sometimes I think of them as the "metropolitan area" of which I, construed narrowly, am just the zone inside the official city limits. Everyone has friends, and in that sense I am no different from anyone else, but this cloud is *my* cloud, and it somehow defines me, and I am proud of it and proud of them all. And so I say to this cloud of friends, with all my heart, "Thank you so very much, one and all!"

❧ ❧ ❧

PREFACE

An Author and His Book

❧ ❧ ❧

Facing the Physicality of Consciousness

FROM an early age onwards, I pondered what my mind was and, by analogy, what all minds are. I remember trying to understand how I came up with the puns I concocted, the mathematical ideas I invented, the speech errors I committed, the curious analogies I dreamt up, and so forth. I wondered what it would be like to be a girl, to be a native speaker of another language, to be Einstein, to be a dog, to be an eagle, even to be a mosquito. By and large, it was a joyous existence.

When I was twelve, a deep shadow fell over our family. My parents, as well as my seven-year-old sister Laura and I, faced the harsh reality that the youngest child in our family, Molly, then only three years old, had something terribly wrong with her. No one knew what it was, but Molly wasn't able to understand language or to speak (nor is she to this day, and we never did find out why). She moved through the world with ease, even with charm and grace, but she used no words at all. It was so sad.

For years, our parents explored every avenue imaginable, including the possibility of some kind of brain surgery, and as their quest for a cure or at least some kind of explanation grew ever more desperate, my own anguished thinking about Molly's plight and the frightening idea of people opening up my tiny sister's head and peering in at the mysterious stuff that filled it (an avenue never explored, in the end) gave me the impetus to read a couple of lay-level books about the human brain. Doing so had a huge impact on my life, since it forced me to consider, for the first time, the physical basis of consciousness and of being — or of having — an "I", which I found disorienting, dizzying, and profoundly eerie.

Right around that time, toward the end of my high-school years, I encountered the mysterious metamathematical revelations of the great Austrian logician Kurt Gödel and I also learned how to program, using Stanford University's only computer, a Burroughs 220, which was located in the deliciously obscure basement of decrepit old Encina Hall. I rapidly became addicted to this "Giant Electronic Brain", whose orange lights flickered in strange magical patterns revealing its "thoughts", and which, at my behest, discovered beautiful abstract mathematical structures and composed whimsical nonsensical passages in various foreign languages that I was studying. I simultaneously grew obsessed with symbolic logic, whose arcane symbols danced in strange magical patterns reflecting truths, falsities, hypotheticals, possibilities, and counterfactualities, and which, I was sure, afforded profound glimpses into the hidden wellsprings of human thought. As a result of these relentlessly churning thoughts about symbols and meanings, patterns and ideas, machines and mentality, neural impulses and mortal souls, all hell broke loose in my adolescent mind/brain.

The Mirage

One day when I was around sixteen or seventeen, musing intensely on these swirling clouds of ideas that gripped me emotionally no less than intellectually, it dawned on me — and it has ever since seemed to me — that what we call "consciousness" was a kind of mirage. It had to be a very peculiar kind of mirage, to be sure, since it was a mirage that perceived itself, and of course it didn't *believe* that it was perceiving a mirage, but no matter — it still *was* a mirage. It was almost as if this slippery phenomenon called "consciousness" lifted itself up by its own bootstraps, almost as if it made itself out of nothing, and then disintegrated back into nothing whenever one looked at it more closely.

So caught up was I in trying to understand what being alive, being human, and being conscious are all about that I felt driven to try to capture my elusive thoughts on paper lest they flit away forever, and so I sat down and wrote a dialogue between two hypothetical contemporary philosophers whom I flippantly named "Plato" and "Socrates" (I knew almost nothing about the real Plato and Socrates). This may have been the first serious piece of writing I ever did; in any case, I was proud of it, and never threw it away. Although I now see my dialogue between these two pseudo-Greek philosophers as pretty immature and awkward, not to mention extremely sketchy, I decided nonetheless to include it herein as my Prologue, because it hints at many of the ideas to come, and I think it sets a pleasing and provocative tone for the rest of the book.

A Shout into a Chasm

When, some ten years or so later, I started working on my first book, whose title I imagined would be "Gödel's Theorem and the Human Brain", my overarching goal was to relate the concept of a human self and the mystery of consciousness to Gödel's stunning discovery of a majestic wraparound self-referential structure (a "strange loop", as I later came to call it) in the very midst of a formidable bastion from which self-reference had been strictly banished by its audacious architects. I found the parallel between Gödel's miraculous manufacture of self-reference out of a substrate of meaningless symbols and the miraculous appearance of selves and souls in substrates consisting of inanimate matter so compelling that I was convinced that here lay the secret of our sense of "I", and thus my book *Gödel, Escher, Bach* came about (and acquired a catchier title).

That book, which appeared in 1979, couldn't have enjoyed a greater success, and indeed yours truly owes much of the pathway of his life since then to its success. And yet, despite the book's popularity, it always troubled me that the fundamental message of *GEB* (as I always call it, and as it is generally called) seemed to go largely unnoticed. People liked the book for all sorts of reasons, but seldom if ever for its most central *raison d'être*! Years went by, and I came out with other books that alluded to and added to that core message, but still there didn't seem to be much understanding out there of what I had really been trying to say in *GEB*.

In 1999, *GEB* celebrated its twentieth anniversary, and the folks at Basic Books suggested that I write a preface for a special new edition. I liked the idea, so I took them up on it. In my preface, I told all sorts of tales about the book and its vicissitudes, and among other things I described my frustration with its reception, ending with the following plaint: "It sometimes feels as if I had shouted a deeply cherished message out into an empty chasm and nobody heard me."

Well, one day in the spring of 2003, I received a very kind email message from two young philosophers named Ken Williford and Uriah Kriegel, inviting me to contribute a chapter to an anthology they were putting together on what they called "the self-referentialist theory (or theories)" of consciousness. They urged me to participate, and they even quoted back to me that very lamentation of mine from my preface, and they suggested that this opportunity would afford me a real chance to change things. I was genuinely gratified by their sincere interest in my core message and moved by their personal warmth, and I saw that indeed, contributing to their volume would be a grand occasion for me to try once again to articulate my ideas about self and consciousness for exactly the right audience of specialists — philosophers of mind. And so it wasn't too hard for me to decide to accept their invitation.

From the Majestic Dolomites to Gentle Bloomington

I started writing my chapter in a quiet and simple hotel room in the beautiful Alpine village of Anterselva di Mezzo, located in the Italian Dolomites, only a few stone's throws from the Austrian border. Inspired by the loveliness of the setting, I quickly dashed off ten or fifteen pages, thinking I might already have reached the halfway point. Then I returned home to Bloomington, Indiana, where I kept on plugging away.

It took me a good deal longer than I had expected to finish it (some of my readers will recognize this as a quintessential example of Hofstadter's Law, which states, "It always takes longer than you think it will take, even when you take into account Hofstadter's Law"), and worse, the chapter wound up being four times longer than the specified limit — a disaster! But when they finally received it, Ken and Uriah were very pleased with what I had written and were most tolerant of my indiscretions; indeed, so keen were they to have a contribution from me in their book that they said they could accept an extra-long chapter, and Ken in particular helped me cut it down to half its length, which was a real labor of love on his part.

In the meantime, I was starting to realize that what I had on my hands could be more than a book chapter — it could become a book unto itself. And so what had begun as a single project fissioned into two. I gave my chapter the title "What is it like to be a strange loop?", alluding to a famous article on the mystery of consciousness called "What is it like to be a bat?" by the philosopher of mind Thomas Nagel, while the book-to-be was given the shorter, sweeter title "I Am a Strange Loop".

In Ken Williford and Uriah Kriegel's anthology, *Self-Representational Approaches to Consciousness,* which appeared in the spring of 2006, my essay was placed at the very end, in a two-chapter section entitled "Beyond Philosophy" (why it qualified as lying "beyond philosophy" is beyond me, but I rather like the idea nonetheless). I don't know if, in that distinguished but rather specialized setting, this set of ideas will have much impact on anyone, but I certainly hope that in this book, its more fully worked-out and more visible incarnation, it will be able to reach all sorts of people, both inside and outside of philosophy, both young and old, both specialists and novices, and will give them new imagery about selves and souls (not to mention loops!). In any case, I owe a great deal to Ken and Uriah for having provided the initial spark that gave rise to this book, as well as for giving me much encouragement along the way.

And so, after just about forty-five years (good grief!), I've come full circle, writing once again about souls, selves, and consciousness, banging up against the same mysteriousness and eeriness that I first experienced when I was a teen-ager horrified and yet riveted by the awful and awesome physicality of that which makes us be what we are.

An Author and His Audience

Despite its title, this book is not about me, but about the concept of "I". It's thus about you, reader, every bit as much as it is about me. I could just as well have called it "You Are a Strange Loop". But the truth of the matter is that, in order to suggest the book's topic and goal more clearly, I should probably have called it "'I' Is a Strange Loop" — but can you imagine a clunkier title? Might as well call it "I Am a Lead Balloon".

In any case, this book is about the venerable topic of what an "I" is. And what is its audience? Well, as always, I write in order to reach a general educated public. I almost never write for specialists, and in a way that's because I'm not really a specialist myself. Oh, I take it back; that's unfair. After all, at this point in my life, I have spent nearly thirty years working with my graduate students on computational models of analogy-making and creativity, observing and cataloguing cognitive errors of all sorts, collecting examples of categorization and analogy, studying the centrality of analogies in physics and math, musing on the mechanisms of humor, pondering how concepts are created and memories are retrieved, exploring all sorts of aspects of words, idioms, languages, and translation, and so on — and over these three decades I have taught seminars on many aspects of thinking and how we perceive the world.

So yes, in the end, I am a kind of specialist — I specialize in thinking about thinking. Indeed, as I stated earlier, this topic has fueled my fire ever since I was a teen-ager. And one of my firmest conclusions is that we always think by seeking and drawing parallels to things we know from our past, and that we therefore communicate best when we exploit examples, analogies, and metaphors galore, when we avoid abstract generalities, when we use very down-to-earth, concrete, and simple language, and when we talk directly about our own experiences.

The Horsies-and-Doggies Religion

Over the years, I have fallen into a style of self-expression that I call the "horsies-and-doggies" style, a phrase inspired by a charming episode in the famous cartoon "Peanuts", which I've reproduced on the following page.

I often feel just the way that Charlie Brown feels in that last frame — like someone whose ideas are anything but "in the clouds", someone who is so down-to-earth as to be embarrassed by it. I realize that some of my readers have gotten an impression of me as someone with a mind that enormously savors and indefatigably pursues the highest of abstractions, but that is a very mistaken image. I'm just the opposite, and I hope that reading this book will make that evident.

I don't have the foggiest idea why I wrongly remembered the poignant phrase that Charlie Brown utters here, but in any case the slight variant "horsies and doggies" long ago became a fixture in my own speech, and so, for better or for worse, that's the standard phrase I always use to describe my teaching style, my speaking style, and my writing style.

In part because of the success of *Gödel, Escher, Bach,* I have had the good fortune of being given a great deal of freedom by the two universities on whose faculties I have served — Indiana University (for roughly twenty-five years) and the University of Michigan (for four years, in the 1980's). Their wonderful generosity has given me the luxury of being able to explore my variegated interests without being under the infamous publish-or-perish pressures, or perhaps even worse, the relentless pressures of grant-chasing.

I have not followed the standard academic route, which involves publishing paper after paper in professional journals. To be sure, I have published some "real" papers, but mostly I have concentrated on expressing myself through books, and these books have always been written with an eye to maximal clarity.

Clarity, simplicity, and concreteness have coalesced into a kind of religion for me — a set of never-forgotten guiding principles. Fortunately, a large number of thoughtful people appreciate analogies, metaphors, and examples, as well as a relative lack of jargon, and last but not least, accounts from a first-person stance. In any case, it is for people who appreciate that way of writing that this book, like all my others, has been written. I believe that this group includes not only outsiders and amateurs, but also many professional philosophers of mind.

If I tell many first-person stories in this book, it is not because I am obsessed with my own life or delude myself about its importance, but simply because it is the life I know best, and it provides all sorts of examples that I suspect are typical of most people's lives. I believe most people understand abstract ideas most clearly if they hear them through stories, and so I try to convey difficult and abstract ideas through the medium of my own life. I wish that more thinkers wrote in a first-person fashion.

Although I hope to reach philosophers with this book's ideas, I don't think that I write very much like a philosopher. It seems to me that many philosophers believe that, like mathematicians, they can actually *prove* the points they believe in, and to that end, they often try to use highly rigorous and technical language, and sometimes they attempt to anticipate and to counter all possible counter-arguments. I admire such self-confidence, but I am a bit less optimistic and a bit more fatalistic. I don't think one can truly prove anything in philosophy; I think one can merely try to convince, and probably one will wind up convincing only those people who started out fairly close to the position one is advocating. As a result of this mild brand of fatalism, my strategy for conveying my points is based more on metaphor and analogy than on attempts at rigor. Indeed, this book is a gigantic salad bowl full of metaphors and analogies. Some will savor my metaphor salad, while others will find it too... well, too metaphorical. But I particularly hope that *you*, dear eater, will find it seasoned to your taste.

A Few Last Random Observations

I take analogies very seriously, so much so that I went to a great deal of trouble to index a large number of the analogies in my "salad". There are thus two main headings in the index for my lists of examples. One is "analogies, serious examples of"; the other is "throwaway analogies, random examples of". I made this droll distinction because whereas many of my analogies play key roles in conveying ideas, some are there just to add spice. There's another point to be made, though: in the final analysis, virtually every thought in this book (or in any book) is an analogy, as it involves recognizing something as being a variety of something else. Thus every time I write "similarly" or "by contrast", there is an implicit analogy, and every time I pick a word or phrase (*e.g.*, "salad", "storehouse", "bottom line"), I am making an analogy to something in my life's storehouse of experiences. The bottom line is, *every* thought herein could be listed under "analogies". However, I refrained from making my index that detailed.

I initially thought this book was just going to be a distilled retelling of the central message of *GEB,* employing little or no formal notation and not indulging in Pushkinian digressions into such variegated topics as Zen Buddhism, molecular biology, recursion, artificial intelligence, and so forth. In other words, I thought I had already fully stated in *GEB* and my other books what I intended to (re)state here, but to my surprise, as I started to write, I saw new ideas sprouting everywhere under foot. That was a relief, and made me feel that my new book was more than just a rehash of an earlier book (or books).

Among the keys to *GEB*'s success was its alternation between chapters and dialogues, but I didn't intend, thirty years later, to copycat myself with another such alternation. I was in a different frame of mind, and I wanted this book to reflect that. But as I was approaching the end, I wanted to try to compare my ideas with well-known ideas in the philosophy of mind, and so I started saying things like, "Skeptics might reply as follows…" After I had written such phrases a few times, I realized I had inadvertently fallen into writing a dialogue between myself and a hypothetical skeptical reader, so I invented a pair of oddly-named characters and let them have at each other for what turned out to be one of the longest chapters in the book. It's not intended to be uproariously funny, although I hope my readers will occasionally smile here and there as they read it. In any case, fans of the dialogue form, take heart — there are two dialogues in this book.

I am a lifelong lover of form–content interplay, and this book is no exception. As with several of my previous books, I have had the chance to typeset it down to the finest level of detail, and my quest for visual elegance on each page has had countless repercussions on how I phrase my ideas. To some this may sound like the tail wagging the dog, but I think that attention to form improves anyone's writing. I hope that reading this book not only is stimulating intellectually but also is a pleasant visual experience.

A Useful Youthfulness

GEB was written by someone pretty young (I was twenty-seven when I started working on it and twenty-eight when I completed the first draft — all written out in pen on lined paper), and although at that tender age I had already experienced my fair or unfair share of suffering, sadness, and moral soul-searching, one doesn't find too much allusion to those aspects of life in the book. In this book, though, written by someone who has known considerably more suffering, sadness, and soul-searching, those hard aspects of life are much more frequently touched on. I think that's one of the things about growing older — one's writing becomes more inward, more reflective, perhaps wiser, or perhaps just sadder.

I have long been struck by the poetic title of André Malraux's famous novel *La Condition humaine.* I guess each of us has a personal sense of what this evocative phrase means, and I would characterize *I Am a Strange Loop* as being my own best shot at describing what "the human condition" is.

One of my favorite blurbs for *GEB* came from the physicist and writer Jeremy Bernstein, and in part it said, "It has a youthful vitality and a wonderful brilliance…" True music to my ears! But unfortunately this flattering phrase got garbled at some point, and as a result there are now thousands of copies of *GEB* floating around on whose back cover Bernstein proclaims, "It has a useful vitality…" What a letdown, compared with a "youthful" vitality! And yet perhaps this new book, in its older, more sober style, will someday be described by someone somewhere as having a "useful" vitality. I guess worse things could be said about a book.

And so now I will stop talking about my book, and will let my book talk for itself. In it I hope you will discover messages imbued with interest and novelty, and even with a useful, if no longer youthful, vitality. I hope that reading this book will make you reflect in fresh ways on what being human is all about — in fact, on what just-plain *being* is all about. And I hope that when you put the book down, you will perhaps be able to imagine that you, too, are a strange loop. Now that would please me no end.

— Bloomington, Indiana
December, MMVI.

❧ ❧ ❧

I AM A
STRANGE
LOOP

PROLOGUE

An Affable Locking of Horns

ɹ ɹ ɹ

[As I stated in the Preface, I wrote this dialogue when I was a teen-ager, and it was my first, youthful attempt at grappling with these difficult ideas.]

Dramatis personæ:

    Plato: a seeker of truth who suspects consciousness is an illusion

    Socrates: a seeker of truth who believes in consciousness' reality

•   •   •

PLATO: But what then do you mean by "life", Socrates? To my mind, a living creature is a body which, after birth, grows, eats, learns how to react to various stimuli, and which is ultimately capable of reproduction.

SOCRATES: I find it interesting, Plato, that you say a living creature *is* a body, rather than *has* a body. For surely, many people today would say that there are at least some living creatures that have souls independent of their bodies.

PLATO: Yes, and with those I would agree. I should have said that living creatures *have* bodies.

SOCRATES: Then you would agree that fleas and mice have souls, however insignificant.

PLATO: My definition does require that, yes.

SOCRATES: And do trees have souls, and blades of grass?

PLATO: You have used words to put me in this situation, Socrates. I will revise what I said — only animals have souls.

SOCRATES: But no, I have not only used words, for there is no distinction to be found between plants and animals, if you examine small enough creatures.

PLATO: You mean there are some creatures sharing the properties of plant and animal? Yes, I guess I can imagine such a thing, myself. Now I suppose you will force me into saying that only humans have souls.

SOCRATES: No, on the contrary, I will ask you, what animals do you usually consider to have souls?

PLATO: Why, all higher animals — those which are able to think.

SOCRATES: Then, at least higher animals are alive. Now can you truly consider a stalk of grass to be a living creature like yourself?

PLATO: Let me put it this way, Socrates: I can only imagine true life with a soul, and so I must discard grass as true life, though I could say it has the symptoms of life.

SOCRATES: I see. So you would classify soulless creatures as only *appearing* alive, and creatures with souls as *true* life. Then am I right if I say that your question "What is true life?" depends on the understanding of the soul?

PLATO: Yes, that is right.

SOCRATES: And you have said that you consider the soul as the ability to think?

PLATO: Yes.

SOCRATES: Then you are really seeking the answer to "What is thinking?"

PLATO: I have followed each step of your argument, Socrates, but this conclusion makes me uneasy.

SOCRATES: It has not been *my* argument, Plato. You have provided all the facts, and I have only drawn logical conclusions from them. It is curious, how one often mistrusts one's own opinions if they are stated by someone else.

PLATO: You are right, Socrates. And surely it is no simple task to explain thinking. It seems to me that the purest thought is the *knowing* of something; for clearly, to know something is more than just to write it down or to assert it. These can be done if one knows something; and one can learn to know something from hearing it asserted or from seeing it written. Yet knowing is more than this — it is conviction — but I am only using a synonym. I find it beyond me to understand what knowing is, Socrates.

SOCRATES: That is an interesting thought, Plato. Do you say that knowing is not so familiar as we think it is?

PLATO: Yes. Because we humans have knowledge, or convictions, we are humans, yet when we try to analyze knowing itself, it recedes, and evades us.

SOCRATES: Then had one not better be suspicious of what we call "knowing", or "conviction", and not take it so much for granted?

PLATO: Precisely. We must be cautious in saying "I know", and we must ponder what it truly means to say "I know" when our minds would have us say it.

SOCRATES: True. If I asked you, "Are you alive?", you would doubtless reply, "Yes, I am alive." And if I asked you, "How do you know that you are alive?", you would say "I *feel* it, I *know* I am alive — indeed, is not knowing and feeling one is alive *being* alive?" Is that not right?

PLATO: Yes, I would certainly say something to that effect.

SOCRATES: Now let us suppose that a machine had been constructed which was capable of constructing English sentences and answering questions. And suppose I asked this English machine, "Are you alive?" and suppose it gave me precisely the same answers as you did. What would you say as to the validity of its answers?

PLATO: I would first of all object that no machine can know what words are, or mean. A machine merely deals with words in an abstract mechanical fashion, much as canning machines put fruit in cans.

SOCRATES: I do not accept your objections for two reasons. Surely you do not contend that the basic unit of human thought is the word? For it is well known that humans have nerve cells, the laws of whose operation are arithmetical. Secondly, you cautioned earlier that we must be wary of the verb "to know", yet here you use it quite nonchalantly. What makes you say that no machine could ever "know" what words are, or mean?

PLATO: Socrates, do you argue that machines can know facts, as we humans do?

SOCRATES: You declared just now that you yourself cannot even explain what knowing is. How did you learn the verb "to know" as a child?

PLATO: Evidently, I assimilated it from hearing it used around me.

SOCRATES: Then it was by automatic action that you gained control of it.

PLATO: No... Well, perhaps I see what you mean. I grew accustomed to hearing it in certain contexts, and thus came to be able to use it myself in those contexts, in a more or less automatic fashion.

SOCRATES: Much as you use language now — without having to reflect on each word?

PLATO: Yes, exactly.

SOCRATES: Thus now, if you say, "I know I am alive", that sentence is merely a reflex coming from your brain, and is not a product of conscious thought.

PLATO: No, no! You or I have used faulty logic. Not all thoughts I utter are simply products of reflex actions. Some thoughts I think about *consciously* before uttering.

SOCRATES: In what sense do you think consciously about them?

PLATO: I don't know. I suppose that I try to find the correct words to describe them.

SOCRATES: What guides you to the correct words?

PLATO: Why, I search logically for synonyms, similar words, and so on, with which I am familiar.

SOCRATES: In other words, *habit* guides your thought.

PLATO: Yes, my thought is guided by the habit of connecting words with one another systematically.

SOCRATES: Then once again, these conscious thoughts are produced by reflex action.

PLATO: I do not see how I can know I am conscious, how I can feel alive, if this is true, yet I have followed your argument.

SOCRATES: But this argument itself shows that your reaction is merely habit, or reflex action, and that no conscious thought is leading you to say you know you are alive. If you stop to consider it, do you really understand what you mean by saying such a sentence? Or does it just come into your mind without your thinking consciously of it?

PLATO: Indeed, I am so confused I scarcely know.

SOCRATES: It becomes interesting to see how one's mind fails when working in new channels. Do you see how little you understand of that sentence "I am alive"?

PLATO: Yes, it is truly a sentence which, I must admit, is not so obvious to understand.

SOCRATES: I think it is in the same way as you fashioned that sentence that many of our actions come about — we think they arise through conscious thought, yet, on careful analysis, each bit of that thought is seen to be automatic and without consciousness.

PLATO: Then feeling one is alive is merely an illusion propagated by a reflex that urges one to utter, without understanding, such a sentence, and a truly living creature is reduced to a collection of complex reflexes. Then you have told me, Socrates, what you think life is.

CHAPTER 1

On Souls and Their Sizes

ન૭ ન૭ ન૭

Soul-Shards

O_{NE} gloomy day in early 1991, a couple of months after my father died, I was standing in the kitchen of my parents' house, and my mother, looking at a sweet and touching photograph of my father taken perhaps fifteen years earlier, said to me, with a note of despair, "What meaning does that photograph have? None at all. It's just a flat piece of paper with dark spots on it here and there. It's useless." The bleakness of my mother's grief-drenched remark set my head spinning because I knew instinctively that I disagreed with her, but I did not quite know how to express to her the way I felt the photograph should be considered.

After a few minutes of emotional pondering — soul-searching, quite literally — I hit upon an analogy that I felt could convey to my mother my point of view, and which I hoped might lend her at least a tiny degree of consolation. What I said to her was along the following lines.

"In the living room we have a book of the Chopin études for piano. All of its pages are just pieces of paper with dark marks on them, just as two-dimensional and flat and foldable as the photograph of Dad — and yet, think of the powerful effect that they have had on people all over the world for 150 years now. Thanks to those black marks on those flat sheets of paper, untold thousands of people have collectively spent millions of hours moving their fingers over the keyboards of pianos in complicated patterns, producing sounds that give them indescribable pleasure and a sense of great meaning. Those pianists in turn have conveyed to many millions of listeners, including you and me, the profound emotions that churned in Frédéric Chopin's heart, thus affording all of us some partial access to

Chopin's interiority — to the experience of living in the head, or rather the soul, of Frédéric Chopin. The marks on those sheets of paper are no less than soul-shards — scattered remnants of the shattered soul of Frédéric Chopin. Each of those strange geometries of notes has a unique power to bring back to life, inside our brains, some tiny fragment of the internal experiences of another human being — his sufferings, his joys, his deepest passions and tensions — and we thereby know, at least in part, what it was like to be that human being, and many people feel intense love for him. In just as potent a fashion, looking at that photograph of Dad brings back, to us who knew him intimately, the clearest memory of his smile and his gentleness, activates inside our living brains some of the most central representations of him that survive in us, makes little fragments of his soul dance again, but in the medium of brains other than his own. Like the score to a Chopin étude, that photograph is a soul-shard of someone departed, and it is something we should cherish as long as we live."

Although the above is a bit more flowery than what I said to my mother, it gives the essence of my message. I don't know what effect it had on her feelings about the picture, but that photo is still there, on a counter in her kitchen, and every time I look at it, I remember that exchange.

What Is It Like to Be a Tomato?

I slice up and devour tomatoes without the slightest sense of guilt. I do not go to bed uneasily after having consumed a fresh tomato. It does not occur to me to ask myself *which* tomato I ate, or whether by eating it I have snuffed an inner light, nor do I believe it is meaningful to try to imagine how the tomato felt as it was sitting on my plate being sliced apart. To me, a tomato is a desireless, soulless, nonconscious entity, and I have no qualms about doing with its "body" as I like. Indeed, a tomato is nothing but its body. There is no "mind–body problem" for tomatoes. (I hope, dear reader, that we agree on this much!)

I also swat mosquitoes without a qualm, though I try to avoid stepping on ants, and when there is an insect other than a mosquito in the house, I usually try to capture it and carry it outside, where I let it go unharmed. I eat chicken and fish sometimes [Note: This is no longer the case — see the Post Scriptum to this chapter], but many years ago I stopped eating the flesh of mammals. No beef, no ham, no bacon, no spam, no pork, no lamb — no thank you, ma'am! Mind you, I would still enjoy the *taste* of a BLT or well-done burger, but for moral reasons, I simply don't partake of them. I don't want to go on a crusade here, but I do need to talk a little bit about my vegetarian leanings, because they have everything to do with souls.

Guinea Pig

When I was fifteen, I had a summer job punching buttons on a Friden mechanical calculator in a physiology lab at Stanford University. (This was back in those days when there was but one computer on the whole Stanford campus and few scientists even knew of its existence, let alone thought about using it for their calculations.) It was pretty grueling work to do such "number-punching" for hours on end, and one day, Nancy, the graduate student for whose research project I was doing all this, asked me if, for relief, I'd like to try my hand at other kinds of tasks around the lab. I said "Sure!", and so that afternoon she escorted me up to the fourth floor of the physiology building and showed me the cages where they kept the animals — literally guinea pigs — that they used in their experiments. I still remember the pungent smell and the scurrying-about of all those little orange-furred rodents.

The next afternoon, Nancy asked me if I would please go up to the top floor and bring down two animals for her next round of experiments. I didn't have a chance to reply, however, for no sooner had I started to imagine myself reaching into one of those cages and selecting two small soft furry beings to be snuffed than my head began spinning, and in a flash I fainted right away, banging my head on the concrete floor. The next thing I knew, I was looking up into the face of the lab's director, George Feigen, a dear old family friend, who was deeply concerned that I might have injured myself in the fall. Luckily I was fine, and I slowly stood up and then rode my bike home for the rest of the day. Nobody ever asked me again to pick animals to be sacrificed for the sake of science.

Pig

Oddly enough, despite that extremely troubling head-on encounter with the concept of taking the life of a living creature, I kept on eating hamburgers and other kinds of meat for several years. I don't think I thought about it very much, since none of my friends did, and certainly no one talked about it. Meat-eating was just a background fact in the life of everyone I knew. Moreover, I admit with shame that in my mind, back in those days, the word "vegetarian" conjured up an image of weird, sternly moralistic nutcases (the movie *The Seven Year Itch* has a terrific scene in a vegetarian restaurant in Manhattan that conveys this stereotype to a tee). But one day when I was twenty-one, I read a short story called "Pig" by the Norwegian–English writer Roald Dahl, and this story had a profound effect on my life — and through me, on the lives of other creatures as well.

"Pig" starts off lightly and amusingly — a naïve young man named Lexington, raised as a strict vegetarian by his Aunt Glosspan ("Pangloss" reversed), discovers after her death that he loves the taste of meat (though he doesn't know what it is that he's eating). Soon, as in all Dahl stories, things take weird twists.

Driven by curiosity about this tasty substance called "pork", Lexington, on the recommendation of a new friend, decides to take a tour of a slaughterhouse. We join him as he sits in the waiting room with other tourists. He idly watches as various waiting parties are called, one by one, to take their tours. Eventually, Lexington's turn comes, and he is escorted from the waiting room into the shackling area where he watches pigs being hoisted by their back legs onto hooks on a moving chain, getting their throats slit, and, with blood gushing out, proceeding head downwards down the "disassembly line" to fall into a cauldron of boiling water where their body hair is removed, after which their heads and limbs are chopped off and they are prepared for being gutted and sent off, in neat little cellophane-wrapped packages, to supermarkets all over the country, where they will sit in glass cases, along with other rose-colored rivals, waiting for purchasers to admire them and hopefully to select them to take home.

As he is observing all this with detached fascination, Lexington himself is suddenly yanked by the leg and flipped upside down, and he realizes that he too is now dangling from the moving chain, just like the pigs he's been watching. His placidity all gone, he yells out, "There has been a frightful mistake!", but the workers ignore his cries. Soon the chain pulls him alongside a friendly-looking chap who Lexington hopes will grasp the situation's absurdity, but instead, the gentle "sticker" grasps Lexington's ear, pulls the dangling lad a bit closer, and then, smiling at him with lovingkindness, deftly slits the boy's jugular vein wide open with a razor-sharp knifeblade. As young Lexington continues his unanticipated inverted journey, his powerful heart pumps his blood out of his throat and onto the concrete floor, and even though he is upside down and losing awareness rapidly, he dimly perceives the pigs ahead of him dropping, one by one, into the steaming cauldron. One of them, oddly enough, seems to have white gloves on its two front trotters, and he is reminded of the glove-clad young woman who had just preceded him from the waiting room into the tour area. And with that curious final thought, Lexington woozily slips out of this, "the best of all possible worlds", into the next.

The closing scene of "Pig" reverberated in my head for a long time. In my mind, I kept on flipping back and forth between being an upside-down oinking pig on a hook and being Lexington, spilling into the cauldron…

Revulsion, Revelation, Revolution

A month or two after reading this haunting story, I accompanied my parents and my sister Laura to the city of Cagliari, at the southern end of the rugged island of Sardinia, where my father was participating in a physics conference. To wind up the meeting in grand local style, the organizers had planned a sumptuous banquet in a park on the outskirts of Cagliari, in which a suckling piglet was to be roasted and then sliced apart in front of all the diners. As honored guests of the conference, we were all expected to take part in this venerated Sardinian tradition. I, however, was deeply under the influence of the Dahl story I had recently read, and I simply could not envision participating in such a ritual. In my new frame of mind, I couldn't even imagine how anybody could wish to be there, let alone partake of the piglet's body. It turned out that my sister Laura was also horrified by the prospect, and so the two of us stayed behind in our hotel and were very happy to eat some pasta and vegetables.

The one–two punch of the Norwegian "Pig" and the Sardinian piglet resulted in my following my sister's lead in completely giving up meat-eating. I also refused to buy leather shoes or belts. Soon I became a fervent proselytizer for my new credo, and I remember how gratified I was that I managed to sway a couple of my friends for a few months, although to my disappointment, they gradually gave up on it.

In those days, I often wondered how some of my personal idols — Albert Einstein, for instance — could have been meat-eaters. I found no explanation, although recently, to my great pleasure, a Web search yielded hints that Einstein's sympathies were, in fact, toward vegetarianism, and not for health reasons but out of compassion towards living beings. But I didn't know that fact back then, and in any case many other heroes of mine were certainly carnivores who knew exactly what they were doing. Such facts saddened and confused me.

Reversion, Re-evolution

The very strange thing is that only a few years later, I, too, found the pressures of daily life in American society so strong that I gave up on my once-passionate vegetarianism, and for a while all my intense ruminations went totally underground. I think that the me of the mid-sixties would have found this reversal totally unfathomable, and yet the two versions of me had both lived in the very same skull. Was I really the same person?

Several years passed this way, almost as if I had never had any epiphany, but then one day, when I was a beginning assistant professor at

Indiana University, I met a highly thoughtful woman who had adopted the same vegetarian philosophy as I once had, and had done so for similar reasons, but she had stuck to it for a longer time than I had. Sue and I became good friends, and I admired the purity of her stance. Our friendship caused me to think it all through once more, and in short order I had swung back to my post-"Pig" stance of no killing at all.

Over the next several years there came a few more oscillations, but by my late thirties I had finally settled into a stable state — a compromise representing my evolving intuition that there are souls of different sizes. Though it was anything but crystal-clear to me, I was willing to accept the vague idea that some souls, provided they were "small enough", could legitimately be sacrificed for the sake of the desires of "larger" souls, such as mine and those of other human beings. Although drawing the dividing line at mammals was clearly somewhat arbitrary (as any such dividing line must be), that became my new credo and I stuck with it for two more decades.

The Mystery of Inanimate Flesh

We English speakers do not eat pig or cow; we eat pork and beef. We do eat chicken — but we don't eat chickens. One time the very young daughter of a friend of mine exclaimed with great mirth to her father that the word for a certain farm bird that clucks and lays eggs was also the word for a substance that she often found on her plate at dinnertime. She found this a most humorous coincidence, similar to the humorous coincidence that "calf" means both a young cow and a part of one's leg. She was upset, needless to say, when she was told that the tasty foodstuff and the clucky egg-layer were one and the same thing.

Presumably we all go through much the same confusion when, as children, we discover we are eating animals that our culture tells us are cute — lambs, bunnies, calves, chicks, and so forth. I remember, albeit dimly, my own genuine childhood confusion at this mystery, but since meat-eating was such a bland commonplace, I usually swept it under the rug and didn't give it much thought.

Nonetheless, grocery stores had an annoying way of bringing the issue up very vividly. There were big display cases with all sorts of slimy-looking blobs of various strange colors, labeled "liver", "tripe", "heart", and "kidney", and sometimes even "tongue" and "brain". Not only did these sound like animal parts, they *looked* like them as well. Fortunately, what was called "ground beef" didn't look terribly much like an animal part, and I say "fortunately" because it tasted so good. Wouldn't want to be talked out of *that*! Bacon tasted great too, and strips of the stuff were so thin and, once cooked, so crunchy, that they hardly conjured up thoughts of an animal at all. How fortunate!

It was the unloading docks at the rear of grocery stores that made the mystery come back with a vengeance. Sometimes a big truck would pull up and when its rear doors swung open, I would see huge hunks of flesh and bones dangling lifelessly on scary-looking metal hooks. I would watch with morbid curiosity as these carcasses were carried into the back of the store and attached to hooks that slid along overhead rails, so that they could be moved around easily. All this made the preadolescent me very uneasy, and as I gazed at a carcass, I could not help musing, "Who was that animal?" I wasn't wondering about its *name*, because I knew that farm animals didn't have names; I was grasping at something more philosophical — how it had felt to be *that* animal as opposed to some *other* one. What was the unique inner light that had suddenly gone off when this animal had been slaughtered?

When I went to Europe as a teenager, the issue was raised more starkly. There, lifeless animal bodies (usually skinned, headless and tailless, but sometimes not) were on display in front of all customers. My most vivid recollection is of one grocery store that, around the Christmas season, featured the severed head of a pig on a table in the middle of an aisle. If you chanced to approach it from the rear, you would see a flat cross-section showing all the inner structures of that pig's neck, exactly as if it had been guillotined. There were all the dense communication lines that had once connected all the far-flung parts of this individual's body to the central "headquarters" in its head. Seen from the other side, this pig had what looked like a smile frozen on its face, and that gave me the creeps.

Once again, I couldn't help wondering, "Who once had been in that head? Who had lived there? Who had looked out through those eyes, heard through those ears? Who had this hunk of flesh really been? Was it a male or a female?" No answers came, of course, and no other customers seemed to pay any attention to this display. It seemed to me that nobody else was facing the intense questions of life, death, and "porcinal identity" that this silent, still head provoked so powerfully and agitatedly inside mine.

I sometimes asked myself the analogous question if I squished an ant or clothes moth or mosquito — but not so often. My instincts told me that there was less meaning to the question "Who is 'in there'?" in such cases. Nonetheless, the sight of a partly squished insect writhing around on the floor would always give rise to some soul-searching.

And indeed, the reason I have raised all these grim images is not in order to crusade for a cause to which probably most of my readers have already given considerable thought; it is, rather, to raise the burning issue of what a "soul" is, and who or what possesses one. It is an issue that concerns everyone throughout their life — implicitly at the very least, and for many people quite explicitly — and it is the core issue of this book.

Give Me Some Men Who Are Stouter-souled Men

I alluded earlier to my deep love for the music of Chopin. In my teens and twenties, I played a lot of Chopin on the piano, often out of the bright yellow editions published by G. Schirmer in New York City. Each of those volumes opened with an essay penned in the early 1900's by the American critic James Huneker. Today, many people would find Huneker's prose overblown, but I did not; its unrestrained emotionality resonated with my perception of Chopin's music, and I still love his style of writing and his rich metaphors. In his preface to the volume of Chopin's études, Huneker asserts of the eleventh étude in Opus 25, in A minor (a titanic outburst often called the "Winter Wind", though that was certainly neither Chopin's title nor his image for it), the following striking thought: "Small-souled men, no matter how agile their fingers, should not attempt it."

I personally can attest to the terrifying technical difficulty of this incredible surging piece of music, having valiantly attempted to learn it when I was around sixteen and having sadly been forced to give it up in mid-stream, since playing just the first page up to speed (which I finally managed to do after several weeks of unbelievably arduous practice) made my right hand throb with pain. But the technical difficulty is, of course, not what Huneker was referring to. Quite rightly, he is saying that the piece is majestic and noble, but more controversially, he is drawing a dividing line between different levels or "sizes" of human souls, suggesting that some people are simply not up to playing this piece, not because of any physical limitations of their bodies, but because their souls are not "large enough". (I won't bother to criticize the sexism of Huneker's words; that was par for the course in those days.)

This kind of sentiment does not go down well in today's egalitarian America. It would not play in Peoria. Quite frankly, it rings terribly elitist, perhaps even repugnant, to our modern democratic ears. And yet I have to admit that I somewhat agree with Huneker, and I can't help wondering if we don't all of us implicitly believe in the validity of something vaguely like the idea of "small-souled" and "large-souled" human beings. In fact, I can't help suggesting that this is indeed the belief of almost all of us, no matter how egalitarian we publicly profess to be.

Small-souled and Large-souled Humans

Some of us believe in capital punishment — the intentional public squelching of a human soul, no matter how ardently that soul would plead for mercy, would tremble, would shake, would shriek, would desperately struggle to escape, on being led down the corridor to the site of their doom.

Some of us, perhaps almost all of us, believe that it is legitimate to kill enemy soldiers in a war, as if war were a special circumstance that shrinks the sizes of enemy souls.

In earlier days, perhaps some of us would have believed (as did George Washington, Thomas Jefferson, and Benjamin Franklin, each in their own way, at least for some period of time) that it was not immoral to own slaves and to buy and sell them, breaking up families willy-nilly, just as we do today with, for example, horses, dogs, and cats.

Some religious people believe that atheists, agnostics, and followers of other faiths — and worst of all, traitors who have abandoned "the" faith — have no souls at all, and are therefore eminently deserving of death.

Some people (including some women) believe that women have no souls — or perhaps, a little more generously, that women have "smaller souls" than men do.

Some of us (myself included) believe that the late President Reagan was essentially "all gone" many years before his body gave up the ghost, and more generally we believe that people in the final stages of Alzheimer's disease are essentially all gone. It strikes us that although there is a human brain couched inside each of those cranial shells, something has gone away from that brain — something essential, something that contains the secrets of that person's soul. The "I" has either wholly or partly vanished, gone down the drain, never to be found again.

Some of us (again, I count myself in this group) believe that neither a just-fertilized egg nor a five-month old fetus possesses a full human soul, and that, in some sense, a potential mother's life counts more than the life of that small creature, alive though it indisputably is.

Hattie the Chocolate Labrador

Kellie: After brunch we're going out to see Lynne's turkey, which we haven't seen yet.
Doug: *Which*, or *whom*?
Kellie: *Which*, I'd say. A turkey's not a *whom*.
Doug: I see… So is Hattie a *whom*, or a *which*?
Kellie: Oh, she's a *whom*, no doubt.

Ollie the Golden Retriever

Doug: So how did Ollie enjoy the outing this afternoon at Lake Griffy?
Danny: Oh, he had a pretty good time, but he didn't play much with the
        other dogs. He liked playing with the people, though.
Doug: Really? How come?
Danny: Ollie's a people person.

Where to Draw that Fateful, Fatal Line?

All human beings — at least all sufficiently large-souled ones — have
to make up their minds about such matters as the swatting of mosquitoes or
flies, the setting of mousetraps, the eating of rabbits or lobsters or turkeys
or pigs, perhaps even of dogs or horses, the purchase of mink stoles or ivory
statues, the usage of leather suitcases or crocodile belts, even the penicillin-
based attack on swarms of bacteria that have invaded their body, and on
and on. The world imposes large and small moral dilemmas on us all the
time — at the very least, meal after meal — and we are all forced to take a
stand. Does a baby lamb have a soul that matters, or is the taste of lamb
chops just too delicious to worry one's head over that? Does a trout that
went for the bait and is now helplessly thrashing about on the end of a
nylon line deserve to survive, or should it just be given one sharp thwack on
the head and "put out of its misery" so that we can savor the indescribable
and yet strangely predictable soft, flaky texture of its white muscles? Do
grasshoppers and mosquitoes and even bacteria have a tiny little "light on"
inside, no matter how dim, or is it all dark "in there"? (In *where*?) Why do I
not eat dogs? Who was the pig whose bacon I am enjoying for breakfast?
Which tomato is it that I am munching on? Should we chop down that
magnificent elm in our front yard? And while I'm at it, shall I yank out the
wild blackberry bush? And all the weeds growing right by it?

What gives us word-users the right to make life-and-death decisions
concerning other living creatures that have no words? Why do we find
ourselves in positions of such anguish (at least for some of us)? In the final
analysis, it is simply because *might makes right,* and we humans, thanks to the
intelligence afforded us by the complexity of our brains and our
embeddedness in rich languages and cultures, are indeed high and mighty,
relative to the "lower" animals (and vegetables). By virtue of our might, we
are forced to establish some sort of ranking of creatures, whether we do so
as a result of long and careful personal reflections or simply go along with
the compelling flow of the masses. Are cows just as comfortably killable as
mosquitoes? Would you feel any less troubled by swatting a fly preening

on a wall than by beheading a chicken quivering on a block? Obviously, such questions can bc endlessly proliferated (note the ironic spelling of this verb), but I will not do so here.

Below, I have inserted my own personal "consciousness cone". It is not meant to be exact; it is merely suggestive, but I submit that some comparable structure exists inside your head, as well as in the head of each language-endowed human being, although in most cases it is seldom if ever subjected to intense scrutiny, because it is not even explicitly formulated.

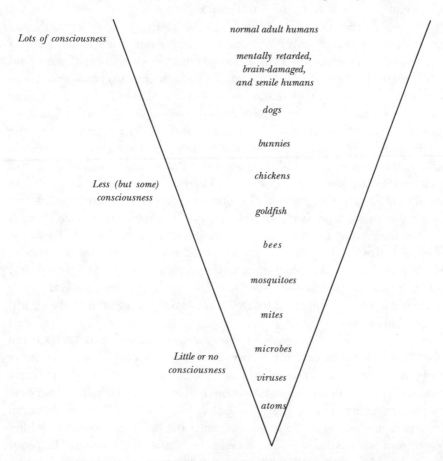

Interiority — What Has it, and to What Degree?

It is most unlikely that you, a reader of this book, have missed all the *Star Wars* movies, with their rather unforgettable characters C-3PO and R2-D2. Absurdly unrealistic though these two robots are, especially as perceived by someone like myself who has worked for decades trying to understand just the most primordial mechanisms of human intelligence by building computational models thereof, they nonetheless serve one very useful purpose — they are mind-openers. Seeing C-3PO and R2-D2 "in flesh and blood" on the screen makes us realize that whenever we look at an entity made of metal or plastic, we are not inherently destined to jump reflexively to the dogmatic conclusion, "That thing is necessarily an inanimate object since it is made of 'the wrong stuff'." Rather, we find, perhaps to our own surprise, that we are easily able to imagine a thinking, feeling entity made of cold, rigid, unfleshlike stuff.

In one of the *Star Wars* films, I recall seeing a huge squadron of hundreds of uniformly marching robots — and when I say "uniformly", I mean *really* uniformly, with all of them strutting in perfect synchrony, and all of them featuring identical, impassive, vacuous, mechanical facial expressions. I suspect that thanks to this unmistakable image of absolute interchangeability, virtually no viewer feels the slightest twinge of sadness when a bomb falls on the charging platoon and all of its members — these factory-made "creatures" — are instantly blown to smithereens. After all, in diametric opposition to C-3PO and R2-D2, *these* robots are not creatures at all — they are just hunks of metal! There is no more *interiority* to these metallic shells than there is to a can-opener or a car or a battleship, a fact revealed to us by their perfect identicality. Or else, if perchance there is inside of them some *tiny* degree of interiority, it is on the same order as the interiority of an ant. These metallic marchers are mere soldier robots, members of a dronelike caste in some larger robot colony, and are merely following out, in their zombie-ish way, the inflexible mechanical drives implanted in their circuitry. *If* there is interiority somewhere in there, it is of a negligible level.

What is it, then, that gives us the undeniable sense that C-3PO and R2-D2 have a "light on" inside, that there is lots of genuine interiority inside their inorganic crania, located somewhere behind their funny circular "eyes"? Where does our undeniable sense of their "I"'s come from? And contrariwise, what was it that was *lacking* in former President Reagan in his last years and in that mass of identical blown-up soldier robots, and what is it that is *not* lacking in Hattie the chocolate labrador and in R2-D2 the robot, that makes all the difference to us?

The Gradual Growth of a Soul

I stated above that I am among those who reject the notion that a full-fledged human soul comes into being the moment that a human sperm joins a human ovum to form a human zygote. By contrast, I believe that a human soul — and, by the way, it is my aim in this book to make clear what I mean by this slippery, shifting word, often rife with religious connotations, but here not having any — comes slowly into being over the course of years of development. It may sound crass to put it this way, but I would like to suggest, at least metaphorically, a numerical scale of "degrees of souledness". We can initially imagine it as running from 0 to 100, and the units of this scale can be called, just for the fun of it, "hunekers". Thus you and I, dear reader, both possess 100 hunekers of souledness, or thereabouts. Shake!

Oops! I just realized that I have committed an error that comes from long years of indoctrination into the admirable egalitarian traditions of my native land — namely, I unconsciously assumed that there is a value at which souledness "maxes out", and that all normal adults reach that ceiling and can go no higher. Why, though, should I make any such assumption? Why could souledness not be like tallness? There is an average tallness for adults, but there is also a considerable spread around that average. Why should there not likewise be an average degree of souledness for adults (100 hunekers, say), plus a wide range around that average, maybe (as for IQ) going as high as 150 or 200 hunekers in rare cases, and down to 50 or lower in others?

If that's how things are, then I retract my reflexive claim that you and I, dear reader, share 100 hunekers of souledness. Instead, I'd like to suggest that we both have considerably *higher* readings than that on the hunekometer! (I hope you agree.) However, this is starting to feel like dangerous moral territory, verging on the suggestion that some people are *worth more* than others — a thought that is anathema in our society (and which troubles me, as well), so I won't spend much time here trying to figure out how to calculate a person's souledness value in hunekers.

It strikes me that when sperm joins ovum, the resulting infinitesimal bio-blob has a soul-value of essentially zero hunekers. What has happened, however, is that a dynamic, snowballing entity has come into existence that over a period of years will be capable of developing a complex set of internal structures or patterns — and the presence, to a higher and higher degree, of those intricate patterns is what would endow that entity (or rather, the enormously more complex entities into which it slowly metamorphoses, step by step) with an ever-larger value along the Huneker soul-scale, homing in on a value somewhere in the vicinity of 100.

The cone shown on the following page gives a crude but vivid sense of how I might attach huneker values to human beings of ages from zero to twenty (or alternatively, to just one human being, but at different stages).

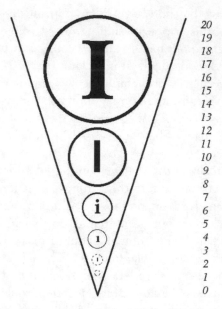

20
19
18
17
16
15
14
13
12
11
10
9
8
7
6
5
4
3
2
1
0

In short, I would here argue, echoing and generalizing the provocative statement by James Huneker, that "souledness" is by no means an off–on, black-and-white, discrete variable having just two possible states like a bit, a pixel, or a light bulb, but rather is a shaded, blurry numerical variable that ranges continuously across different species and varieties of object, and that also can rise or fall over time as a result of the growth or decay, within the entity in question, of a special kind of subtle pattern (the elucidation of whose nature will keep us busy for much of this book). I would also argue that most people's largely unconscious prejudices about whether to eat or not to eat this or that food, whether to buy or not to buy this or that article of clothing, whether to swat or not to swat this or that insect, whether to root or not to root for this or that species of robot in a sci-fi film, whether to be sad or not to be sad if a human character in a film or a novel meets with a violent end, whether to claim or not to claim that a particular senescent person "is no longer there", and so forth, reflect precisely this kind of numerical continuum in their minds, whether they admit it or not.

You might wonder whether my having drawn a cone that impenitently depicts "degrees of souledness" during the development of a given human being implies that I would be more willing, if placed under enormous pressure (as in the film *Sophie's Choice*), to extinguish the life of a two-year-old child than the life of a twenty-year-old adult. The answer is, "No, it does not." Even though I sincerely believe there is much more of a soul in the twenty-year-old than in the two-year-old (a view that will no doubt dismay many readers), I nonetheless have enormous respect for the *potential* of the two-year-old to *develop* a much larger soul over the course of a dozen or so years. In addition, I have been built, by the mechanisms of billions of years of evolution, to perceive in the two-year-old what, for lack of a better word, I will call "cuteness", and the perceived presence of that quality grants the two-year-old an amazingly strong shell of protectedness against attacks not just by me, but by humans of all ages, sexes, and persuasions.

Lights On?

The central aim of this book is to try to pinpoint the nature of that "special kind of subtle pattern" that I have come to believe underlies, or gives rise to, what I have here been calling a "soul" or an "I". I could just as well have spoken of "having a light on inside", "possessing interiority", or that old standby, "being conscious".

Philosophers of mind often use the terms "possessing intentionality" (which means having beliefs and desires and fears and so forth) or "having semantics" (which means the ability to genuinely think *about* things, as contrasted with the "mere" ability to juggle meaningless tokens in complicated patterns — a distinction that I raised in the dialogue between my versions of Socrates and Plato).

Although each of these terms puts the focus on a slightly different aspect of the elusive abstraction that concerns us, they are all, from my perspective, pretty much interchangeable. And for all of these terms, I reiterate that they have to be understood as coming in *degrees* along a sliding scale, rather than as on/off, black/white, yes/no switches.

Post Scriptum

The first draft of this chapter was written two years ago, and although it discussed meat-eating and vegetarianism, it had far less on the topic than this final version does. Some months later, while I was "fleshing it out" by summarizing the short story "Pig", I suddenly found myself questioning the dividing line that I had carefully drawn two decades earlier and had lived with ever since (although occasionally somewhat uneasily) — namely, the line between mammals and other animals.

All at once, I started feeling distinctly uncomfortable with the idea of eating chicken and fish, even though I had done so for some twenty years, and so, catching myself by surprise, I stopped "cold turkey". And by a remarkable coincidence, my two children independently came to similar conclusions at almost exactly the same time, so that over a period of just a couple of weeks our family's diet was transmuted into a completely vegetarian one. I've returned to the same spot as I was in when I was twenty-one in Sardinia, and it's the spot I plan to stay in.

Writing this chapter thus gave rise to a totally unexpected boomerang effect on its author — and as we shall see in later chapters, such an unpredictable bouncing-back of choices one has just made, followed by the incorporation of their repercussions into one's self-model, serves as an excellent example of the meaning of the motto "I am a strange loop."

CHAPTER 2
_____

This Teetering Bulb
of Dread and Dream

&ε; &ε; &ε;

What Is a "Brain Structure"?

I HAVE often been asked, when people hear that my research amounts to a quest after the hidden machinery of human thought, "Oh, so that means that you study the brain?"

One part of me wants to reply, "No, no — I think about *thinking*. I think about how concepts and words are related, what 'thinking in French' is, what underlies slips of the tongue and other types of errors, how one event effortlessly reminds us of another, how we recognize written letters and words, how we understand sloppily spoken, slurred, slangy speech, how we toss off untold numbers of utterly bland-seeming yet never-before-made analogies and occasionally come up with sparklingly original ones, how each of our concepts grows in subtlety and fluidity over our lifetime, and so forth. I don't think *in the least* about the brain — I leave the wet, messy, tangled web of the brain to the neurophysiologists."

Another part of me, however, wants to reply, "*Of course* I think about the human brain. *By definition*, I think about the brain, since the human brain is precisely the machinery that carries out human thinking."

This amusing contradiction has forced me to ask myself, "What do I mean, and what do other people mean, by 'brain research'?", and this leads naturally to the question, "What are the structures in the brain that someone could in principle study?" Most neuroscientists, if they were asked such a question, would make a list that would include (at least some of) the following items (listed roughly in order of physical size):

amino acids
neurotransmitters
DNA and RNA
synapses
dendrites
neurons
Hebbian neural assemblies
columns in the visual cortex
area 19 of the visual cortex
the entire visual cortex
the left hemisphere

Although these are all legitimate and important objects of neurological study, to me this list betrays a limited point of view. Saying that studying the brain is limited to the study of physical entities such as these would be like saying that literary criticism must focus on paper and bookbinding, ink and its chemistry, page sizes and margin widths, typefaces and paragraph lengths, and so forth. But what about the high abstractions that are the heart of literature — plot and character, style and point of view, irony and humor, allusion and metaphor, empathy and distance, and so on? Where did these crucial essences disappear in the list of topics for literary critics?

My point is simple: abstractions are central, whether in the study of literature or in the study of the brain. Accordingly, I herewith propose a list of abstractions that "researchers of the brain" should be just as concerned with:

the concept "dog"
the associative link between the concepts "dog" and "bark"
object files (as proposed by Anne Treisman)
frames (as proposed by Marvin Minsky)
memory organization packets (as proposed by Roger Schank)
long-term memory and short-term memory
episodic memory and melodic memory
analogical bridges (as proposed by my own research group)
mental spaces (as proposed by Gilles Fauconnier)
memes (as proposed by Richard Dawkins)
the ego, id, and superego (as proposed by Sigmund Freud)
the grammar of one's native language
sense of humor
"I"

I could extend this list arbitrarily. It is merely suggestive, intended to convey my thesis that the term "brain structure" should include items of this general sort. It goes without saying that some of the above-listed theoretical notions are unlikely to have lasting validity, while others may be increasingly confirmed by various types of research. Just as the notion of "gene" as an invisible entity that enabled the passing-on of traits from parents to progeny was proposed and studied scientifically long before any physical object could be identified as an actual carrier of such traits, and just as the notion of "atoms" as the building blocks of all physical objects was proposed and studied scientifically long before individual atoms were isolated and internally probed, so any of the notions listed above might legitimately be considered as invisible structures for brain researchers to try to pinpoint physically in the human brain.

Although I'm convinced that finding the exact physical incarnation of any such structure in "the human brain" (is there only one?) would be an amazing stride forward, I nonetheless don't see why physical mapping should constitute the be-all and end-all of neurological inquiry. Why couldn't the establishment of various sorts of precise relationships among the above-listed kinds of entities, prior to (or after) physical identification, be just as validly considered brain research? This is how scientific research on genes and atoms went on for many decades before genes and atoms were confirmed as physical objects and their inner structure was probed.

A Simple Analogy between Heart and Brain

I wish to offer a simple but crucial analogy between the study of the brain and the study of the heart. In our day, we all take for granted that bodies and their organs are made of cells. Thus a heart is made of many billions of cells. But concentrating on a heart at that microscopic scale, though obviously important, risks missing the big picture, which is that *a heart is a pump*. Analogously, *a brain is a thinking machine*, and if we're interested in understanding what thinking is, we don't want to focus on the trees (or their leaves!) at the expense of the forest. The big picture will become clear only when we focus on the brain's large-scale architecture, rather than doing ever more fine-grained analyses of its building blocks.

At some point a billion years or so ago, natural selection, in its usual random-walk fashion, bumped into cells that contracted rhythmically, and little beings possessing such cells did well for themselves because the cells' contractions helped send useful stuff here and there inside the being itself. Thus, by accident, were pumps born, and in the abstract design space of all such proto-pumps, nature favored designs that were more efficient. The inner workings of the pulsating cells making up those pumps had been found, in essence, and the cells' innards thus ceased being the crucial variables that were selected for. It was a brand-new game, in which rival *architectures* of hearts became the chief contenders for selection by nature, and on that new level, ever more complex patterns quickly evolved.

For this reason, heart surgeons don't think about the details of heart cells but concentrate instead on large architectural structures in the heart, just as car buyers don't think about the physics of protons and neutrons or the chemistry of alloys, but concentrate instead on high abstractions such as comfort, safety, fuel efficiency, maneuverability, sexiness, and so forth. And thus, to close out my heart–brain analogy, the bottom line is simply that the microscopic level may well be — or rather, almost certainly is — the wrong level in the brain on which to look, if we are seeking to explain such enormously abstract phenomena as concepts, ideas, prototypes, stereotypes, analogies, abstraction, remembering, forgetting, confusing, comparing, creativity, consciousness, sympathy, empathy, and the like.

Can Toilet Paper Think?

Simple though this analogy is, its bottom line seems sadly to sail right by many philosophers, brain researchers, psychologists, and others interested in the relationship between brain and mind. For instance, consider the case of John Searle, a philosopher who has spent much of his career heaping scorn on artificial-intelligence research and computational models of thinking, taking special delight in mocking Turing machines.

A momentary digression… Turing machines are extremely simple idealized computers whose memory consists of an infinitely long (*i.e.,* arbitrarily extensible) "tape" of so-called "cells", each of which is just a square that either is blank or has a dot inside it. A Turing machine comes with a movable "head", which looks at any one square at a time, and can "read" the cell (*i.e.,* tell if it has a dot or not) and "write" on it (*i.e.,* put a dot there, or erase a dot). Lastly, a Turing machine has, stored in its "head", a fixed list of instructions telling the head under which conditions to move left one cell or right one cell, or to make a new dot or to erase an old dot. Though the basic operations of all Turing machines are supremely trivial, any computation of any sort can be carried out by an appropriate Turing machine (numbers being represented by adjacent dot-filled cells, so that "•••" flanked by blanks would represent the integer 3).

Back now to philosopher John Searle. He has gotten a lot of mileage out of the fact that a Turing machine is an abstract machine, and therefore could, in principle, be built out of any materials whatsoever. In a ploy that, in my opinion, should fool only third-graders but that unfortunately takes in great multitudes of his professional colleagues, he pokes merciless fun at the idea that *thinking* could ever be implemented in a system made of such far-fetched physical substrates as *toilet paper and pebbles* (the tape would be an infinite roll of toilet paper, and a pebble on a square of paper would act as the dot in a cell), or *Tinkertoys,* or a vast assemblage of *beer cans and ping-pong balls* bashing together.

In his vivid writings, Searle gives the appearance of tossing off these humorous images light-heartedly and spontaneously, but in fact he is carefully and premeditatedly instilling in his readers a profound prejudice, or perhaps merely profiting from a preexistent prejudice. After all, it *does* sound preposterous to propose "thinking toilet paper" (no matter how long the roll might be, and regardless of whether pebbles are thrown in for good measure), or "thinking beer cans", "thinking Tinkertoys", and so forth. The light-hearted, apparently spontaneous images that Searle puts up for mockery are in reality skillfully calculated to make his readers scoff at such notions without giving them further thought — and sadly, they often work.

The Terribly Thirsty Beer Can

Indeed, Searle goes very far in his attempt to ridicule the systems that he portrays in this humorous fashion. For example, to ridicule the notion that a gigantic system of interacting beer cans might "have experiences" (yet another term for consciousness), he takes *thirst* as the experience in question, and then, in what seems like a casual allusion to something obvious to everyone, he drops the idea that in such a system there would have to be *one particular can* that would "pop up" (whatever that might mean, since he conveniently leaves out all description of how these beer cans might interact) on which the English words "I am thirsty" are written. The popping-up of this single beer can (a micro-element of a vast system, and thus comparable to, say, one neuron or one synapse in a brain) is meant to constitute the system's experience of thirst. In fact, Searle has chosen this silly image very deliberately, because he knows that no one would attribute it the slightest amount of plausibility. How could a metallic beer can possibly experience thirst? And how would its "popping up" *constitute* thirst? And why should the words "I am thirsty" written on a beer can be taken any more seriously than the words "I want to be washed" scribbled on a truck caked in mud?

The sad truth is that this image is the most ludicrous possible distortion of computer-based research aimed at understanding how cognition and sensation take place in minds. It could be criticized in any number of ways, but the key sleight of hand that I would like to focus on here is how Searle casually states that the experience claimed for this beer-can brain model is localized to *one single beer can,* and how he carefully avoids any suggestion that one might instead seek the system's experience of thirst in a more complex, more global, high-level property of the beer cans' configuration.

When one seriously tries to think of how a beer-can model of thinking or sensation might be implemented, the "thinking" and the "feeling", no matter how superficial they might be, would not be localized phenomena associated with a single beer can. They would be vast processes involving millions or billions or trillions of beer cans, and the state of "experiencing thirst" would not reside in three English words pre-painted on the side of a single beer can that popped up, but in a very intricate pattern involving huge numbers of beer cans. In short, Searle is merely mocking a trivial target of his own invention. No serious modeler of mental processes would ever propose the idea of one lonely beer can (or neuron) for each sensation or concept, and so Searle's cheap shot misses the mark by a wide margin.

It's also worth noting that Searle's image of the "single beer can as thirst-experiencer" is but a distorted replay of a long-discredited idea in neurology — that of the "grandmother cell". This is the idea that your visual recognition of your grandmother would take place if and only if one special cell in your brain were activated, that cell constituting your brain's physical representation of your grandmother. What significant difference is there between a grandmother cell and a thirst can? None at all. And yet, because John Searle has a gift for catchy imagery, his specious ideas have, over the years, had a great deal of impact on many professional colleagues, graduate students, and lay people.

It's not my aim here to attack Searle in detail (that would take a whole dreary chapter), but to point out how widespread is the tacit assumption that the level of the most primordial physical components of a brain must *also* be the level at which the brain's most complex and elusive mental properties reside. Just as many aspects of a mineral (its density, its color, its magnetism or lack thereof, its optical reflectivity, its thermal and electrical conductivity, its elasticity, its heat capacity, how fast sound spreads through it, and on and on) are properties that come from how its billions of atomic constituents interact and form high-level patterns, so mental properties of the brain reside not on the level of a single tiny constituent but on the level of *vast abstract patterns* involving those constituents.

Dealing with brains as multi-level systems is essential if we are to make even the slightest progress in analyzing elusive mental phenomena such as perception, concepts, thinking, consciousness, "I", free will, and so forth.

Trying to localize a concept or a sensation or a memory (etc.) down to a single neuron makes no sense at all. Even localization to a higher level of structure, such as a column in the cerebral cortex (these are small structures containing on the order of forty neurons, and they exhibit a more complex collective behavior than single neurons do), makes no sense when it comes to aspects of thinking like analogy-making or the spontaneous bubbling-up of episodes from long ago.

Levels and Forces in the Brain

I once saw a book whose title was "Molecular Gods: How Molecules Determine Our Behavior". Although I didn't buy it, its title stimulated many thoughts in my brain. (What is *a thought in a brain*? Is a thought really *inside* a brain? Is a thought made of molecules?) Indeed, the very fact that I soon placed the book back up on the shelf is a perfect example of the kinds of thoughts that its title triggered in my brain. What exactly determined my behavior that day (*e.g.*, my interest in the book, my pondering about its title, my decision not to buy it)? Was it some *molecules* inside my brain that made me reshelve it? Or was it some *ideas* in my brain? What is the proper way to talk about what was going on in my head as I first flipped through that book and then put it back?

At the time, I was reading books by many different writers on the brain, and in one of them I came across a chapter by the neurologist Roger Sperry, which not only was written with a special zest but also expressed a point of view that resonated strongly with my own intuitions. I would like to quote here a short passage from Sperry's essay "Mind, Brain, and Humanist Values", which I find particularly provocative.

> In my own hypothetical brain model, conscious awareness does get representation as a very real causal agent and rates an important place in the causal sequence and chain of control in brain events, in which it appears as an active, operational force....
>
> To put it very simply, it comes down to the issue of who pushes whom around in the population of causal forces that occupy the cranium. It is a matter, in other words, of straightening out the peck-order hierarchy among intracranial control agents. There exists within the cranium a whole world of diverse causal forces; what is more, there are forces within forces within forces, as in no other cubic half-foot of universe that we know....
>
> To make a long story short, if one keeps climbing upward in the chain of command within the brain, one finds at the very top those over-all organizational forces and dynamic properties of the large

patterns of cerebral excitation that are correlated with mental states or psychic activity.... Near the apex of this command system in the brain.... we find ideas.

Man over the chimpanzee has ideas and ideals. In the brain model proposed here, the causal potency of an idea, or an ideal, becomes just as real as that of a molecule, a cell, or a nerve impulse. Ideas cause ideas and help evolve new ideas. They interact with each other and with other mental forces in the same brain, in neighboring brains, and, thanks to global communication, in far distant, foreign brains. And they also interact with the external surroundings to produce *in toto* a burstwise advance in evolution that is far beyond anything to hit the evolutionary scene yet, including the emergence of the living cell.

Who Shoves Whom Around Inside the Cranium?

Yes, reader, I ask you: Who shoves whom around in the tangled mega-ganglion that is your brain, and who shoves whom around in "this teetering bulb of dread and dream" that is mine? (The marvelously evocative phrase in quotes, serving also as this chapter's title, is taken from "The Floor" by American poet Russell Edson.)

Sperry's pecking-order query puts its finger on what we need to know about ourselves — or, more pointedly, about our *selves*. What was *really* going on in that fine brain on that fine day when, allegedly, something calling itself "I" did something called "deciding", after which a jointed appendage moved in a fluid fashion and a book found itself back where it had been just a few seconds before? Was there truly something referable-to as "I" that was "shoving around" various physical brain structures, resulting in the sending of certain carefully coordinated messages through nerve fibers and the consequent moving of shoulder, elbow, wrist, and fingers in a certain complex pattern that left the book upright in its original spot — or, contrariwise, were there merely myriads of microscopic physical processes (quantum-mechanical collisions involving electrons, photons, gluons, quarks, and so forth) taking place in that localized region of the spatiotemporal continuum that poet Edson dubbed a "teetering bulb"?

Do dreads and dreams, hopes and griefs, ideas and beliefs, interests and doubts, infatuations and envies, memories and ambitions, bouts of nostalgia and floods of empathy, flashes of guilt and sparks of genius, play any role in the world of physical objects? Do such pure abstractions have causal powers? Can they shove massive things around, or are they just impotent fictions? Can a blurry, intangible "I" dictate to concrete physical objects such as electrons or muscles (or for that matter, books) what to do?

Have religious beliefs caused any wars, or have all wars just been caused by the interactions of quintillions (to underestimate the truth absurdly) of infinitesimal particles according to the laws of physics? Does fire cause smoke? Do cars cause smog? Do drones cause boredom? Do jokes cause laughter? Do smiles cause swoons? Does love cause marriage? Or, in the end, are there just myriads of particles pushing each other around according to the laws of physics — leaving, in the end, no room for selves or souls, dreads or dreams, love or marriage, smiles or swoons, jokes or laughter, drones or boredom, cars or smog, or even smoke or fire?

Thermodynamics and Statistical Mechanics

I grew up with a physicist for a father, and to me it was natural to see physics as underlying every last thing that happened in the universe. Even as a very young boy, I knew from popular science books that chemical reactions were a consequence of the physics of interacting atoms, and when I became more sophisticated, I saw molecular biology as the result of the laws of physics acting on complex molecules. In short, I grew up seeing no room for "extra" forces in the world, over and above the four basic forces that physicists had identified (gravity, electromagnetism, and two types of nuclear force — strong and weak).

But how, as I grew older, did I reconcile that rock-solid belief with my additional convictions that evolution caused hearts to evolve, that religious dogmas have caused wars, that nostalgia inspired Chopin to write a certain étude, that intense professional jealousy has caused the writing of many a nasty book review, and so forth and so on? These easily graspable macroscopic causal forces seem radically different from the four ineffable forces of physics that I was sure caused every event in the universe.

The answer is simple: I conceived of these "macroscopic forces" as being merely *ways of describing* complex patterns engendered by basic physical forces, much as physicists came to realize that such macroscopic phenomena as friction, viscosity, translucency, pressure, and temperature could be understood as highly predictable regularities determined by the statistics of astronomical numbers of invisible microscopic constituents careening about in spacetime and colliding with each other, with everything dictated by only the four basic forces of physics.

I also realized that this kind of shift in levels of description yielded something very precious to living beings: *comprehensibility*. To describe a gas's behavior by writing a gigantic piece of text having Avogadro's number of equations in it (assuming such a herculean feat were possible) would not lead to anyone's understanding of anything. But throwing away huge amounts of information and making a statistical summary could do a lot for comprehensibility. Just as I feel comfortable referring to "a pile of autumn leaves" without specifying the exact shape and orientation and color of each leaf, so I feel comfortable referring to a gas by specifying just its temperature, pressure, and volume, and nothing else.

All of this, to be sure, is very old hat to all physicists and to most philosophers as well, and can be summarized by the unoriginal maxim *Thermodynamics is explained by statistical mechanics,* but perhaps the idea becomes somewhat clearer when it is turned around, as follows: *Statistical mechanics can be bypassed by talking at the level of thermodynamics.*

Our existence as animals whose perception is limited to the world of everyday macroscopic objects forces us, quite obviously, to function without any reference to entities and processes at microscopic levels. No one really knew the slightest thing about atoms until only about a hundred years ago, and yet people got along perfectly well. Ferdinand Magellan circumnavigated the globe, William Shakespeare wrote some plays, J. S. Bach composed some cantatas, and Joan of Arc got herself burned at the stake, all for their own good (or bad) reasons, none of which, from their point of view, had the least thing to do with DNA, RNA, and proteins, or with carbon, oxygen, hydrogen, and nitrogen, or with photons, electrons, protons, and neutrons, let alone with quarks, gluons, W and Z bosons, gravitons, and Higgs particles.

Thinkodynamics and Statistical Mentalics

It thus comes as no news to anyone that different levels of description have different kinds of utility, depending on the purpose and the context, and I have accordingly summarized my view of this simple truth as it applies to the world of thinking and the brain: *Thinkodynamics is explained by statistical mentalics,* as well as its flipped-around version: *Statistical mentalics can be bypassed by talking at the level of thinkodynamics.*

What do I mean by these two terms, "thinkodynamics" and "statistical mentalics"? It is pretty straightforward. Thinkodynamics is analogous to thermodynamics; it involves large-scale structures and patterns in the brain, and makes no reference to microscopic events such as neural firings. Thinkodynamics is what psychologists study: how people make choices, commit errors, perceive patterns, experience novel remindings, and so on.

By contrast, by "mentalics" I mean the small-scale phenomena that neurologists traditionally study: how neurotransmitters cross synapses, how cells are wired together, how cell assemblies reverberate in synchrony, and so forth. And by "statistical mentalics", I mean the averaged-out, collective behavior of these very small entities — in other words, the behavior of a huge swarm as a whole, as opposed to a tiny buzz inside it.

However, as neurologist Sperry made very clear in the passage cited above, there is not, in the brain, just one single natural upward jump, as there is in a gas, all the way from the basic constituents to the whole thing; rather, there are many way-stations in the upward passage from mentalics to thinkodynamics, and this means that it is particularly hard for us to see, or even to imagine, the ground-level, neural-level explanation for why a certain professor of cognitive science once chose to reshelve a certain book on the brain, or once refrained from swatting a certain fly, or once broke out in giggles during a solemn ceremony, or once exclaimed, lamenting the departure of a cherished co-worker, "She'll be hard shoes to fill!"

The pressures of daily life require us, force us, to talk about events *at the level on which we directly perceive them.* Access at that level is what our sensory organs, our language, and our culture provide us with. From earliest childhood on, we are handed concepts such as "milk", "finger", "wall", "mosquito", "sting", "itch", "swat", and so on, on a silver platter. We perceive the world in terms of such notions, not in terms of microscopic notions like "proboscis" and "hair follicle", let alone "cytoplasm", "ribosome", "peptide bond", or "carbon atom". We can of course acquire such notions later, and some of us master them profoundly, but they can never replace the silver-platter ones we grew up with. In sum, then, we are victims of our macroscopicness, and cannot escape from the trap of using everyday words to describe the events that we witness, and perceive as *real.*

This is why it is much more natural for us to say that a war was triggered for religious or economic reasons than to try to imagine a war as a vast pattern of interacting elementary particles and to think of what triggered it in similar terms — even though physicists may insist that that is the only "true" level of explanation for it, in the sense that no information would be thrown away if we were to speak at that level. But having such phenomenal accuracy is, alas (or rather, "Thank God!"), not our fate.

We mortals are condemned *not* to speak at that level of no information loss. We *necessarily* simplify, and indeed, vastly so. But that sacrifice is also our glory. Drastic simplification is what allows us to reduce situations to their bare bones, to discover abstract essences, to put our fingers on what matters, to understand phenomena at amazingly high levels, to survive reliably in this world, and to formulate literature, art, music, and science.

❧ ❧ ❧

CHAPTER 3

The Causal Potency of Patterns

ॐ ॐ ॐ

The Prime Mover

AS THE rest of this book depends on having a clear sense for the interrelationships between different levels of description of entities that think, I would like to introduce here a few concrete metaphors that have helped me a great deal in developing my intuitions on this elusive subject.

My first example involves the familiar notion of a chain of falling dominos. However, I'll jazz up the standard image a bit by stipulating that each domino is spring-loaded in a clever fashion (details do not concern us) so that whenever it gets knocked down by its neighbor, after a short "refractory" period it flips back up to its vertical state, all set to be knocked down once more. With such a system, we can implement a mechanical computer that works by sending signals down stretches of dominos that can bifurcate or join together; thus signals can propagate in loops, jointly trigger other signals, and so forth. Relative timing, of course, will be of the essence, but once again, details do not concern us. The basic idea is just that we can imagine a network of precisely timed domino chains that amounts to a computer program for carrying out a particular computation, such as determining if a given input is a prime number or not. (John Searle, so fond of unusual substrates for computation, should like this "domino chainium" thought experiment!)

Let us thus imagine that we can give a specific numerical "input" to the chainium by taking any positive integer we are interested in — 641, say — and placing exactly that many dominos end to end in a "reserved" stretch of the network. Now, when we tip over the chainium's first domino, a Rube Goldberg–type series of events will take place in which domino after domino will fall, including, shortly after the outset, all 641 of the dominos constituting our input stretch, and as a consequence various loops will be triggered, with some loop presumably testing the input number for divisibility by 2, another for divisibility by 3, and so forth. If ever a divisor is found, then a signal will be sent down one particular stretch — let's call it the "divisor stretch" — and when we see that stretch falling, we will know that the input number has some divisor and thus is not prime. By contrast, if the input has no divisor, then the divisor stretch will never be triggered and we will know the input is prime.

Suppose an observer is standing by when the domino chainium is given 641 as input. The observer, who has not been told what the chainium was made for, watches keenly for while, then points at one of the dominos in the divisor stretch and asks with curiosity, "How come that domino there is never falling?"

Let me contrast two very different types of answer that someone might give. The first type of answer — myopic to the point of silliness — would be, "Because its predecessor never falls, you dummy!" To be sure, this is correct as far as it goes, but it doesn't go very far. It just pushes the buck to a different domino, and thus begs the question.

The second type of answer would be, "Because 641 is prime." Now this answer, while just as correct (indeed, in some sense it is far more on the mark), has the curious property of not talking about anything physical at all. Not only has the focus moved upwards to collective properties of the chainium, but those properties somehow transcend the physical and have to do with pure abstractions, such as primality.

The second answer bypasses all the physics of gravity and domino chains and makes reference only to concepts that belong to a completely different domain of discourse. The domain of prime numbers is as remote from the physics of toppling dominos as is the physics of quarks and gluons from the Cold War's "domino theory" of how communism would inevitably topple country after neighboring country in Southeast Asia. In both cases, the two domains of discourse are many levels apart, and one is purely local and physical, while the other is global and organizational.

Before passing on to other metaphors, I'd just like to point out that although here, 641's primality was used as an explanation for why a certain domino did *not* fall, it could equally well serve as the explanation for why a different domino *did* fall. In particular, in the domino chainium, there could be a stretch called the "prime stretch" whose dominos all topple when the set of potential divisors has been exhausted, which means that the input has been determined to be prime.

The point of this example is that 641's primality is the best explanation, perhaps even the *only* explanation, for why certain dominos *did* fall and certain other ones *did not* fall. In a word, 641 is the prime mover. So I ask: Who shoves whom around inside the domino chainium?

The Causal Potency of Collective Phenomena

My next metaphor was dreamt up on an afternoon not long ago when I was caught in a horrendous traffic jam on some freeway out in the countryside, with several lanes of nearly touching cars all sitting stock still. For some reason I was reminded of big-city traffic jams where you often hear people honking angrily at each other, and I imagined myself suddenly starting to honk my horn over and over again at the car in front of me, as if to say, "Get out of my way, lunkhead!"

The thought of myself (or anyone) taking such an outrageously childish action made me smile, but when I considered it a bit longer, I saw that there might be a slim rationale for honking that way. After all, if the next car were magically to poof right out of existence, I could fill the gap and thus make one car-length's worth of progress. Now a car poofing out of existence is not too terribly likely, and one car-length is not much progress, but somehow, through this image, the idea of honking became just barely comprehensible to me. And then I remembered my domino chainium and the silly superlocal answer, "That domino didn't fall because its neighbor didn't fall, you dummy!" This myopic answer and my fleeting thought of honking at the car just ahead of me seemed to be cut from the same cloth.

As I continued to sit in this traffic jam, twiddling my thumbs instead of honking, I let these thoughts continue, in their bully-like fashion, to push my helpless neurons around. I imagined a counterfactual situation in which the highway was shrouded in the densest pea-soup fog imaginable, so that I could barely make out the rear of the car ahead of me. In such a case, honking my horn wouldn't be quite so blockheaded. For all I know, that car alone might well be the entire cause of my being stuck, and if only it would just get out of the way, I could go sailing down the highway!

If you're totally fog-bound like that, or if you're incredibly myopic, then you might think to yourself, "It's all my neighbor's fault!", and there's at least a small chance that you're right. But if you have a larger field of view and can see hordes of immobilized cars on all sides, then honking at your immediate predecessor is an absurdity, for it's obvious that the problem is not local. The root problem lies at some level of discourse other than that of cars. Though you may not know its nature, some higher-level, more abstract reason must lie behind this traffic jam.

Perhaps a very critical baseball game just finished three miles up the road. Perhaps it's 7:30 on a weekday morning and you're heading towards Silicon Valley. Perhaps there's a huge blizzard ten miles ahead. Or it may be something else, but it's surely some social or natural event of the type that induces large numbers of people all to do the same thing as one another. No amount of expertise in car mechanics will help you to grasp the essence of such a situation; what is needed is knowledge of the abstract forces that can act on freeways and traffic. Cars are just pawns in the bigger game and, aside from the fact that they can't pass through each other and emerge intact post-crossing (as do ripples and other waves), their physical nature plays no significant role in traffic jams. We are in a situation analogous to that in which the global, abstract, math-level answer "641 is prime" is far superior to a local, physical, domino-level answer.

Neurons and Dominos

The foregoing down-to-earth images provide us with helpful metaphors for talking about the many levels of causality inside a human brain. Suppose it were possible to monitor any selected neuron in my brain. In that case, someone might ask, as I listened to some piece of music, "How come neuron #45826493842 never seems to fire?" A local, myopic answer might be, "Because the neurons that feed into it never fire jointly", and this answer would be just as correct but also just as useless and uninformative as the myopic answers in the other situations. On the other hand, the global, organizational answer "Because Doug Hofstadter doesn't care for the style of Fats Domino" would be much more on target.

Of course we should not fall into the trap of thinking that neuron #45826493842 is the sole neuron designated to fire whenever I resonate to some piece of music I'm listening to. It's just one of many neurons that participate in the high-level process, like voters in a national election. Just as no special voter makes the decision, so no special neuron is privileged. As long as we avoid simplistic notions such as a privileged "grand-music neuron", we can use the domino-chainium metaphor to think about brains, and especially to remind ourselves of how, for a given phenomenon in a brain, there can be vastly different explanations belonging to vastly different domains of discourse at vastly different levels of abstraction.

Patterns as Causes

I hope that in light of these images, Roger Sperry's comments about "the population of causal forces" and "overall organizational forces and dynamic properties" in a complex system like the brain or the chainium have become clearer. For instance, let us try to answer the question, "Can the primality of 641 really play a causal role in a physical system?" Although 641's primality is obviously not a physical force, the answer nonetheless has to be, "Yes, it does play a causal role, because the most efficient and most insight-affording explanation of the chainium's behavior depends crucially on that notion." Deep understanding of causality sometimes requires the understanding of very large patterns and their abstract relationships and interactions, not just the understanding of microscopic objects interacting in microscopic time intervals.

I have to emphasize that there's no "extra" physical (or extra-physical) force here; the local, myopic laws of physics take care of everything on their own, but the global *arrangement* of the dominos is what determines what happens, and if you notice (and understand) that arrangement, then an insight-giving shortcut to the answer of the non-falling domino in the divisor stretch (as well as the falling domino in the prime stretch) is served to you on a silver platter. On the other hand, if you don't pay attention to that arrangement, then you are doomed to taking the long way around, to understanding things only locally and without insight. In short, considering 641's primality as a physical cause in our domino chainium is analogous to considering a gas's temperature as a physical cause (*e.g.*, of the amount of pressure it exerts against the walls of its container).

Indeed, let us think for a moment about such a gas — a gas in a cylinder with a movable piston. If the gas suddenly heats up (as occurs in any cylinder in your car engine when its spark plug fires), then its pressure suddenly increases and *therefore* (note the causal word) the piston is suddenly shoved outwards. Thus combustion engines can be built.

What I just told is the story at a gross (thermodynamic) level. Nobody who designs combustion engines worries about the fine-grained level — that of molecules. No engineer tries to figure out the exact trajectories of 10^{23} molecules banging into each other! The locations and velocities of individual molecules are simply irrelevant. All that matters is that they can be counted on to *collectively* push the piston out. Indeed, it doesn't matter whether they are molecules of type X or type Y or type Z — pressure is pressure, and that's all that matters. The explosion — a high-level event — will do its job in heating the gas, and the gas will do its job in pushing the piston. This high-level description of what happens is the *only* level of description that is relevant, because all the microdetails could be changed and exactly the same thing (at least from the human engineer's point of view) would still happen.

The Strange Irrelevance of Lower Levels

This idea — that the bottom level, though 100 percent *responsible* for what is happening, is nonetheless *irrelevant* to what happens — sounds almost paradoxical, and yet it is an everyday truism. Since I want this to be crystal-clear, let me illustrate it with one more example.

Consider the day when, at age eight, I first heard the fourth étude of Chopin's Opus 25 on my parents' record player, and instantly fell in love with it. Now suppose that my mother had placed the needle in the groove a millisecond later. One thing for sure is that all the molecules in the room would have moved completely differently. If you had been one of those molecules, you would have had a wildly different life story. Thanks to that millisecond delay, you would have careened and bashed into completely different molecules in utterly different places, spun off in totally different directions, and on and on, *ad infinitum*. No matter which molecule you were in the room, your life story would have turned out unimaginably different. But would any of that have made an iota of difference to the life story of the kid listening to the music? No — not the teensiest, tiniest iota of difference. All that would have mattered was that Opus 25, number 4 got transmitted faithfully through the air, and *that* would most surely have happened. *My* life story would not have been changed in any way, shape, or form if my mother had put the needle down in the groove a millisecond earlier or later. Or a second earlier or later.

Although the air molecules were crucial mediating agents for a series of high-level events involving a certain kid and a certain piece of music, their precise behavior was not crucial. Indeed, saying it was "not crucial" is a ridiculous understatement. Those air molecules could have done exactly the same kid–music job in an astronomical number of different but humanly indistinguishable fashions. The lower-level laws of their collisions played a role only in that they gave rise to predictable high-level events (propagation of the notes in the Chopin étude to little Douggie's ear). But the positions, speeds, directions, even the chemical identity of the molecules — all of this was changeable, and the high-level events would have been the same. It would have been the same music to my ears. One can even imagine that the microscopic laws of physics could have been different — what matters is not the detailed laws but merely the fact that they reliably give rise to stable statistical consequences.

Flip a quarter a million times and you'll very reliably get within one percent of 500,000 heads. Flip a penny the same number of times, and the same statement holds. Use a different coin on every flip — dimes, quarters, new pennies, old pennies, buffalo nickels, silver dollars, you name it — and still you'll get the same result. Shave your penny so that its outline is hexagonal instead of circular — no difference. Replace the hexagonal outline by an elephant shape. Dip the penny in apple butter before each flip. Bat the penny high into the air with a baseball bat instead of tossing it up. Flip the penny in helium gas instead of air. Do the experiment on Mars instead of Earth. These and countless other variations on the theme will not have any effect on the fact that out of a million tosses, within one percent of 500,000 will wind up heads. That high-level statistical outcome is robust and invariant against the details of the substrate and the microscopic laws governing the flips and bounces; the high-level outcome is insulated and sealed off from the microscopic level. It is a fact in its own right, at its own level.

That is what it means to say that although what happens on the lower level is *responsible* for what happens on the higher level, it is nonetheless *irrelevant* to the higher level. The higher level can blithely ignore the processes on the lower level. As I put it in Chapter 2, "Our existence as animals whose perception is limited to the world of everyday macroscopic objects forces us, quite obviously, to function without any reference to entities and processes at microscopic levels. No one really knew the slightest thing about atoms until only about a hundred years ago, and yet people got along perfectly well."

A Hat-tip to the Spectrum of Unpredictability

I am not suggesting that the invisible, swarming, chaotic, microscopic level of the world can be totally swept under the rug and forgotten. Although in many circumstances we rely on the familiar macroworld to be completely predictable to us, there are many other circumstances where we are very aware of not being able to predict what will happen. Let me first, however, make a little list of some sample predictables that we rely on unthinkingly all the time.

When we turn our car's steering wheel, we know for sure where our car will go; we don't worry that a band of recalcitrant little molecules might mutiny and sabotage our turn. When we turn a burner to "high" under a saucepan filled with water, we know that the water will boil within a few minutes. We can't predict the pattern of bubbles inside the boiling water, but we really don't give a hoot about that. When we take a soup can down from the shelf in the grocery store and place it in our cart, we know for sure that it will not turn into a bag of potato chips, will not burn our hand, will not be so heavy that we cannot lift it, will not slip through the grill of the cart, will sit still if placed vertically, and so forth. To be sure, if we lay the soup can down horizontally and start wheeling the cart around the store, the can will roll about in the cart in ways that are not predictable to us, though they lie completely within the bounds of our expectations and have little interest or import to us, aside from being mildly annoying.

When we speak words, we know that they will reach the ears of our listeners without being changed by the intermediary pressure waves into other words, will even come through with the exact intonations that we impart to them. When we pour milk into a glass, we know just how far to tilt the milk container to get the desired amount of flow without spilling a drop. We control the milk and we get exactly the result we want.

There is no surprise in any of this! And I could extend this list forever, and it would soon grow very boring, because you know it all instinctively and take it totally for granted. Every day of our lives, we all depend in a million tacit ways on innumerable rock-solid predictabilities about how things happen in the visible, tangible world (the solidity of rocks being yet another of those countless rock-solid predictabilities).

On the other hand, there's also plenty of unpredictability "up here" in the macroworld. How about a second list, giving typical unpredictables?

When we toss a basketball towards a basket, we don't have any idea whether it will go through or not. It might bounce off the backboard and then teeter for a couple of seconds on the rim, keeping us in suspense and perhaps even holding an entire crowd in tremendous, tingling tension. A championship basketball game could go one way or the other, depending on a microscopic difference in the position of the pinky of the player who makes a desperate last-second shot.

When we begin to utter a thought, we have no idea what words we will wind up using nor which grammatical pathways we will wind up following, nor can we predict the speech errors or the facts about our unconscious mind that our little slips will reveal. Usually such revelations will make little difference, but once in a while — in a job interview, say — they can have huge repercussions. Think of how people jump on a politician whose unconscious mind chooses a word loaded with political undertones (*e.g.,* "the crusade against terrorism").

When we ski down a slope, we don't know if we're going to fall on our next turn or not. Every turn is a risk — slight for some, large for others. A broken bone can come from an event whose cause we will never fathom, because it is so deeply hidden in detailed interactions between the snow and our ski. And the tiniest detail about the manner in which we fall can make all the difference as to whether we suffer a life-changing multiple break or a just a trivial hairline fracture.

The macroscopic world as experienced by humans is, in short, an intimate mixture ranging from the most predictable events all the way to wildly unpredictable ones. Our first few years of life familiarize us with this spectrum, and the degree of predictability of most types of actions that we undertake becomes second nature to us. By the time we emerge from childhood, we have acquired a reflex-level intuition for where most of our everyday world's loci of unpredictability lie, and the more unpredictable end of this spectrum simultaneously beckons to us and frightens us. We're pulled by but fearful of risk-taking. That is the nature of life.

The Careenium

I now move to a somewhat more complex metaphor for thinking about the multiple levels of causality in our brains and minds (and eventually, if you will indulge me in this terminology, in our souls). Imagine an elaborate frictionless pool table with not just sixteen balls on it, but myriads of extremely tiny marbles, called "sims" (an acronym for "small interacting marbles"). These sims bash into each other and also bounce off the walls, careening about rather wildly in their perfectly flat world — and since it is frictionless, they just keep on careening and careening, never stopping.

So far our setup sounds like a two-dimensional ideal gas, but now we'll posit a little extra complexity. The sims are also magnetic (so let's switch to "simms", with the extra "m" for "magnetic"), and when they hit each other at lowish velocities, they can stick together to form clusters, which I hope you will pardon me for calling "simmballs". A simmball consists of a very large number of simms (a thousand, a million, I don't care), and on its periphery it frequently loses a few simms while gaining others. There are thus two extremely different types of denizen of this system: tiny, light, zipping simms, and giant, ponderous, nearly-immobile simmballs.

The dynamics taking place on this pool table — hereinafter called the "careenium" — thus involves simms crashing into each other and also into simmballs. To be sure, the details of the physics involve transfers of momentum, angular momentum, kinetic energy, and rotational energy, just as in a standard gas, but we won't even think about that, because this is just a *thought* experiment (in two senses of the term). All that matters for our purposes is that there are these collisions taking place all the time.

Simmballism

Why the corny pun on "symbol"? Because I now add a little more complexity to our system. The vertical walls that constitute the system's boundaries react sensitively to outside events (*e.g.*, someone touching the outside of the table, or even a breeze) by momentarily flexing inward a bit. This flexing, whose nature retains some traces of the external causing event, of course affects the motions of the simms that bounce internally off that section of wall, and indirectly this will be registered in the slow motions of the nearest simmballs as well, thus allowing the simmballs to *internalize* the event. We can posit that one particular simmball always reacts in some standard fashion to breezes, another to sharp blows, and so forth. Without going into details, we can even posit that the configurations of simmballs *reflect the history* of the impinging outer-world events. In short, for someone who looked at the simmballs and knew how to read their configuration, the simmballs would be *symbolic*, in the sense of *encoding events*. That's why the corny pun.

Of course this image is far-fetched, but remember that the careenium is merely intended as a useful metaphor for understanding our brains, and the fact is that our brains, too, are rather far-fetched, in the sense that they too contain tiny events (neuron firings) and larger events (patterns of neuron firings), and the latter presumably somehow have *representational* qualities, allowing us to register and also to remember things that happen outside of our crania. Such internalization of the outer world in symbolic patterns in a brain is a pretty far-fetched idea, when you think about it, and yet we know it somehow came to exist, thanks to the pressures of evolution. If you wish, then, feel free to imagine that careenia, too, evolved. You can think of them as emerging as the end result of billions of more primitive systems fighting for survival in the world. But the evolutionary origins of our careenium need not concern us here. The key idea is that whereas no simm on its own encodes anything or plays a symbolic role, the simmballs, on their far more macroscopic level, *do* encode and *are* symbolic.

Taking the Reductionistic View of the Careenium

The first inclination of a modern physicist who heard this story might be reductionistic, in the sense of pooh-poohing the large simmballs as mere *epiphenomena*, meaning that although they are undeniably *there*, they are not essential to an understanding of the system, since they are composed of simms. Everything that happens in the careenium is explainable in terms of simms alone. And there's no doubt that this is true. A volcano, too, is undeniably *there*, but who needs to talk about mountains and subterranean pressures and eruptions and lava and such things? We can dispense with such epiphenomenal concepts altogether by shifting to the deeper level of atoms or elementary particles. The bottom line, at least for our physicist, is that epiphenomena are just convenient shorthands that summarize a large number of deeper, lower-level phenomena; they are never essential to any explanation. Reductionism ho!

The only problem is the enormous escalation in complexity when we drop all macroscopic terms and ways of looking at things. If we refuse to use any language that involves epiphenomena, then we are condemned to seeing only untold myriads of particles, and that is certainly not a very welcoming thought. Moreover, when one perceives only myriads of particles, there are no natural sharp borders in the world. One cannot draw a line around the volcano and declare, "Only particles in this zone are involved", because particles won't respect any such macroscopic line — no more than ants respect the property lines carefully surveyed and precisely drawn by human beings. No fixed portion of the universe can be tightly fenced off from interacting with the rest — not even approximately. To a reductionist, the idea of carving the universe up into zones with inviolable macroscopic spatiotemporal boundary lines makes no sense.

Here is a striking example of the senselessness of local spatiotemporal boundaries. In November of 1993, I read several newspaper articles about a comet that was "slowly" making its way towards Jupiter. It was still some eight months from t-zero but astrophysicists had already predicted to the minute, if not to the second, when it would strike Jupiter, and where. This fact about some invisible comet that was billions of miles away from earth had already had enormous impacts on the surface of our planet, where teams of scientists were already calculating its Jovian arrival time, where newspapers and magazines were already printing front-page stories about it, and where millions of people like me were already reading about it. Some of these people were possibly missing planes because of being engrossed in the story, or possibly striking up a new friendship with someone because of a common interest in it, or possibly arriving at a traffic light one second later than otherwise because of having reread one phrase in the article, and so on. As t-zero approached and finally the comet hit Jupiter's far side exactly as predicted, denizens of the Earth paid enormous attention to this remote cosmic event. There is no doubt that many months before the comet hit Jupiter, certain fender-benders took place on our planet that wouldn't have taken place if the comet hadn't been coming, certain babies were conceived that wouldn't have been conceived otherwise, certain flies were swatted, certain coffee cups were chipped, and so on. All of this crazy stuff happening on our tiny planet was due to a comet coasting through silent space billions of miles away and nearly half a million minutes in advance of its encounter with the huge planet.

The point is that one gets into very hot water if one goes the fully reductionistic route; not only do all the objects in "the system" become microscopic and uncountably numerous, but also the system itself grows beyond bounds in space and time and becomes, in the end, the entire universe taken over all of time. There is no comprehensibility left, since everything is shattered into a trillion trillion trillion invisible pieces that are scattered hither and yon. Reductionism is merciless.

Taking a Higher-level View of the Careenium

If, on the other hand, there is a perceptible and comprehensible "logic" to events at the level of epiphenomena, then we humans are eager to jump to that level. In fact, we have no choice. And so we *do* talk of volcanoes and eruptions and lava and so forth. Likewise, we talk of bitten fingernails and rye bread and wry smiles and Jewish senses of humor rather than of cells and proteins, let alone of atoms and photons. After all, we ourselves are pretty big epiphenomena, and as I've already observed many times in this book, this fact dooms us to talking about the world in terms of other epiphenomena at about our size level (*e.g.,* our mothers and fathers, our cats and cars and cakes, our sailboats and saxophones and sassafras trees).

Now let's return to the careenium and talk about what happens in it. The way I've portrayed it so far focuses on the simms and their dashing and bashing. The simmballs are also present, but they serve a similar function to the walls — they are just big stationary objects off of which the simms bounce. In my mind's eye, I often see the simms as acting like the silver marbles in a pinball machine, with the simmballs acting like the "pins" — that is, the larger stationary cylindrical objects which the marbles strike and ricochet off of as they roll down the sloped board of play.

But now I'm going to describe a different way of looking at the careenium, which is characterized by two perceptual shifts. First, we shift to time-lapse photography, meaning that imperceptibly slow motions get speeded up so as to become perceptible, while fast motions become so fast that they are not even seen as blurs — they become imperceptible, like the spinning blades of an electric fan. The second shift is that we spatially back away or zoom out, thus rendering simms too small to be seen, and so the simmballs alone necessarily become our focus of attention.

Now we see a completely different type of dynamics on the table. Instead of seeing simms bashing into what look like large stationary blobs, we realize that these blobs are not stationary at all but have a lively life of their own, moving back and forth across the table and interacting with each other, as if there were nothing else on the table but them. Of course we know that deep down, this is all happening thanks to the teeny-weeny simms' bashing-about, *but we cannot see the simms any more.* In our new way of seeing things, their frenetic careening-about on the table forms nothing but a stationary gray background.

Think of how the water in a glass sitting on a table seems completely still to us. If our eyes could shift levels (think of the twist that zooms binoculars in or out) and allow us to peer at the water at the micro-level, we would realize that it is not peaceful at all, but a crazy tumult of bashings of water molecules. In fact, if colloidal particles are added to a glass of water, then it becomes a locus of Brownian motion, which is an incessant random jiggling of the colloidal particles, due to a myriad of imperceptible collisions with the water molecules, which are far tinier. (The colloidal particles here play the role of simmballs, and the water molecules play the role of simms.) The effect, which is visible under a microscope, was explained in great detail in 1905 by Albert Einstein using the theory of molecules, which at the time were only hypothetical entities, but Einstein's explanation was so far-reaching (and, most crucially, consistent with experimental data) that it became one of the most important confirmations that molecules do exist.

Who Shoves Whom Around inside the Careenium?

And so we finally have come to the crux of the matter: *Which of these two views of the careenium is the truth?* Or, to echo the key question posed by Roger Sperry, *Who shoves whom around in the population of causal forces that occupy the careenium?* In one view, the meaningless tiny simms are the primary entities, zipping around like mad, and in so doing they very slowly push the heavy, passive simmballs about, hither and thither. In this view, it is the tiny simms that shove the big simmballs around, and that is all there is to it. In fact, in this view the simmballs are not even recognized as separate entities, since anything we might say about their actions is just a shorthand way of talking about what simms do. From this perspective, there are no simmballs, no symbols, no ideas, no thoughts going on — just a great deal of tumultuous, pointless careening-about of tiny, shiny, magnetic spheres.

In the other view, speeded up and zoomed out, all that is left of the shiny tiny simms is a featureless gray soup, and the interest resides solely in the simmballs, which give every appearance of richly interacting with each other. One sees groups of simmballs triggering other simmballs in a kind of "logic" that has nothing to do with the soup churning around them, except in the rather pedestrian sense that the simmballs derive their *energy* from that omnipresent soup. Indeed, the simmballs' logic, not surprisingly, has to do with the *concepts* that the simmballs symbolize.

The Dance of the Simmballs

From our higher-level macroscopic vantage point as we hover above the table, we can see *ideas* giving rise to other *ideas,* we can see one symbolic event *reminding* the system of another symbolic event, we can see elaborate patterns of simmballs coming together and forming even larger patterns that constitute *analogies* — in short, we can visually eavesdrop on the logic of a thinking mind taking place in the patterned dance of the simmballs. And in this latter view, *it is the simmballs that shove each other about,* at their own isolated symbolic level.

The simms are still there, to be sure, but they are simply serving the simmballs' dance, allowing it to happen, with the microdetails of their bashings being no more relevant to the ongoing process of cognition than the microdetails of the bashings of air molecules are relevant to the turning of the blades of a windmill. Any old air-molecule bashings will do — the windmill will turn no matter what, thanks to the aerodynamic nature of its blades. Likewise, any old simm-bashings will do — the "thoughtmill" will churn no matter what, thanks to the symbolic nature of its simmballs.

If any of this strikes you as too far-fetched to be plausible, just return to the human brain and consider what must be going on inside it in order to allow our thinking's logic to take place. What else is going on inside every human cranium but some story like this?

Of course we have come back to the question that that long-ago-shelved book's title made me ask, and the question that Roger Sperry also asked: Who is shoving whom about in here? And the answer is that it all depends on what level you choose to focus on. Just as, on one level, the primality of 641 could legitimately be said to be shoving about dominos in the domino-chain network, so here there is a level on which the meanings attached to various simmballs can legitimately be said to be shoving other simmballs about. If this all seems topsy-turvy, it certainly is — but it is nonetheless completely consistent with the fundamental causality of the laws of physics.

CHAPTER 4

Loops, Goals, and Loopholes

☙ ☙ ☙

The First Flushes of Desire

WHEN the first mechanical systems with feedback in them were designed, a set of radically new ideas began coming into focus for humanity. Among the earliest of such systems was James Watt's steam-engine governor; subsequent ones, which are numberless, include the float-ball mechanism governing the refilling of a flush toilet, the technology inside a heat-seeking missile, and the thermostat. Since the flush toilet is probably the most familiar and the easiest to understand, let's consider it for a moment.

A flush toilet has a pipe that feeds water into the tank, and as the water level rises, it lifts a hollow float. Attached to the rising float is a rigid rod whose far end is fixed, so that the rod's angle of tilt reflects the amount of water in the tank. This variable angle controls a valve that regulates the flow of water in the pipe. Thus at a critical level of filling, the angle reaches a critical value and the valve closes totally, thereby shutting off all flow in the pipe. However, if there is leakage from the tank, the water level gradually falls, and of course the float falls with it, the valve opens, and the inflow of water is thereby turned back on. Thus one sometimes gets into cyclic situations where, because a little rubber gizmo didn't land exactly centered on the tank's drain right after a flush, the tank slowly leaks for a few minutes, then suddenly fills for a few seconds, then again slowly leaks for a few minutes, then again fills for a few seconds, and so on, in a cyclic pattern that somewhat resembles breathing, and that never stops — that is, not until someone jiggles the toilet handle, thus jiggling the rubber gizmo, hopefully making it land properly on the drain, thus fixing the leak.

Once a friend of mine who was watching my house while I was away for a few weeks' vacation flushed the toilet on the first day and, by chance, the little rubber gizmo didn't fall centered, so this cycle was entered. My friend diligently returned a few times to check out the house but he never noticed anything untoward, so the toilet tank kept on leaking and refilling periodically for my entire absence, and as a result I had a $300 water bill. No wonder people are suspicious of feedback loops!

We might anthropomorphically describe a flush toilet as a system that is "trying" to make the water reach and stay at a certain level. Of course, it's easy to bypass such anthropomorphic language since we effortlessly see how the mechanism works, and it's pretty clear that such a simple system has no desires; even so, when working on a toilet whose tank has sprung a leak, one might be tempted to say the toilet is "trying" get the water up to the mark but "can't". One doesn't *truly* impute desires or frustrations to the device — it's just a manner of speaking — but it is a convenient shorthand.

A Soccer Ball Named Desire

Why does this move to a goal-oriented — that is, *teleological* — shorthand seem appealing to us for a system endowed with feedback, but not so appealing for a less structured system? It all has to do with the way the system's "perceptions" feed back (so to speak) into its behavior. When the system always moves towards a certain state, we see that state as the system's "goal". It is the self-monitoring, self-controlling nature of such a system that tempts us to use teleological language.

But what kinds of systems have feedback, have goals, have desires? Does a soccer ball rolling down a grassy hill "want" to get to the bottom? Most of us, reflexively recoiling at such a primitive Aristotelian conception of why things move, would answer *no* without hesitation. But let's modify the situation just a tiny amount and ask the question again.

What about a soccer ball zipping down a long, narrow roadside gutter having a U-shaped cross-section — is it seeking any goal? Such a ball, as it speeds along, will first roll up one side of the gutter and then fall back to the center, cross it and then roll up the other side, then again back down, and so forth, gradually converging from a sinusoidal pathway wavering about the gutter's central groove to a straight pathway at the bottom of the gutter. Is there "feedback" here or not? Is this soccer ball "seeking" the gutter's mid-line? Does it "want" to be rolling along the gutter's valley? Well, as this example and the previous one of the ball rolling down a hill show, the presence or absence of feedback, goals, or desires is not a black-and-white matter; such things are judgment calls.

The Slippery Slope of Teleology

As we move to systems where the feedback is more sophisticated and its mechanisms are more hidden, our tendency to shift to teleological terms — first the language of goals and then the language of "wishing", "desiring", "trying" — becomes ever more seductive, ever harder to resist. The feedback doesn't even need to be very sophisticated, as long as it is hidden.

In San Francisco's Exploratorium museum, there is an enclosure where people can stand and watch a spot of red light dancing about on the walls and floor. If anyone tries to touch the little spot, it darts away at the last moment. In fact, it dances about in a way that seems to be teasing the people chasing it — sometimes stopping completely, taunting them, daring them to approach, and then flitting away just barely in time. However, despite appearances, there is no hidden person guiding it — just some simple feedback mechanisms in some circuitry monitoring the objects in the enclosure and controlling the light beam. But the red spot *seems* for all the world to have a personality, an impish desire to tease people, even a sense of humor! The Exploratorium's red dot seems more alive than, say, a mosquito or a fly, both of which attempt to avoid being swatted but certainly don't have any detectable sense of humor.

In the video called "Virtual Creatures" by Karl Sims, there are virtual objects made out of a few (virtual) tubes hinged together, and these objects can "flap" their limbs and thus locomote across a (virtual) flat plane. When they are given a rudimentary sort of perception and a simple feedback loop is set up that causes them to pursue certain kinds of resources, then the driven manner in which they pursue what looks like food and frantically struggle with "rivals" to reach this resource gives viewers an eerie sensation of witnessing primitive living creatures engaged in life-and-death battles.

On a more familiar level, there are plants — consider a sunflower or a growing vine — which, when observed at normal speed, seem as immobile as rocks and thus patently devoid of goals, but when observed in time-lapse photography, seem all of a sudden to be highly aware of their surroundings and to possess clear goals as well as strategies to reach them. The question is whether such systems, despite their lack of brains, are nonetheless imbued with goals and desires. Do they have hopes and aspirations? Do they have dreads and dreams? Beliefs and griefs?

The presence of a feedback loop, even a rather simple one, constitutes for us humans a strong pressure to shift levels of description from the goalless level of mechanics (in which *forces* make things move) to the goal-oriented level of cybernetics (in which, to put it very bluntly, *desires* make things move). The latter is, as I have stressed, nothing but a more efficient rewording of the former; nonetheless, with systems that possess increasingly subtle and sophisticated types of feedback loops, that shorthand's efficiency becomes well-nigh irresistible. And eventually, not only does teleological language become indispensable, but we cease to realize that there could be any other perspective. At that point, it is locked into our worldview.

Feedback Loops and Exponential Growth

The type of feedback with which we are all most familiar, and probably the case that gave it its name, is audio feedback, which typically takes place in an auditorium when a microphone gets too close to a loudspeaker that is emitting, with amplification, the sounds picked up by the microphone. In goes some sound (any sound — it makes no difference), out it comes louder, then *that* sound goes back in, comes out yet louder, then back in again, and all of a sudden, almost out of nowhere, you have a loop, a vicious circle, producing a terrible high-pitched screech that makes the audience clap their hands over their ears.

This phenomenon is so familiar that it seems to need no comment, but in fact there are a couple of things worth pointing out. One is that each cycling-around of any input sound would theoretically amplify its volume by a fixed factor, say k — thus, two loops would amplify by a factor of k^2, three loops by k^3, and so on. Well, we all know the power of exponential growth from hearing horror stories about exponential growth of the earth's population or some such disaster. (In my childhood, the power of exponentials was more innocently but no less indelibly imprinted on me by the story of a sultan who commanded that on each square of a chessboard there be placed twice as many grains of rice as on the previous square — and after less than half the board was full, it was clear there was not nearly enough rice in the sultanate or even the whole world to get anywhere close to the end.) In theory, then, the softest whisper would soon grow to a roar, which would continue growing without limit, first rendering everyone in the auditorium deaf, shortly thereafter violently shaking the building's rafters till it collapsed upon the now-deaf audience, and then, only a few loops later, vibrating the planet apart and finishing up by annihilating the entire universe. What is specious about this apocalyptic scenario?

Fallacy the First

The primary fallacy in this scenario is that we have not taken into account the actual device carrying out the exponential process — the sound system itself, and in particular the amplifier. To make my point in the most blatant manner, I need merely remind you that the moment the auditorium's roof collapsed, it would land on the amplifier and smash it to bits, thus bringing the out-of-control feedback loop to a swift halt. The little system contains the seeds of its own destruction!

But there is something specious about this scenario, too, because as we all know, things never get that far. The auditorium never collapses, nor are the audience members deafened by the din. Something slows down the runaway process far earlier. What is that thing?

Fallacy the Second

The other fallacy in our reasoning also involves a type of self-destruction of the sound system, but it is subtler than being smashed to smithereens. It is that as the sound gets louder and louder, the amplifier stops amplifying with that constant factor of k. At a certain level it starts to fail. Just as a floored car will not continue accelerating at a constant rate (reaching 100 miles per hour, then 200, 300, 400, soon breaking the sound barrier, etc.) but eventually levels out at some peak velocity (which is a function of road friction, air resistance, the motor's internal limits, and so forth), so an amplifier will not uniformly amplify sounds of any volume but will eventually saturate, giving less and less amplification until at some volume level the output sound has the same volume as the input sound, and that is where things stabilize. The volume at which the amplification factor becomes equal to 1 is that of the familiar screech that drives you mad but doesn't deafen you, much less brings the auditorium crashing down on your head.

And why does it always give off that same high-pitched screeching sound? Why not a low roar? Why not the sound of a waterfall or a jet engine or long low thunder? This has to do with the natural resonance frequency of the system — an acoustic analogue of the natural oscillation frequency of a playground swing, roughly once every couple of seconds. An amplifier's feedback loop has a natural oscillation frequency, too, and for reasons that need not concern us, it usually has a pitch close to that of a high-frequency scream. However, the system does not instantly settle down precisely on its final pitch. If you could drastically slow down the process, you would hear it homing in on that squealing pitch much as the rolling soccer ball seeks the bottom of the gutter — namely, by means of a very rapid series of back-and-forth swings in frequency, almost as if it "wanted" to reach that natural spot in the sonic spectrum.

What we have seen here is that even the simplest imaginable feedback loop has levels of subtlety and complexity that are seldom given any thought, but that turn out to be rich and full of surprise. Imagine, then, what happens in the case of more complex feedback loops.

Feedback and Its Bad Rap

The first time my parents wanted to buy a video camera, sometime in the 1970's, I went to the store with them and we asked to see what they had. We were escorted to an area of the store that had several TV screens on a shelf, and a video camera was plugged into the back of one of them, thus allowing us to see what the camera was looking at and to gauge its color accuracy and such things. I took the camera and pointed it at my father, and we saw his amused smile jump right up onto the screen. Next I pointed the camera at my own face and presto, there was I, up on the screen, replacing my father. But then, inevitably, I felt compelled to try pointing the camera at the TV screen itself.

Now comes the really curious fact, which I will forever remember with some degree of shame: I was *hesitant* to close the loop! Instead of just going ahead and doing it, I balked and timidly asked the salesperson for *permission* to do so. Now why on earth would I have done such a thing? Well, perhaps it will help if I relate how he replied to my request. What he said was this: "No, no, *no!* Don't do *that* — you'll break the camera!"

And how did I react to his sudden panic? With scorn? With laughter? Did I just go ahead and follow my whim anyway? No. The truth is, I wasn't quite sure of myself, and his panicky outburst reinforced my vague uneasiness, so I held my desire in check and didn't do it. Later, though, as we were driving home with our brand-new video camera, I reflected carefully on the matter, and I just couldn't see where in the world there would have been any danger to the system — either to the camera or to the TV — if I had closed the loop (though *a priori* either one of them would seem vulnerable to a meltdown). And so when we got home, I gingerly tried pointing the camera at the screen and, *mirabile dictu*, nothing terrible happened at all.

The danger I suppose one could fear is something analogous to audio feedback: perhaps one particular spot on the screen (the spot the camera is pointing straight at, of course) would grow brighter and brighter and brighter, and soon the screen would melt down right there. But why might this happen? As in audio feedback, it would have to come from some kind of amplification of the light's intensity; however, we know that video cameras are not designed to *amplify* an image in any way, but simply to *transmit* it to a different place. Just as I had figured out in the calm of the drive home, there is no danger at all in standard video feedback (by the way, I don't know when the term "video feedback" was invented, nor by whom; certainly I had never heard it back then). But danger or no danger, I remember well my hesitation at the store, and so I can easily imagine the salesperson's panic, irrational though it was. Feedback — making a system turn back or twist back on itself, thus forming some kind of mystically taboo loop — seems to be dangerous, seems to be tempting fate, perhaps even to be intrinsically *wrong*, whatever that might mean.

These are primal, irrational intuitions, and who knows where they come from. One might speculate that fear of any kind of feedback is just a simple, natural generalization from one's experience with audio feedback, but I somehow doubt that the explanation is that simple. We all know that some tribes are fearful of mirrors, many societies are suspicious of cameras, certain religions prohibit making drawings of people, and so forth. Making representations of one's own self is seen as suspicious, weird, and perhaps ultimately fatal. This suspicion of loops just runs in our human grain, it would seem. However, as with many daring activities such as hang-gliding or parachute jumping, some of us are powerfully drawn to it, while others are frightened to death by the mere thought of it.

God, Gödel, Umlauts, and Mystery

When I was fourteen years old, browsing in a bookstore, I stumbled upon a little paperback entitled "Gödel's Proof". I had no idea who this Gödel person was or what he (I'm sure I didn't think "he or she" at that early age and stage of my life) might have proven, but the idea of a whole book about just one mathematical proof — any mathematical proof — intrigued me. I must also confess that what doubtlessly added a dash of spice to the dish was the word "God" blatantly lurking inside "Gödel", as well as the mysterious-looking umlaut perched atop the center of "God". My brain's molecules, having been tickled in the proper fashion, sent signals down to my arms and fingers, and accordingly I picked up the umlaut-decorated book, flipped through its pages, and saw tantalizing words like "meta-mathematics", "meta-language", and "undecidability". And then, to my delight, I saw that this book discussed paradoxical self-referential sentences like "I am lying" and more complicated cousins. I could see that whatever Gödel had proved wasn't focused on numbers *per se*, but on reasoning itself, and that, most amazingly, *numbers* were being put to use in reasoning about the nature of mathematics.

Although to some readers this next may sound implausible, I remember being particularly drawn in by a long footnote about the proper use of quotation marks to distinguish between use and mention. The authors — Ernest Nagel and James R. Newman — took the two sentences "Chicago is a populous city" and "Chicago is trisyllabic" and asserted that the former is true but the latter is false, explaining that if one wishes to talk about properties of a *word*, one must use its *name*, which is the expression resulting from putting it inside quotes. Thus, the sentence "'Chicago' is trisyllabic" does not concern a city but its name, and states a truth. The authors went on to talk about the necessity of taking great care in making such distinctions inside formal reasoning, and pointed out that names themselves have names (made using quote marks), and so on, *ad infinitum*. So here was a book talking about how language can talk about itself talking about itself (etc.), and about how reasoning can reason about itself (etc.). I was hooked! I still didn't have a clue what Gödel's theorem was, but I knew I had to read this book. The molecules constituting the book had managed to get the molecules in my head to get the molecules in my hands to get the molecules in my wallet to… Well, you get the idea.

Savoring Circularity and Self-application

What seemed to me most magical, as I read through Nagel and Newman's compelling booklet, was the way in which mathematics seemed to be doubling back on itself, engulfing itself, twisting itself up inside itself. I had always been powerfully drawn to loopy phenomena of this sort. For instance, from early childhood, I had loved the idea of closing a cardboard box by tucking its four flaps over each other in a kind of "circular" fashion — A on top of B, B on top of C, C on top of D, and then D on top of A. Such grazing of paradoxicality enchanted and fascinated me.

Also, I had always loved standing between two mirrors and seeing the implied infinitude of images as they faded off into the distance. (The photo was taken by Kellie Gutman.) A mirror mirroring a mirror — what idea could be more provocative? And I loved the picture of the Morton Salt girl holding a box of Morton Salt, with herself drawn on it, holding the box, and on and on, by implication, in ever-tinier copies, without any end, ever.

Years later, when I took my children to Holland and we visited the park called "Madurodam" (those quote marks, by the way, are a testimony to the lifelong effect on me of Nagel and Newman's insistence on the importance of distinguishing between use and mention), which contains dozens of beautifully constructed miniature replicas of famous buildings from all over Holland, I was most disappointed to see that there was no miniature replica of Madurodam itself, containing, of course, a yet tinier replica, and so on... I was particularly surprised that this lacuna existed in Holland, of all places — not only the native land of M. C. Escher, but also the home of Droste's famous hot chocolate, whose box, much like the Morton's Salt box, implicated itself in an infinite regress, something that all Dutch people grow up knowing very well.

The roots of my fascination with such loops go very far back. When I was but a tyke, around four or five years old, I figured out, or was told, that two twos made four. This catchy phrase — "two twos" — sent thrills up and down my spine, because I realized that it involved applying the notion of "two" *to itself.* It was a kind of self-referential operation, the twisting-back of a concept on itself. Just like a daredevil pilot or rock-climber, I craved more such experiences and riskier ones as well, so I quite naturally asked myself what *three threes* made. Being too small to figure this mystery out for myself (by making a square with three rows of three dots each, for instance), I asked my mother, that Font of Wisdom, for the answer, and she calmly informed me that it was nine.

At first I was delighted, but it didn't take long before vague worries started setting in that I hadn't asked her the right question. I was troubled that both my new phrase and the old phrase contained only *two* copies of the number in question, whereas my goal had been to *transcend* twoness. So I pushed my luck and invented the more threeful phrase "three three threes" — but unfortunately, I didn't know what I meant by it. And so I naturally turned once again to the All-Wise One for help. I remember we had a conversation about this matter (which, at that tender age, I was convinced was surely beyond the grasp of anyone on earth), and I remember she assured me that she fully understood my idea, and she even told me the answer, but I've forgotten what it was — surely 9 or 27.

But the answer is not the point. The point is that among my earliest memories is a relishing of loopy structures, of self-applied operations, of circularity, of paradoxical acts, of implied infinities. This, for me, was the cat's meow and the bee's knees rolled into one.

The Timid Theory of Types

The foregoing vignette reveals a personality trait that I share with many people, but by no means with everyone. I first encountered this split in people's instincts when I read about Bertrand Russell's invention of the so-called "theory of types" in *Principia Mathematica,* his famous *magnum opus* written jointly with his former professor Alfred North Whitehead, which was published in the years 1910–1913.

Some years earlier, Russell had been struggling to ground mathematics in the theory of sets, which he was convinced constituted the deepest bedrock of human thought, but just when he thought he was within sight of his goal, he unexpectedly discovered a terrible loophole in set theory. This loophole (the word fits perfectly here) was based on the notion of "the set of all sets that don't contain themselves", a notion that was legitimate in set theory, but that turned out to be deeply self-contradictory. In order to convey the fatal nature of his discovery to a wide audience, Russell made it more vivid by translating it into the analogous notion of the hypothetical village barber "who shaves all those in the village who don't shave themselves". The stipulation of such a barber's existence is paradoxical, and for exactly the same reason.

When set theory turned out to allow self-contradictory entities like this, Russell's dream of solidly grounding mathematics came crashing down on him. This trauma instilled in him a terror of theories that permitted loops of self-containment or of self-reference, since he attributed the intellectual devastation he had experienced to loopiness and to loopiness alone.

In trying to recover, then, Russell, working with his old mentor and new colleague Whitehead, invented a novel kind of set theory in which a definition of a set could never invoke that set, and moreover, in which a strict linguistic hierarchy was set up, rigidly preventing any sentence from referring to itself. In *Principia Mathematica,* there was to be no twisting-back of sets on themselves, no turning-back of language upon itself. If some formal language had a word like "word", that word could not refer to or apply to itself, but only to entities on the levels *below* itself.

When I read about this "theory of types", it struck me as a pathological retreat from common sense, as well as from the fascination of loops. What on earth could be wrong with the word "word" being a member of the category "word"? What could be wrong with such innocent sentences as "I started writing this book in a picturesque village in the Italian Dolomites", "The main typeface in this chapter is Baskerville", or "This carton is made of recyclable cardboard"? Do such declarations put anyone or anything in danger? I can't see how.

What about "This sentence contains eleven syllables" or "The last word in this sentence is a four-letter noun"? They are both very easy to understand, they are clearly true, and certainly they are not paradoxical. Even silly sentences such as "The ninth word in this sentence contains ten letters" or "The tenth word in this sentence contains nine letters" are no more problematical than the sentence "Two plus two equals five". All three are false or at worst meaningless assertions (the second one refers to something that doesn't exist), but there is nothing paradoxical about any of them. Categorically banishing all loops of reference struck me as such a paranoid maneuver that I was disappointed for a lifetime with the once-bitten twice-shy mind of Bertrand Russell.

Intellectuals Who Dread Feedback Loops

Many years thereafter, when I was writing a monthly column called "Metamagical Themas" for *Scientific American* magazine, I devoted a couple of my pieces to the topic of self-reference in language, and in them I featured a cornucopia of sentences invented by myself, a few friends, and quite a few readers, including some remarkable and provocative flights of fancy, such as these:

> If the meanings of "true" and "false" were switched, this sentence wouldn't be false.

> I am going two-level with you.

> The following sentence is totally identical with this one, except that the words "following" and "preceding" have been exchanged, as have the words "except" and "in", and the phrases "identical with" and "different from".

> The preceding sentence is totally different from this one, in that the words "preceding" and "following" have been exchanged, as have the words "in" and "except", and the phrases "different from" and "identical with".

> This analogy is like lifting yourself up by your own bootstraps.

> Thit sentence it not self-referential because "thit" it not a word.

> If wishes were horses, the antecedent clause in this conditional sentence would be true.

> This sentence every third, but it still comprehensible.

> If you think this sentence is confusing, then change one pig.

> How come *this* noun phrase doesn't denote the same thing as *this* noun phrase does?

> I eee oai o ooa a e ooi eee o oe.

> Ths sntnc cntns n vwls nd th prcdng sntnc n cnsnnts.

> This pangram tallies five a's, one b, one c, two d's, twenty-eight e's, eight f's, six g's, eight h's, thirteen i's, one j, one k, three l's, two m's, eighteen n's, fifteen o's, two p's, one q, seven r's, twenty-five s's, twenty-two t's, four u's, four v's, nine w's, two x's, four y's, and one z.

Although I received from readers a good deal of positive feedback (if you'll excuse the term), I also received some extremely negative feedback concerning what certain readers considered sheer frivolity in an otherwise respectable journal. One of the most vehement objectors was a professor of education at the University of Delaware, who quoted the famous behavioral psychologist B. F. Skinner on the topic of self-referring sentences:

> Perhaps there is no harm in playing with sentences in this way or in analyzing the kinds of transformations which do or do not make sentences acceptable to the ordinary reader, but it is still a waste of time, particularly when the sentences thus generated could not have been emitted as verbal behavior. A classical example is a paradox, such as "This sentence is false", which appears to be true if false and false if true. The important thing to consider is that no one could ever have emitted the sentence as verbal behavior. A sentence must be in existence before a speaker can say, "This sentence is false", and the response itself will not serve, since it did not exist until it was emitted.

This kind of knee-jerk reaction against even the *possibility* that someone might meaningfully utter a self-referential sentence was new to me, and caught me off guard. I reflected long and hard on the education professor's lament, and for the next issue of the magazine I wrote a lengthy reply to it, citing case after case of flagrant and often useful, even indispensable, self-reference in ordinary human communication as well as in humor, art, literature, psychotherapy, mathematics, computer science, and so forth. I have no idea how he or other objectors to self-reference took it. What remained with me, however, was the realization that some highly educated and otherwise sensible people are irrationally allergic to the idea of self-reference, or of structures or systems that fold back upon themselves.

I suspect that such people's allergy stems, in the final analysis, from a deep-seated fear of paradox or of the universe exploding (metaphorically), something like the panic that the television sales clerk evinced when I threatened to point the video camera at the TV screen. The contrast between my lifelong savoring of such loops and the allergic recoiling from them on the part of such people as Bertrand Russell, B. F. Skinner, this education professor, and the TV salesperson taught me a lifelong lesson in the "theory of types" — namely, that there are indeed "two types" of people in this world.

❧ ❧ ❧

CHAPTER 5

On Video Feedback

ờ ờ ờ

Two Video Voyages, Three Decades Apart

THE loop of video feedback is rich, as I found out in my first explorations with our family's new video camera in the mid-1970s. A few months later, my appreciation of the phenomenon deepened considerably when I decided to explore it in detail as a visual study for my book *Gödel, Escher, Bach.* I made an appointment at the Stanford University television studios, and upon arriving I found that the very friendly fellow there had already set up a TV and a camera on a tripod for me to play around with. It was a piece of cake to point the camera at the screen, zoom in and out, tilt the camera, change angles, regulate brightness and contrast, and so on. He told me I was free to use the system as long as I wanted, and so I spent several hours that afternoon navigating around in the ocean of "taboo" possibilities opened up by this video loop. Like any curious tourist, I snapped dozens of photos (just black-and-white stills) during my exotic trip, and later I selected twelve of my favorites to use in one of *GEB*'s dialogues.

Since that first adventure in video feedback, three decades have passed and technology has advanced a bit, so for my new book I decided to give it another shot. This time I was aided and abetted by Bill Frucht, who, because of (or in spite of) being my editor at Basic Books for a dozen years or so, has become a good friend, and who flew in from New York just for this purpose. Together in my kids' old "playroom", Bill and I spent many delightful hours sailing the same old seas but in a somewhat newer craft, and we wound up with several hundred color snapshots that archived our voyage superbly. Aside from the cover illustration, sixteen of my favorites, covering a wide range, can be found in the color insert.

Although both video voyages were vivid and variegated, I decided for this chapter to write up a "diary" of the earlier one, undertaken long ago at Stanford, since that's when I first explored the phenomenon and learned about it step by step. So the story below involves a different television, a different TV camera, and in general an older technology than was used in making this book's color insert. Nonetheless, as you will see, much of the old diary still pertains to the newer voyage, though there are a few small discrepancies that I'll mention when I come to them.

Diary of a Video Trip

There happened to be a shiny metallic strip running down the right side of the TV set I was given, and the presence of this random object had the fortuitous effect of making the various layers of screens-within-screens easily distinguishable. The first thing I discovered, then, was that there was a critical angle that determined whether the regress of nested screens was finite or infinite. If I pointed the camera at the metal strip instead of the center of the screen, this gave me what looked like a snapshot of the right wall of a long corridor, showing a few evenly spaced "doorways" (which actually were images of that metal strip), moving away from where I was "standing". But I was not able to peer all the way down to the end of this "corridor". I'll therefore call what was visible on the screen in such a case a *truncated* corridor.

If I slowly panned the camera leftwards, thus towards the center of the screen and perforce further down the apparent corridor, more and more doorways would come into view along the right wall, smaller and smaller and farther and farther away — and all of a sudden, at a critical moment, there was a wonderful, dizzying sense of infinity as I would find myself peering *all the way* down the corridor toward a gaping emptiness, stretching arbitrarily far away toward a single point of convergence (the "vanishing point", as it is called in the theory of perspective). I'll call this an *endless* corridor. (Note that essentially this same kind of corridor is also visible in the photo of the self-reflecting mirrors in Chapter 4.)

Of course my impression of seeing an infinite number of doorways was illusory, since the graininess of the TV screen and the speed of light set a limit as to how many nestings could occur. Nevertheless, peering down what looked like a magically endless corridor was much more enticing and provocative than merely peering down a mundanely truncated corridor.

My next set of experiments involved tilting the camera. When I did this, each screen obediently tilted at exactly the same angle with respect to its containing screen, which instantly gave rise to a receding *helical* corridor

— a corridor that twisted like a corkscrew. Though quite attractive to the eye, this was not terribly surprising to the mind.

An unanticipated surprise, however, was that at certain angles of camera twist, instead of peering down a helical corridor punctuated by doorways, I seemed to be looking at a flat spiral resembling a galaxy as seen through a telescope. The edges of this spiral were smooth, continuous curves of light rather than jagged sets of straight lines (coming from the edges of the TV screen), and such smoothness mystified me; I saw no reason why a sudden jump from jagged corners to graceful curves should take place. I also noticed that at the very core of each "galaxy", there was nearly always a beautiful circular "black hole". (On our more recent video voyage, Bill and I were unable to reproduce this "black hole" phenomenon, to our puzzlement and chagrin, so you won't see any black holes in the photos in the insert.)

Enigmatic, Emergent Reverberation

At some point during the session, I accidentally stuck my hand momentarily in front of the camera's lens. Of course the screen went all dark, but when I removed my hand, the previous pattern did not just pop right back onto the screen, as I expected. Instead, I saw a different pattern on the screen, but this pattern, unlike anything I'd seen before, was not stationary. Instead, it was throbbing, like a heart! Its "pulse rate" was about one cycle per second, and over the course of each short "heartbeat", the shapes before my eyes metamorphosed greatly. Where, then, had this mysterious periodic pulsation come from, given that there was nothing in the room that was moving?

Whoops — I'm sorry! What I just wrote is a patent falsity — there *was* something in the room that was moving. Do you know what it was, dear reader? Well, the image *itself* was moving. Now that may strike you as a fatuous, trivial, or smart-alecky answer, but since the image was *of itself* (albeit at a slight delay), it is in fact quite to the point. A faithful image of something changing will itself necessarily keep changing! In this case, motion begat motion endlessly because I was dealing with a cyclic setup — a loop. And the original motion that had set things going — the prime mover — had been my hand's motion, of which this video reverberation now constituted a stable, self-sustaining visible memory trace!

This situation reminds me of another loopy phenomenon that I call "reverberant barking", which one sometimes can hear in a neighborhood where many dogs live. If a jogger passes one house and triggers one dog's bark, then neighbor dogs may pick up the barking and a chain reaction involving a dozen dogs may ensue. Soon the barking party has taken on a life of its own, and in the meantime its unwitting instigator has long since exited the neighborhood. If dogs were a bit more like robots and didn't eventually grow tired of doing the same thing over and over again, their reverberant barking could become a stable, self-sustaining audible memory trace of the jogger's fleeting passage through their street.

The dynamically pulsating patterns that I encountered in my video voyage were completely unlike the unwavering "steady-state universes" that I had observed up till then. Stable, periodic video reverberation was a strange and unanticipated phenomenon that I'd bumped into by accident while exploring the possibilities lurking in video feedback.

Even today, all these years later, the origins of such pulsation remain quite unclear, even mysterious, to me; for that reason, it is an *emergent* phenomenon, otherwise known as an *epiphenomenon,* as discussed in Chapter 3. In general, an emergent phenomenon *somehow* emerges quite naturally and automatically from rigid rules operating at a lower, more basic level, but *exactly how* that emergence happens is not at all clear to the observer.

I admit to feeling a little dense for not having fully fathomed what lies behind video reverberation, but at this point I am so accustomed to it that it "makes sense" to me. That is, I have a clear intuition for how to induce it on the screen, and I know that once it starts, it is a robust phenomenon that will continue unabated probably for hours, perhaps even forever, if I don't interfere with it. Rather than trying to figure out how to account precisely for video reverberation in terms of phenomena at lower levels, I have come to just accept it as a fact, and I deal with it at as a phenomenon that exists at its own level. This should sound familiar to you, since it's how we deal with almost everything in our physical and biological world.

Feeding "Content" to the Loop

As I mentioned at the outset, one lucky thing about the Stanford setup was the seemingly random metallic strip on one side of the television set I'd been given to use. That strip — a kind of interloper — added a key note of "spice" to the image that was being cycled round and round, and in that sense it was a crucial ingredient of Video Voyage I.

While Bill and I were conducting Video Voyage II, there were times, to our surprise, when the seas we were sailing seemed a bit too placid for our taste, and we longed for a bit more action, more visual excitement. This brought to my mind the crucial "spicy" role played by the interloping metal strip during Voyage I, so on a lark we decided to introduce something that would play an analogous role in our system. I picked up various objects around the room and dangled them in front of the camera without any idea of what would happen when the image was cycled round and round the video loop. Usually we got marvelous results that were (once again) unanticipatable. For instance, when I dangled a chain of beads in front of the screen, what emerged (the choice of verb is not accidental) was a random-looking swirl of pockmarked bluish-white globs that reminded me a bit of some kind of exotic cheese.

Of course each such interloping object opened up a whole new universe of possibilities, since we could vary its position as well as all the other standard variables (the amount of zoom, the angle of tilt, the direction of the camera, the brightness, the contrast, and others). I tried such things as a glass vase, a compact disk, and, eventually, my own hands. The results were quite fantastic, as you can see in the color insert, but alas, Bill and I didn't have infinite amounts of time to explore the manifold universes we had uncovered and sampled. We played with the possibilities for perhaps a dozen hours and from that we got a 400-photo memory album, and that's all. Like any excursion to a wondrous and exotic place, our trip had to end earlier than we would have preferred, but we were very glad to have taken it and to have savored it together.

A Mathematical Analogue

As might be expected, all the unexpected phenomena that I observed depended on the nesting of screens being (theoretically) infinite — that is, on the apparent corridor being endless, not truncated. This was the case because the most unpredictable of the visual phenomena always seemed to happen right in the vicinity of that central point where the infinite regress converges down to a magical dot.

My explorations did not teach me that *any* shape whatsoever can arise as a result of video feedback, but they did show me that I had entered a far richer universe of possibilities than I had expected. Today, this visual richness reminds me of the amazing visual universe discovered around 1980 by mathematician Benoit Mandelbrot when he studied the properties of the simple iteration defined by $z \rightarrow z^2 + c$, where c is a fixed complex number and z is a variable complex number whose initial value is 0. This is a mathematical feedback loop where one value of z goes in and a new value comes out, ready to be fed back in again, just as in audio or video feedback. The key question is this: If you, playing the role of microphone and loudspeaker (or camera and TV), do this over and over again, will the z values you get grow unboundedly, sailing off into the wild blue (or wild yellow or wild red) yonder, or will they instead home in on a finite value?

The details need not concern us here; the basic point is that the answer to the question depends in a very subtle way on the value of the parameter c, and if you make a map by color-coding different values of c according to the rate of z's divergence, you get amazing pictures. (This is why I joked about the "wild yellow" and "wild red" yonders.) Both in video feedback and in this mathematical system, a very simple looping process gives rise to a family of truly unanticipated and incredibly intricate swirling patterns.

The Phenomenon of "Locking-in"

The mysterious and strangely robust phenomena that emerge out of looping processes such as video feedback will serve from here on out as one of the main metaphors in this book, as I broach the central questions of consciousness and self.

From my video voyages I have gained a sense of the immense richness of the phenomenon of video feedback. More specifically, I have learned that very often, wonderfully complex structures and patterns come to exist on the screen whose origins are, to human viewers, utterly opaque. I have been struck by the fact that it is the circularity — the loopiness — of the system that brings these patterns into existence and makes them persist. Once a pattern is *on* the screen, then all that is needed to justify its *staying* up there is George Mallory's classic quip about why he felt compelled to scale Mount Everest: "Because it's there!" When loops are involved, circular justifications are the name of the game.

To put it another way, feedback gives rise to a new kind of abstract phenomenon that can be called "locking-in". From just the barest hint (the very first image sent to the TV screen in the first tiny fraction of a second) comes, almost instantly (after perhaps twenty or thirty iterations), the full realization of all the implications of this hint — and this new higher-level structure, this emergent pattern on the screen, this epiphenomenon, is then "locked in", thanks to the loop. It will not go away because it is forever refreshing itself, feeding on itself, giving rebirth to itself. Otherwise put, the emergent output pattern is a self-stabilizing structure whose origins, despite the simplicity of the feedback loop itself, are nearly impenetrable because the loop is cycled through so many times.

Emergent New Realities of Video Feedback

Coming up with vivid and helpful nicknames for unexpected visual patterns had certainly not figured in my initial plans for my video voyage at Stanford, but this little game soon became necessary. At the outset, I had thought I was undertaking a project that would involve straightforward terms like "screen inside screen", "silver strip", "angle of tilt", "zooming in", and so forth — but soon I found myself forced, willy-nilly, to use completely unexpected descriptive terms for what I was observing. As you have seen, I started talking about "corridors" and "walls", "doorways" and "galaxies", "spirals" and "black holes", "hubs" and "spokes", "petals" and "pulsations", and so forth. In the second video voyage with Bill, many of these same terms were once again needed, and some new ones were called for, such as "starfish", "cheese", "fire", "foam", and others.

Such words are hardly the kind of language I had thought I would be dealing with when I first broached the idea of video feedback. Although the system to which I was applying these terms was mechanical and deterministic, the patterns that emerged as a consequence of the loop were unpredictable, and therefore it turned out that words were needed that no one could have predicted in advance.

Simple but evocative metaphors like "corridor", "galaxy", and others turned out to be *indispensable* in describing the abstract shapes and events I witnessed on the screen. The initial terms I had tacitly assumed I would use wound up getting mostly ignored, because they yielded little insight. Of course, in principle, everything could be explained in terms of them, in a rigorous and incomprehensibly verbose fashion (like explaining a gas's temperature and pressure by writing out Avogadro's number of equations) — but such a boringly reductionistic, nearly pixel-by-pixel explanation would entirely leave out the wonderful higher-level visual phenomena to which a human eye and mind intuitively resonate.

In short, there are surprising new structures that looping gives rise to that constitute a new level of reality that could *in principle* be deduced from the basic loop and its detailed properties, but that *in practice* have a different kind of "life of their own" and that demand — at least when it comes to extremely finite, simplicity-seeking, pattern-loving creatures like us — a new vocabulary and a new level of description that transcend the basic level out of which they emerge.

CHAPTER 6

Of Selves and Symbols

❧ ❧ ❧

Perceptual Looping as the Germ of "I"-ness

I FIND it curious that, other than proper nouns and adjectives, the only word in the English tongue that is always capitalized is the first-person pronoun (nominative case) with which this sentence most flamboyantly sets sail. The convention is striking and strange, hinting that the word must designate something very important. Indeed, to some people — perhaps to most, perhaps even to us all — the ineffable sense of being an "I" or a "first person", the intuitive sense of "being there" or simply "existing", the powerful sense of "having experience" and of "having raw sensations" (what some philosophers refer to as "qualia"), seem to be the realest things in their lives, and an insistent inner voice bridles furiously at any proposal that all this might be an illusion, or merely the outcome of some kind of physical processes taking place among "third-person" (*i.e.,* inanimate) objects. My goal here is to combat this strident inner voice.

I begin with the simple fact that living beings, having been shaped by evolution, have survival as their most fundamental, automatic, and built-in goal. To enhance the chances of its survival, any living being must be able to react flexibly to events that take place in its environment. This means it must develop the ability to sense and to categorize, however rudimentarily, the goings-on in its immediate environment (most earthbound beings can pretty safely ignore comets crashing on Jupiter). Once the ability to sense external goings-on has developed, however, there ensues a curious side effect that will have vital and radical consequences. This is the fact that the living being's ability to sense certain aspects of its environment flips around and endows the being with the ability to sense certain aspects of *itself*.

That this flipping-around takes place is not in the least amazing or miraculous; rather, it is a quite unremarkable, indeed trivial, consequence of the being's ability to perceive. It is no more surprising than the fact that audio feedback can take place or that a TV camera can be pointed at a screen to which its image is being sent. Some people may find the notion of such self-perception peculiar, pointless, or even perverse, but such a prejudice does not make self-perception a complex or subtle idea, let alone paradoxical. After all, in the case of a being struggling to survive, the one thing that is *always* in its environment is… itself. So why, of all things, should the being be perceptually immune to the most salient item in its world? Now *that* would seem perverse!

Such a lacuna would be reminiscent of a language whose vocabulary kept growing and growing yet without ever developing words for such common concepts as are named by the English words "say", "speak", "word", "language", "understand", "ask", "question", "answer", "talk", "converse", "claim", "deny", "argue", "tell", "sentence", "story", "book", "read", "insist", "describe", "translate", "paraphrase", "repeat", "lie", "hedge", "noun", "verb", "tense", "letter", "syllable", "plural", "meaning", "grammar", "emphasize", "refer", "pronounce", "exaggerate", "bluster", and so forth. If such a peculiarly self-ignorant language existed, then as it grew in flexibility and sophistication, its speakers would engage ever more in talking, arguing, blustering, and so forth, but without ever referring to these activities, and such entities as questions, answers, and lies would become (even while remaining unnamed) ever more salient and numerous. Like the hobbled formalisms that came out of Bertrand Russell's timid theory of types, this language would have a gaping hole at its core — the lack of any mechanism for a word or utterance or book (etc.) to refer to itself. Analogously, for a living creature to have evolved rich capabilities of perception and categorization but to be constitutionally incapable of focusing any of that apparatus onto itself would be highly anomalous. Its selective neglect would be pathological, and would threaten its survival.

Varieties of Looping

To be sure, the most primitive living creatures have little or no self-perception. By analogy, we can think of a TV camera rigidly bolted on top of a TV set and facing away from the screen, like a flashlight tightly attached to a miner's helmet, always pointing away from the miner's eyes, never into them. In such a TV setup, obviously, a self-turned loop is out of the question. No matter how you turn it, the camera and the TV set turn in synchrony, preventing the closing of a loop.

We next imagine a more "evolved", hence more flexible, setup; this time the camera, rather than being bolted onto its TV set, is attached to it by a "short leash". Here, depending on the length and flexibility of the cord, it may be possible for the camera to twist around sufficiently to capture at least part of the TV screen in its viewfinder, giving rise to a truncated corridor. The biological counterpart to feedback of this level of sophistication may be the way our pet animals or even young children are slightly self-aware.

The next stage, obviously, is where the "leash" is sufficiently long and flexible that the video camera can point straight at the center of the screen. This will allow an endless corridor, which is far richer than a truncated one. Even so, the possibility of closing the self-watching loop does not pin down the system's richness, because there still are many options open. Can the camera tilt or not, and if so, by how much? Can it zoom in or out? Is its image in color, or just in black and white? Can brightness and contrast be tweaked? What degree of resolution does the image have? What percentage of time is spent in self-observation as opposed to observation of the environment? Is there some way for the video camera itself to appear on the screen? And on and on. There are still many parameters to play with, so the potential loop has many open dimensions of sophistication.

Reception versus Perception

Despite the richness afforded by all these options, a self-watching television system will always lack one crucial aspect: the capacity of *perception*, as opposed to mere *reception*, or image-receiving. Perception takes as its starting point some kind of input (possibly but not necessarily a two-dimensional image) composed of a vast number of tiny signals, but then it goes much further, eventually winding up in the selective triggering of a small subset of a large repertoire of dormant *symbols* — discrete structures that have representational quality. That is to say, a symbol inside a cranium, just like a simmball in the hypothetical careenium, should be thought of as a triggerable physical structure that constitutes the brain's way of implementing a particular *category* or *concept*.

I should offer a quick caveat concerning the word "symbol" in this new sense, since the word comes laden with many prior associations, some of which I definitely want to avoid. We often refer to written tokens (letters of the alphabet, numerals, musical notes on paper, Chinese characters, and so forth) as "symbols". That's not the meaning I have in mind here. We also sometimes talk of objects in a myth, dream, or allegory (for example, a key, a flame, a ring, a sword, an eagle, a cigar, a tunnel) as being "symbols" standing for something else. This is not the meaning I have in mind, either. The idea I want to convey by the phrase "a symbol in the brain" is that some specific structure inside your cranium (or your careenium, depending on what species you belong to) gets activated whenever you think of, say, the Eiffel Tower. That brain structure, whatever it might be, is what I would call your "Eiffel Tower symbol".

You also have an "Albert Einstein" symbol, an "Antarctica" symbol, and a "penguin" symbol, the latter being some kind of structure inside your brain that gets triggered when you perceive one or more penguins, or even when you are just thinking about penguins without perceiving any. There are also, in your brain, symbols for action concepts like "kick", "kiss", and "kill", for relational concepts like "before", "behind", and "between", and so on. In this book, then, symbols in a brain are the neurological entities that correspond to concepts, just as genes are the chemical entities that correspond to hereditary traits. Each symbol is dormant most of the time (after all, most of us seldom think about cotton candy, egg-drop soup, St. Thomas Aquinas, Fermat's last theorem, Jupiter's Great Red Spot, or dental-floss dispensers), but on the other hand, every symbol in our brain's repertoire is potentially triggerable at any time.

The passage leading from vast numbers of received *signals* to a handful of triggered *symbols* is a kind of funneling process in which initial input signals are manipulated or "massaged", the results of which selectively trigger further (*i.e.*, more "internal") signals, and so forth. This baton-passing by squads of signals traces out an ever-narrowing pathway in the brain, which winds up triggering a small set of symbols whose identities are of course a subtle function of the original input signals.

Thus, to give a hopefully amusing example, myriads of microscopic olfactory twitchings in the nostrils of a voyager walking down an airport concourse can lead, depending on the voyager's state of hunger and past experiences, to a joint triggering of the two symbols "sweet" and "smell", or a triggering of the symbols "gooey" and "fattening", or of the symbols "Cinnabon" and "nearby", or of the symbols "wafting", "advertising", "subliminal", "sly", and "gimmick" — or perhaps a triggering of all eleven of these symbols in the brain, in some sequence or other. Each of these examples of symbol-triggering constitutes an act of *perception,* as opposed to the mere *reception* of a gigantic number of microscopic signals arriving from some source, like a million raindrops landing on a roof.

In the interests of clarity, I have painted too simple a picture of the process of perception, for in reality, there is a great deal of two-way flow. Signals don't propagate solely from the outside inwards, towards symbols;

expectations from past experiences simultaneously give rise to signals propagating outwards from certain symbols. There takes place a kind of negotiation between inward-bound and outward-bound signals, and the result is the locking-in of a pathway connecting raw input to symbolic interpretation. This mixture of directions of flow in the brain makes perception a truly complex process. For the present purposes, though, it suffices to say that perception means that, thanks to a rapid two-way flurry of signal-passing, impinging torrents of input signals wind up triggering a small set of symbols, or in less biological words, activating a few concepts.

In summary, the missing ingredient in a video system, no matter how high its visual fidelity, is a *repertoire of symbols* that can be selectively triggered. Only if such a repertoire existed and were accessed could we say that the system was actually *perceiving* anything. Still, nothing prevents us from imagining augmenting a vanilla video system with additional circuitry of great sophistication that supports a cascade of signal-massaging processes that lead toward a repertoire of potentially triggerable symbols. Indeed, thinking about how one might tackle such an engineering challenge is a helpful way of simultaneously envisioning the process of perception in the brain of a living creature and its counterpart in the cognitive system of an artificial mind (or an alien creature, for that matter). However, quite obviously, not all realizations of such an architecture, whether earthbound, alien, or artificial, will possess equally rich repertoires of symbols to be potentially triggered by incoming stimuli. As I have done earlier in this book, I wish once again to consider sliding up the scale of sophistication.

Mosquito Symbols

Suppose we begin with a humble mosquito (not that I know any arrogant ones). What kind of representation of the outside world does such a primitive creature have? In other words, what kind of symbol repertoire is housed inside its brain, available for tapping into by perceptual processes? Does a mosquito even know or believe that there are objects "out there"? Suppose the answer is yes, though I am skeptical about that. Does it assign the objects it registers as such to any kind of categories? Do words like "know" or "believe" apply in any sense to a mosquito?

Let's be a little more concrete. Does a mosquito (of course without using words) divide the external world up into mental categories like "chair", "curtain", "wall", "ceiling", "person", "dog", "fur", "leg", "head", or "tail"? In other words, does a mosquito's brain incorporate symbols — discrete triggerable structures — for such relatively high abstractions? This seems pretty unlikely; after all, to do its mosquito thing, a mosquito could do perfectly well without such "intellectual" luxuries. Who cares if I'm biting a dog, a cat, a mouse, or a human — and who cares if it's an arm, an ear, a tail, or a leg — as long as I'm drawing blood?

What kinds of categories, then, does a mosquito need to have? Something like "potential source of food" (a "goodie", for short) and "potential place to land" (a "port", for short) seem about as rich as I expect its category system to be. It may also be dimly aware of something that we humans would call a "potential threat" — a certain kind of rapidly moving shadow or visual contrast (a "baddie", for short). But then again, "aware", even with the modifier "dimly", may be too strong a word. The key issue here is whether a mosquito has *symbols* for such categories, or could instead get away with a simpler type of machinery not involving any kind of perceptual cascade of signals that culminates in the triggering of symbols.

If this talk of bypassing symbols and managing with a very austere substitute for perception strikes you as a bit blurry, then consider the following questions. Is a toilet aware, no matter how slightly, of its water level? Is a thermostat aware, albeit extremely feebly, of the temperature it is controlling? Is a heat-seeking missile aware, be it ever so minimally, of the heat emanating from the airplane that it is pursuing? Is the Exploratorium's jovially jumping red spot aware, though only terribly rudimentarily, of the people from whom it is forever so gaily darting away? If you answered "no" to these questions, then imagine similarly unaware mechanisms inside a mosquito's head, enabling it to find blood and to avoid getting bashed, yet to accomplish these feats without using any *ideas.*

Mosquito Selves

Having considered mosquito symbols, we now inch closer to the core of our quest. What is the nature of a mosquito's interiority? That is, what is a mosquito's experience of "I"-ness? How rich a sense of self is a mosquito endowed with? These questions are very ambitious, so let's try something a little simpler. Does a mosquito have a visual image of how it looks? I hope you share my skepticism on this score. Does a mosquito know that it has wings or legs or a head? Where on earth would it get ideas like "wings" or "head"? Does it know that it has eyes or a proboscis? The mere suggestion seems ludicrous. How would it ever find such things out? Let's instead speculate a bit about our mosquito's knowledge of its own *internal* state. Does it have a sense of being hot or cold? Of being tuckered out or full of pep? Hungry or starved? Happy or sad? Hopeful or frightened? I'm sorry, but even these strike me as lying well beyond the pale, for an entity as humble as a mosquito.

Well then, how about more basic things like "in pain" and "not in pain"? I am still skeptical. On the other hand, I can easily imagine signals sent from a mosquito's eye to its brain and causing other signals to bounce back to its wings, amounting to a reflex verbalizable to us humans as "Flee threat on left" or simply "Outta here!" — but putting it into telegraphic English words in this fashion *still* makes the mosquito sound too aware, I am afraid. I would be quite happy to compare a mosquito's inner life to that of a flush toilet or a thermostat, but that's about as far as I personally would go. Mosquito behavior strikes me as perfectly comprehensible without recourse to anything that deserves the name "symbol". In other words, a mosquito's wordless and conceptless danger-fleeing behavior may be less like perception as we humans know it, and more like the wordless and conceptless hammer-fleeing behavior of your knee when the doctor's hammer hits it and you reflexively kick. Does a mosquito have more of an inner life than your knee does?

Does a mosquito have even the tiniest glimmering of itself as being a moving part in a vast world? Once again, I suspect not, because this would require all sorts of abstract symbols to reside in its microscopic brain — symbols for such notions as "big", "small", "part", "place", "move", and so on, not to mention "myself". Why would a mosquito need such luxuries? How would they help it find blood or a mate more efficiently? A hypothetical mosquito that had enough brainpower to house fancy symbols like these would be an egghead with a lot more neurons to carry around than its more streamlined and simpleminded cousins, and it would thereby be heavier and slower than they are, meaning that it wouldn't be able to compete with them in the quests for blood and reproduction, and so it would lose out in the evolutionary race.

My intuition, at any rate, is that a mosquito's very efficient teeny little nervous system lacks perceptual categories (and hence symbols) altogether. If I am not mistaken, this reduces the kind of self-perception loops that can exist in a mosquito's brain to an exceedingly low level, thus rendering a mosquito a very "small-souled man" indeed. I hope it doesn't sound too blasphemous or crazy if I suggest that a mosquito's "soul" might be roughly the same "size" as that of the little red spot of light that bounces around on the wall at the Exploratorium — let's say, one ten-billionth of one huneker (*i.e.*., roughly one trillionth of a human soul).

To be sure, I'm being flippant in making this numerical estimate, but I am quite serious in presenting my subjective guess about whether symbols are present or absent in a mosquito's brain. Nevertheless, it is just a subjective guess, and you may not agree with it, but disputes about such fine points are not germane here. The key point is much simpler and cruder: merely that there is *some* kind of creature to which essentially this level of complexity, and no greater level, would apply. If you disagree with my judgment, then I invite you to slide up or down the scale of various animal intellects until you feel you have hit the appropriate level.

One last reflection on all this. Some readers might protest, with what sounds like great sincerity, about all these questions about a mosquito's-eye view on the world: "How could we ever know? You and I can't get inside a mosquito's brain or mind — no one can. For all I know, mosquitoes are every bit as conscious as I am!" Well, I would respectfully suggest that such claims cannot be sincere, because here's ten bucks that say such readers would swat a mosquito perched on their arm without giving it a second thought. Now if they truly believe that mosquitoes are quite possibly every bit as sentient as themselves, then how come they're willing to snuff mosquito lives in an instant? Are these people not vile monsters if they are untroubled by executing living creatures who, they claim, may well enjoy just as much consciousness as do humans? I think you have to judge people's opinions not by their words, but by their deeds.

An Interlude on Robot Vehicles

Before moving on to consider higher animal species, I wish to insert a brief discussion of cars that drive themselves down smooth highways or across rocky deserts. Aboard any such vehicle are one or more television cameras (and laser rangefinders and other kinds of sensors) equipped with extra processors that allow the vehicle to make sense of its environment. No amount of simplistic analysis of just the colors or the raw shapes on the screen is going to provide good advice as to how to get around obstacles without toppling or getting stuck. Such a system, in order to drive itself successfully, has to have a nontrivial storehouse of prepackaged knowledge structures that can be selectively triggered by the scene outside. Thus, some knowledge of such abstractions as "road", "hill", "gulley", "mud", "rock", "tree", "sand", and many others will be needed if the vehicle is going to avoid getting stuck in mud, trapped in a gulley, or wedged between two boulders. The television cameras and the rangefinders (etc.) provide only the simplest *initial* stages of the vehicle's "perceptual process", and the triggering of various knowledge structures of the sort that were just mentioned corresponds to the far end, the *symbolic* end, of the process.

I slightly hesitated about putting quotation marks around the words "perceptual process" in the previous sentence, but I made an arbitrary choice, figuring that I was damned if I did and damned if I didn't. That is, if I left them off, I would be implicitly suggesting that what is going on in such a robot vehicle's processing of its visual input is truly like our own perception, whereas if I put them on, I would be implicitly suggesting that there is some kind of unbridgeable gulf between what "mere machines" can do and what living creatures do. Either choice is too black-and-white a position. Quotation marks, regrettably, don't come in shades of gray; if they did, I would have used some intermediate shade to suggest a more nuanced position.

The self-navigation of today's robot vehicles, though very impressive, is still a far cry from the level of mammalian perception, and yet I think it is fair to say that such a vehicle's "perception" (sorry for the unshaded quotation marks!) of its environment is just as sophisticated as a mosquito's "perception" (there — I hope to have somewhat evened the score), and perhaps considerably more so. (A beautiful treatment of this concept of robot vehicles and what different levels of "perception" will buy them is given by Valentino Braitenberg in his book *Vehicles*.)

Without going into more detail, let me simply say that it makes perfect sense to discuss living animals and self-guiding robots in the same part of this book, for today's technological achievements are bringing us ever closer to understanding what goes on in living systems that survive in complex environments. Such successes give the lie to the tired dogma endlessly repeated by John Searle that computers are forever doomed to mere "simulation" of the processes of life. If an automaton can drive itself a distance of two hundred miles across a tremendously forbidding desert terrain, how can this feat be called merely a "simulation"? It is certainly as genuine an act of survival in a hostile environment as that of a mosquito flying about a room and avoiding being swatted.

Pondering Dogthink

Let us return to our climb up the purely biological ladder of perceptual sophistication, rising from viruses to bacteria to mosquitoes to frogs to dogs to people (I've skipped a few rungs in there, I know). As we move higher and higher, the repertoire of triggerable symbols of course becomes richer and richer — indeed, what else could "climbing up the ladder" mean? Simply judging from their behavior, no one could doubt that pet dogs develop a respectable repertoire of categories, including such examples as "my paw", "my tail", "my food", "my water", "my dish", "indoors", "outdoors", "dog door", "human door", "open", "closed", "hot", "cold", "nighttime", "daytime", "sidewalk", "road", "bush", "grass", "leash", "take a walk", "the park", "car", "car door", "my big owner", "my little owner", "the cat", "the friendly neighbor dog", "the mean neighbor dog", "UPS truck", "the vet", "ball", "eat", "lick", "drink", "play", "sit", "sofa", "climb onto", "bad behavior", "punishment", and on and on. Guide dogs often learn a hundred or more words and respond to highly variegated instances of these concepts in many different contexts, thus demonstrating something of the richness of their internal category systems (*i.e.*, their repertoires of triggerable symbols).

I used a set of English words and phrases in order to suggest the nature of a canine repertoire of categories, but of course I am not claiming that human words are involved when a dog reacts to a neighbor dog or to the UPS truck. But one word bears special mention, and that is the word "my", as in "my tail" or "my dish". I suspect most readers would agree that a pet dog realizes that a particular paw belongs to itself, as opposed to being merely a random physical object in the environment or a part of some other animal. Likewise, when a dog chases its tail, even though it is surely unaware of the loopy irony of the act, it must know that *that* tail is part of its *own* body. I am thus suggesting that a dog has some kind of rudimentary self-model, some kind of sense of itself. In addition to its symbols for "car", "ball", and "leash", and its symbols for other animals and human beings, it has some kind of internal cerebral structure that represents itself (*i.e.*, the dog itself, not the symbol itself!).

If you doubt dogs have this, then what about chimpanzees? What about two-year-old humans? In any case, the emergence of this kind of reflexive symbolic structure, at whatever level of sentience it first enters the picture, constitutes the central germ, the initial spark, of "I"-ness, the tiny core to which more complex senses of "I"-ness will then accrete over a lifetime, like the snowflake that grows around a tiny initial speck of dust.

Given that most grown dogs have a symbol for *dog*, does a dog know, in some sense or other, that it, too, belongs to the category *dog*? When it looks at a mirror and sees its master standing next to "some dog", does it realize that that dog is itself? These are interesting questions, but I will not attempt to answer them. I suspect that this kind of realization lies near the fringes of canine mental ability, but for my purposes in this essay, it doesn't really matter on which side dogs fall. After all, this book is not about dogs. The key point here is that there is *some* level of complexity at which a creature starts applying some of its categories to itself, starts building mental structures that represent itself, starts placing itself in some kind of "intellectual perspective" in relationship to the rest of the world. In this respect, I think dogs are hugely more advanced than mosquitoes, and I suspect you agree.

On the other hand, I suspect that you also agree with me that a dog's soul is considerably "smaller" than a human one — otherwise, why wouldn't we both be out vehemently demonstrating at our respective animal shelters against the daily putting to "sleep" of stray hounds and helpless puppies? Would you condone the execution of homeless people and abandoned babies? What makes you draw a distinction between dogs and humans? Could it be the relative sizes of their souls? How many hunekers would dogs have to have, on the average, for you to decide to organize a protest demonstration at an animal shelter?

Creatures at the sophistication level of dogs, thanks to the inevitable flipping-around of their perceptual apparatus and their modest but nontrivial repertoire of categories, cannot help developing an approximate sense of themselves as physical entities in a larger world. (Robot vehicles in desert-crossing contests don't spend their precious time looking at themselves — it would be as useless as spinning their wheels — so their sense of self is considerably less sophisticated than that of a dog.) Although a dog will never know a thing about its kidneys or its cerebral cortex, it will develop some notion of its paws, mouth, and tail, and perhaps of its tongue or its teeth. It may have seen itself in a mirror and perhaps realized that "that dog over there by my master" is in fact itself. Or it may have seen itself in a home video with its master, recognized the recording of its master's voice, and realized that the barking on the video was its own.

And yet all of this, though in many ways impressive, is still extremely limited in comparison to the sense of self and "I"-ness that continually grows over the course of a normal human being's lifetime. Why is this the case? What's missing in Fido, Rover, Spot, Blackie, and Old Dog Tray?

The Radically Different Conceptual Repertoire of Human Beings

A spectacular evolutionary gulf opened up at some point as human beings were gradually separating from other primates: their category systems became *arbitrarily extensible.* Into our mental lives there entered a dramatic quality of open-endedness, an essentially unlimited extensibility, as compared with a very palpable limitedness in other species.

Concepts in the brains of humans acquired the property that they could get rolled together with other concepts into larger packets, and any such larger packet could then become a new concept in its own right. In other words, concepts could *nest* inside each other hierarchically, and such nesting could go on to arbitrary degrees. This reminds me — and I do not think it is a pure coincidence — of the huge difference, in video feedback, between an infinite corridor and a truncated one.

For instance, the phenomenon of having offspring gave rise to concepts such as "mother", "father", and "child". These concepts gave rise to the nested concept of "parent" — nested because forming it depends upon having three prior concepts: "mother", "father", and the abstract idea of "either/or". (Do dogs have the concept "either/or"? Do mosquitoes?) Once the concept of "parent" existed, that opened the door to the concepts of "grandmother" ("mother of a parent") and "grandchild" ("child of a child"), and then of "great-grandmother" and "great-grandchild". All of these concepts came to us courtesy of nesting. With the addition of "sister" and "brother", then further notions having greater levels of nesting, such as "uncle", "aunt", and "cousin", could come into being. And then a yet more nested notion such as "family" could arise. ("Family" is more nested because it takes for granted and builds on all these prior concepts.)

In the collective human ideosphere, the buildup of concepts through such acts of composition started to snowball, and it turns out to know no limits. Our species would soon find itself leapfrogging upwards to concepts such as "love affair", "love triangle", "fidelity", "temptation", "revenge", "despair", "insanity", "nervous breakdown", "hallucination", "illusion", "reality", "fantasy", "abstraction", "dream", and of course, at the grand pinnacle of it all, "soap opera" (in which are also nested the concepts of "commercial break", "ring around the collar", and "Brand X").

Consider the mundane-seeming concept of "grocery store checkout stand", which I would be willing to bet is a member in good standing of your personal conceptual repertoire. It already sounds like a nested entity, being compounded from four words; thus it tells us straightforwardly that it symbolizes a stand for checking out in a store that deals in groceries. But looking at its visible lexical structure barely scratches the surface. In truth, this concept involves dozens and dozens of other concepts, among which are the following: "grocery cart", "line", "customers", "to wait", "candy rack", "candy bar", "tabloid newspaper", "movie stars", "trashy headlines", "sordid scandals", "weekly TV schedule", "soap opera", "teen-ager", "apron", "nametag", "cashier", "mindless greeting", "cash register", "keyboard", "prices", "numbers", "addition", "scanner", "bar code", "beep", "laser", "moving belt", "frozen food", "tin can", "vegetable bag", "weight", "scale", "discount coupon", "rubber separator bar", "to slide", "bagger", "plastic bag", "paper bag", "plastic money", "paper money", "to load", "to pay", "credit card", "debit card", "to swipe", "receipt", "ballpoint pen", "to sign", and on and on. The list starts to seem endless, and yet we are merely talking about the internal richness of one extremely ordinary human concept.

Not all of these component concepts need be activated when we think about a grocery store checkout stand, to be sure — there is a central nucleus of concepts all of which are reliably activated, while many of these more peripheral components may not be activated — but all of the foregoing, and considerably more, is what constitutes the full concept in our minds. Moreover, this concept, like every other one in our minds, is perfectly capable of being incorporated inside other concepts, such as "grocery store checkout stand romance" or "toy grocery store checkout stand". You can invent your own variations on the theme.

Episodic Memory

When we sit around a table and shoot the breeze with friends, we are inevitably reminded of episodes that happened to us some time back, often many years ago. The time our dog got lost in the neighborhood. The time our neighbor's kid got lost in the airport. The time we missed a plane by a hair. The time we made it onto the train but our friend missed it by a hair. The time it was sweltering hot in the train and we had to stand up in the corridor all the way for four hours. The time we got onto the wrong train and couldn't get off for an hour and a half. The time when nobody could speak a word of English except "Ma-ree-leen Mon-roe!", spoken with lurid grinning gestures tracing out an hourglass figure in the air. The time when we got utterly lost driving in rural Slovenia at midnight and were nearly out of gas and yet somehow managed to find our way to the Italian border using a handful of words of pidgin Slovenian. And on and on.

Episodes are concepts of a sort, but they take place over time and each one is presumably one-of-a-kind, a bit like a proper noun but lacking a name, and linked to a particular moment in time. Although each one is "unique", episodes also fall into their own categories, as the previous paragraph, with its winking "You know what I mean!" tone, suggests. (Missing a plane by a hair is not unique, and even if it has happened to you only once in your life, you most likely know of several members of this category, and can easily imagine an unlimited number of others.)

Episodic memory is our private storehouse of episodes that have happened to us and to our friends and to characters in novels we've read and movies we've seen and newspaper stories and TV news clips, and so on, and it forms a major component of the long-term memory that makes us so human. Obviously, memories of episodes can be triggered by external events that we witness or by other episodes that have been triggered, and equally obviously, nearly all memories of specific episodes are dormant almost all the time (otherwise we would go stark-raving mad).

Do dogs or cats have episodic memories? Do they remember specific events that happened years or months ago, or just yesterday, or even ten minutes ago? When I take our dog Ollie running, does he recall how he strained at the leash the day before, trying to get to say "hi" to that cute Dalmatian across the street (who also was tugging at her leash)? Does he remember how we took a different route from the usual one three days ago? When I take Ollie to the kennel to board over Thanksgiving vacation, he seems to remember the kennel as a *place,* but does he remember anything specific that *happened* there the last time (or any time) he was there? If a dog is frightened of a particular place, does it recall a specific trauma that took place there, or is there just a generalized sense of badness associated with that place?

I do not need answers to these questions here, fascinating though they are to me. I am not writing a scholarly treatise on animal awareness. All I want is that readers think about these questions and then agree with me that some of them merit a "yes" answer, some merit a "no", and for some we simply can't say one way or the other. My overall point, though, is that we humans, unlike other animals, have all these kinds of memories; indeed, we have them all in spades. We recall in great detail certain episodes from vacations we took fifteen or twenty years ago. We know exactly why we are frightened of certain places and people. We can replay in detail the time we ran into so-and-so totally out of the blue in Venice or Paris or London. The depth and complexity of human memory is staggeringly rich. Little wonder, then, that when a human being, possessed of such a rich armamentarium of concepts and memories with which to work, turns its attention to itself, as it inevitably must, it produces a self-model that is extraordinarily deep and tangled. That deep and tangled self-model is what "I"-ness is all about.

CHAPTER 7

The Epi Phenomenon

&? &? &?

As Real as it Gets

THANKS to the funneling-down processes of perception, which lead eventually — that is, in a matter of milliseconds — to the activation of certain discrete symbols in its brain, an animal (and let's not forget robot vehicles!) can relate intimately and reliably to its physical environment. A mature human animal not only does a fine job of not slipping on banana peels and not banging into thorn-bristling rosebushes, it also reacts in a flash to strong odors, strange accents, cute babies, loud crashes, titillating headlines, terrific skiers, garish clothes, and on and on. It even occasionally hits curve balls coming at it at 80 miles an hour. Because an animal's internal mirroring of the world must be highly reliable (the symbol *elephant* should not get triggered by the whine of a mosquito, nor should the symbol *mosquito* get triggered if an elephant ambles into view), its mirroring of the world via its private cache of symbols becomes an unquestioned pillar of stability. The things and patterns it perceives are what define its reality — but not all perceived things and patterns are *equally* real to it.

Of course, in nonverbal animals, a question such as "Which things that I perceive are the most real of all to me?" is never raised, explicitly or implicitly. But in human lives, questions about what is and what is not real inevitably bubble up sooner or later, sometimes getting uttered consciously and carefully, other times remaining unexpressed and inchoate, just quietly simmering in the background. As children and teen-agers, we see directly, or we see on television, or we read about, or we are told about many things that supposedly exist, things that vie intensely with each other for our attention and for acceptance by our reality evaluators — for instance, God,

Godzilla, Godiva, Godot, Gödel, gods, goddesses, ghosts, ghouls, goblins, gremlins, golems, golliwogs, griffins, gryphons, gluons, and grinches. It takes a child a few years to sort out the reality of some of these; indeed, it takes many people a full lifetime to do so (and occasionally a bit longer).

By "sorting out the reality of X", I mean coming to a stable conclusion about how much you believe in X and whether you would feel comfortable relying on the notion of X in explaining things to yourself and others. If you are willing to use griffins in your explanations and don't flinch at other people's doing so in theirs, then it would seem that griffins are a seriously real concept to you. If you had already pretty much sorted out for yourself the reality of griffins and then heard there was going to be a TV special on griffins, you wouldn't feel a need to catch the show in order to help you decide whether or not griffins exist. Perhaps you believe strongly in griffins, perhaps you think of them as a childish fantasy or a joke — but your mind is made up one way or the other. Or perhaps you haven't yet sorted out the reality of griffins; if it were to come up in a dinner-party conversation, you would feel unsure, confused, ignorant, skeptical, or on the fence.

Another way of thinking about "how real X is to you" is how much you would trust a newspaper article that took for granted the existence of X (for example, a living dinosaur, a sighting of Hitler, insects discovered on Mars, a perpetual-motion machine, UFO abductions, God's omniscience, out-of-body experiences, alternate universes, superstrings, quarks, Bigfoot, Big Brother, the Big Bang, Atlantis, the gold in Fort Knox, the South Pole, cold fusion, Einstein's tongue, Holden Caulfield's brain, Bill Gates' checkbook, or the proverbial twenty-mile "wall" for marathon runners). If you stop reading an article the moment you see X's existence being taken for granted, then it would seem that you consider X's "reality" highly dubious.

Pick any of the concepts mentioned above. Almost surely, there are plenty of people who believe fervently in it, others who believe in it just a little, others not at all (whether out of ignorance, cynicism, poor education, or excellent education). Some of these concepts, we are repeatedly told by authorities, are not real, and yet we hear about them over and over again in television shows, books, and newspapers, and so we are left with a curious blurry sense as to whether they do exist, or could exist, or might exist. Others, we are told by authorities, are absolutely real, but somehow we never see them. Others we are told *were* real but are real no longer, and that places them in a kind of limbo as far as reality is concerned. Yet others we are told are real but are utterly beyond our capacity to imagine. Others are said to be real, but only metaphorically or only approximately so — and so on. Sorting all this out is not in the least easy.

Concrete Walls and Abstract Ceilings

To be more concrete about all this, how real is the marathoners' twenty-mile wall, mentioned above? If you're a marathoner, you almost surely have a well-worked-out set of thoughts about it. Perhaps you have experienced it personally, or know people who have. Or perhaps you think the notion is greatly exaggerated. I've never hit the wall myself, but then my longest run ever was only fifteen miles. What I know is that "they say" that most runners, if they haven't trained properly, will bang up against a brutal wall at around twenty miles, in which their body, having used up all of its glycogen, starts burning fat instead (I've heard it described as "your body eating its own muscles"). It comes out of the blue and is extremely painful ("like an elephant falling out of a tree onto my shoulders", said marathoner Dick Beardsley), and many runners simply cannot go any further at that point, and drop out. But is this a universal phenomenon? Is it the same for all people? Do some marathoners never experience it at all? And even if it is scientifically explicable, is it as real and as palpable a phenomenon as a concrete wall into which one bangs?

When I entered math graduate school at Berkeley in 1966, I had the self-image of being quite a math whiz. After all, as an undergraduate math major at Stanford, not only had I coasted through most of my courses without too much work, but I had done lots of original research, and on graduating I was awarded the citation "with Distinction" by the Math Department. I was expecting to become a mathematician and to do great things. Well, at Berkeley two courses were required of all first-year students — abstract algebra and topology — and so I took them. To my shock, both were very hard for me — like nothing I'd ever encountered before. I got good grades in them but only by memorizing and then regurgitating ideas on the finals. For the entire year, my head kept on hurting from a severe lack of imagery such as I had never before experienced. It was like climbing a very high peak and getting piercing headaches as the air grows ever thinner. Abstraction piled on abstraction and the further I plowed, the slower my pace, and the less I grasped. Finally, after a year and a half, I recognized the situation's hopelessness, and with a flood of bitter tears and a crushing loss of self-confidence, I jettisoned my dream of myself as a mathematician and bailed out of the field forever. This hated, rigid "abstraction ceiling" against which I had metaphorically banged my head without any advance warning was a searingly painful, life-changing trauma. And so… how concrete, how genuine, how real a thing was this abstract "abstraction ceiling"? As real as a marathoner's wall? As real as a wooden joist against which my skull could audibly crash? What is really real?

Although nobody planned it that way, most of us wind up emerging from adolescence with a deeply nuanced sense of what is real, with shades of gray all over the place. (However, I have known, and probably you have too, reader, a few adults for whom every issue that strikes me as subtle seems to them to be totally black-and-white — no messy shades of gray at all to deal with. That must make life easy!) Actually, to suggest that for most of us life is filled with "shades of gray" is far too simple, because that phrase conjures up the image of a straightforward one-dimensional continuum with many degrees of grayness running between white and black, while in fact the story is much more multidimensional than that.

All of this is disturbing, because the word "real", like so many words, seems to imply a sharp, clear-cut dichotomy. Surely it ought to be the case that some things simply *are* real while other things simply are *not* real. Surely there should be nothing that is *partly* real — that wouldn't make sense! And yet, though we try very hard to force the world to match this ideal black-and-white dichotomy, things unfortunately get terribly blurry.

The Many-faceted Intellectual Grounding of Reality

That marble over there in that little cardboard box on my desk is certainly real because I *see* the cardboard box sitting there and because I can go over and open it and can *squeeze* the marble, hefting it and feeling its solidity. I hope that makes sense to you.

The upper edge of that 75-foot-tall Shell sign near the freeway exit is real, I am convinced, because every road sign is a solid object and every solid object has a top; also because I can see the sign's bottom edge and its sides and so, by analogy, I can imagine seeing its top; also because, even though I'll certainly never touch it, I could at least theoretically climb up to it or be lowered down onto it from a helicopter. Then again, the sign could topple in an earthquake and I could rush over to it and touch what had once been its upper edge, and so forth.

Antarctica, too, is real because, although I've never been there and almost surely will never go there, I've seen hundreds of photos of it, I've seen photos of the whole earth from space including all of Antarctica, and also I once met someone who told me he went there, and on and on.

Why do I believe what certain people tell me more than I believe what others tell me? Why do I believe in (some) photos as evidence of reality? Why do I trust certain photos in certain books? Why do I trust certain newspapers, and why only up to a certain point? Why do I not trust all newspapers equally? Why do I not trust all book publishers equally? Why do I not trust all authors equally?

Through many types of abstraction and analogy-making and inductive reasoning, and through many long and tortuous chains of citations of all sorts of authorities (which constitute an indispensable pillar supporting every adult's belief system, despite the insistence of high-school teachers who year after year teach that "arguments by authority" are spurious and are convinced that *they* ought to be believed because they are, after all, authority figures), we build up an intricate, interlocked set of beliefs as to *what exists* "out there" — and then, once again, that set of beliefs folds back, inevitably and seamlessly, to apply to our own selves.

Just as we believe in other peoples' kidneys and brains (thanks almost entirely to arguments from analogy and authority), so we come to believe in our own kidneys and brains. Just as we believe in everyone else's mortality (again, thanks primarily to arguments from analogy and authority), so we come, eventually, to believe in our own mortality, as well as in the reality of the obituary notices about us that will appear in local papers even though we know we will never be able to flip those pages and read those notices.

What makes for our sense of utter sureness about such abstract things? It comes firstly from the reliability of our internal symbols to directly mirror the concrete environment (*e.g.*, we purchase a cup of coffee and instantly, somewhere inside our cranium, God only knows where, there springs into existence a physical record reflecting this coffee, tracking where it is on the table or in our hand, constantly updating its color, bitterness level, warmth, and how much there is left of it). It comes secondly from the reliability of our thinking mechanisms to tell us about more abstract entities that we cannot directly perceive (*e.g.*, the role of Napoleon in French history, the impact of Wagner on late-romantic French composers, or the unsolvability by radicals, such as Évariste Galois, of the quintic equation). All of this more abstract stuff is rooted in the constant reinforcement, moment by moment, of the symbols that are haphazardly triggered out of dormancy by events in the world that we perceive first-hand. These immediate mental events constitute the bedrock underlying our broader sense of reality.

Inevitably, what seems realest to us is what gets activated most often. Our hangnails are incredibly real to us (by coincidence, I found myself idly picking at a hangnail while I was reworking this paragraph), whereas to most of us, the English village of Nether Wallop and the high Himalayan country of Bhutan, not to mention the slowly swirling spiral galaxy in Andromeda, are considerably less real, even though our intellectual selves might wish to insist that since the latter are much bigger and longer-lasting than our hangnails, they ought therefore to be far realer to us than our hangnails are. We can say this to ourselves till we're blue in the face, but few of us act as if we really believed it. A slight slippage of subterranean stone that obliterates 20,000 people in some far-off land, the ceaseless plundering of virgin jungles in the Amazon basin, a swarm of helpless stars being swallowed up one after another by a ravenous black hole, even an ongoing collision between two huge galaxies each of which contains a hundred billion stars — such colossal events are so abstract to someone like me that they can't even touch the sense of urgency and importance, and thus the *reality*, of some measly little hangnail on my left hand's pinky.

We are all egocentric, and what is realest to each of us, in the end, is *ourself*. The realest things of all are *my knee, my nose, my anger, my hunger, my toothache, my sideache, my sadness, my joy, my love for math, my abstraction ceiling*, and so forth. What all these things have in common, what binds them together, is the concept of "my", which comes out of the concept of "I" or "me", and therefore, although it is less concrete than a nose or even a toothache, this "I" thing is what ultimately seems to each of us to constitute the most solid rock of undeniability of all. Could it possibly be an illusion? Or if not a total illusion, could it possibly be less real and less solid than we think it is? Could an "I" be more like an elusive, receding, shimmering rainbow than like a tangible, heftable, transportable pot of gold?

No Luck, No Soap, No Dice

One day, many years ago, I wanted to pull out all the envelopes from a small cardboard box lying on the floor of my study and stick them as a group into one of my desk drawers. Accordingly, I picked up the box, reached into it, clasped my right hand around the pack of envelopes inside it (about a hundred in number), and squeezed tightly down on them in order to pull them all out of the box as a unit. Nothing at all surprising in any of this. But all of a sudden I felt, between my thumb and fingers, something very surprising. Oddly enough, there was a *marble* sitting (or floating?) right in the middle of that flimsy little cardboard box!

Like most Americans of my generation, I had held marbles hundreds of times, and I knew without any doubt what I was feeling. Like you, dear reader, I was an "old marble hand". But how had a marble somehow found its way into this box that I usually kept on my desk? At the time I didn't have any kids, so that couldn't be the explanation. And anyway, how could it be hovering in the very *middle* of the box, rather than sitting at the bottom? Why wasn't gravity working?

I peered in between the envelopes, looking for a small, smooth, colored glass sphere. No luck. Then I fumbled about with my fingers between the envelopes, feeling for it. Again no soap. But then, as soon as I grasped the whole set of envelopes as before, there it was again, as solid as ever! Where was this little devil of a marble hiding?

I looked more carefully, and of course took the envelopes out and tried to shake it out from between them, but still no dice. And finally, on checking, I found that each envelope on its own was empty. So what in tarnation was going on?

An Out-of-the-Blue Ode to My Old Friend Epi

To you, my astute reader (and surely an old envelope hand, to boot), it is probably already obvious, but believe me, I was baffled for a minute or two. Eventually it dawned on me that there wasn't any marble in there at all, but that there was something that *felt* for all the world exactly like a marble to this old marble hand. It was an *epiphenomenon* caused by the fact that, for each envelope, at the vertex of the "V" made by its flap, there is a triple layer of paper as well as a thin layer of glue. An unintended consequence of this innocent design decision is that when you squeeze down on a hundred such envelopes all precisely aligned with each other, you can't compress that little zone as much as the other zones — it resists compression. The hardness that you feel at your fingertips has an uncanny resemblance to a more familiar (dare I say "a more real"?) hardness.

An epiphenomenon, as you probably recall from earlier chapters, is a collective and unitary-seeming outcome of many small, often invisible or unperceived, quite possibly utterly unsuspected, events. In other words, an epiphenomenon could be said to be a large-scale *illusion* created by the collusion of many small and indisputably non-illusory events.

Well, I was so charmed and captivated by this epiphenomenal illusion of the marble in the box that I nicknamed the box of envelopes "Epi", and I have kept it ever since — three decades or more, now. (Unfortunately, the box is falling apart after such a long time.) And sometimes, when I take a trip somewhere to give a lecture on the concepts of self and "I", I'll carry Epi along with me and I'll let members of the audience reach in and feel it for themselves, so that the concept of an epiphenomenon — in this case, the Epi phenomenon — becomes very real and vivid for them.

Recently I headed off to give such a lecture in Tucson, Arizona, and I took Epi along with me. One of the audience members, Jeannel King, was so taken with my Epi saga that she wrote a poem about it, translating it with poetic license into her own life, and a few days later she sent it to me. I in turn was so taken with her poem that I asked her for permission to reprint it here, and she generously said she'd be pleased if I did so. So without further ado, here is Jeannel King's delightful poem inspired by Epi.

Ode to a Box of Envelopes
(For all who have lost their marbles…)

by Jeannel King

A box of env'lopes on the floor —
I want to shift them to my drawer.
I squeeze inside — there's something there!
I look inside — there's naught but air.

I squeeze again and marble find.
Is this a marble of my mind?
Determined now, and one by one,
out come the env'lopes — still no plum!

For closer views of each, I must
brave paper cuts and motes of dust.
In tips? Or env'lope forty-six?
My marble, whole, does not exist.

Then coarse-grained Mother whispers, "Nell,
you keep this up, you'll go to hell!"
To which Dad counters, "Mind yer mopes!
Let Nell seek God in envelopes!"

So envelopes lie all around
as I sit, vexed, upon the ground.
My marble's lost, but in my core
could there, perhaps, be something more?

For more than parts this whole has grown:
No single part doth stand alone.
In parts, the marble simply mocks.
Intact, I think, I'll keep this box.

No Sphere, No Radius, No Mass

Perhaps the most bizarre aspect of my epiphenomenal marble was how sure I was that this "object" in the box was *spherical* and how confidently I would have provided an estimate of its *diameter* (about half an inch, like most marbles), as well as described *how hard* it was (as compared with, say, an egg yolk or a ball of clay). Many aspects of this nonexistent object were clear and familiar tactile phenomena. In a word, I had been sucked in by a tactile illusion. There was no marble anywhere in there — there was just a statistical epiphenomenon.

And yet, it's undeniable that the phrase "it felt just like a marble" gets across my experience far more clearly to my readers than if I had written, "I experienced the collective effect of the precise alignment of a hundred triple layers of paper and a hundred layers of glue." It is only because I called it a "marble" that you have a clear impression of how it felt to me. If I hadn't used the word "marble", would you have been able to predict that a thick pack of envelopes would give rise, in its middle, to something (some *thing?*) that felt perfectly *spherical,* felt like it had a *size,* felt extremely *solid* — in short, that this collective effect would feel like a very simple, very familiar physical object? I strongly doubt it. And thus there is something to be gained by not rejecting the term "marble", even if there is no *real* marble in the box. There is something that feels remarkably *like* a marble, and that fact is crucial to my portraying and to your grasping of the situation, just as the concepts of "corridor", "galaxy", and "black hole" were crucial in allowing me to perceive and describe the phenomena on the screen of the self-watching television — even if, strictly speaking, no corridor, no galaxy, and no black hole were there to be seen.

Where the Buck Seems to Stop

I have recounted the story of the half-real, half-unreal marble inside the box of envelopes to suggest a metaphor for the type of reality that applies to our undeniable feeling that something "solid" or "real" resides at the core of ourselves, a powerful feeling that makes the pronoun "I" indispensable and central to our existence. The thesis of this book is that in a non-embryonic, non-infantile human brain, there is a special type of abstract structure or pattern that plays the same role as does that precise alignment of layers of paper and glue — an abstract pattern that gives rise to what *feels* like a self. I intend to talk a great deal about the nature of that abstract pattern, but before I do so, I have to say what I mean by the term "a self", or perhaps more specifically, why we seem to need a notion of that sort.

Each living being, no matter how simple, has a set of innate goals embedded in it, thanks to the feedback loops that evolved over time and that characterize its species. These feedback loops are the familiar, almost clichéd activities of life, such as seeking certain types of food, seeking a certain temperature range, seeking a mate, and so forth. Some creatures additionally develop their own individual goals, such as playing certain pieces of music or visiting certain museums or owning certain types of cars. Whatever a creature's goals are, we are used to saying that it *pursues* those goals, and — at least if it is sufficiently complicated or sophisticated — we often add that it does so because it *wants* certain things.

"Why did you ride your bike to that building?" "I wanted to practice the piano." "And why did you want to practice the piano?" "Because I want to learn that piece by Bach." "And why do you want to learn that piece?" "I don't know, I just do — it's beautiful." "But what is it about this particular piece that is so beautiful?" "I can't say, exactly — it just hits me in some special way."

This creature ascribes its behavior to things it refers to as its *desires* or its *wants,* but it can't say exactly why it has those desires. At a certain point there is no further possibility of analysis or articulation; those desires simply are there, and to the creature, they seem to be the root causes for its decisions, actions, motions. And always, inside the sentences that express why it does what it does, there is the pronoun "I" (or its cousins "me", "my", etc.). It seems that the buck stops there — with the so-called "I".

The Prime Mover, Redux

Late one autumn afternoon, the red, orange, and yellow leaves are so alluring, and the fall weather so mild, compared to the just-finished muggy summer, that I decide to take a good long run. I go into my bedroom, search around for my running shorts and shoes and T-shirt, change my clothes in eagerness, and soon enough, my body finds itself out on the pavement, with my feet pounding the ground and my heart beginning to thump away. Before I know it, I've taken a hundred steps, and moments later it's been three hundred. Then it's been a thousand, then three thousand, and I'm still charging on, breathing hard, sweating, and thinking to myself, "Why do I always tell myself that I *like* running? I *hate* it!" And yet my body doesn't stop for a split second, and no matter how tired my muscles are, my *self* just says to them, like a sadistic drill sergeant sneering at a bunch of new recruits, "Don't be quitters!" — and lo and behold, my poor, huffing, heaving, protesting body unquestioningly obeys my self, even charging up steep hills against its will. In shorts, my rebelling physical body is being quite mercilessly pushed around by my intangible I's equally intangible determination to take this autumn run.

So who is pushing whom around here? Where are the particles of physics in this picture of what makes us do things? They are invisible, and even if you remember that they exist, they seem to be just secondary players. It is this "I", a coherent collection of desires and beliefs, that sets everything in motion. It is this "I" that is the prime mover, the mysterious entity that lies behind, and that launches, all the creature's behaviors. If I *want* something to happen, I just *will* it to happen, and unless it is out of my control, it generally *does* happen. The body's molecules, whether in the fingers, the arm, the legs, the throat, the tongue, or wherever, obediently follow the supreme bidding of the Grand "I" on high.

Thus it is that I push various pedals down and sure enough, my one-ton automobile obediently goes right where *I want* it to go. The ethereal "I" has pushed this huge physical object around. I twiddle my chopsticks and sure enough, the string beans obediently jump on board and I receive the sensory joy that I covet and the nourishment I need. I push certain keys on my Macintosh's keyboard and sure enough, sentences obediently emerge on its screen, and they pretty much express the thoughts that the ethereal "I" hoped to express. And in all of this, where are the particles? Nowhere to be seen. All there seems to be is this "I" making it all happen.

Well then, if this "I" thing is causing everything that a creature does, if this "I" thing is responsible for the creature's decisions and plans and actions and movements, then surely this "I" thing must at least *exist*. How could it be so all-powerful and yet not exist?

God's Eye versus the Careenium's Eye

I'd like to return, at this point, to the image of the careenium. At the heart of my discussion of the tiny zipping simms and the far larger, more sluggish simmballs in the careenium was the fact that this system can be seen on two very distant levels, yielding widely discrepant interpretations.

From the higher-level "thinkodynamics" viewpoint, there is symbolic activity in which simmballs interact with each other, taking advantage of the "heat energy" provided by the churning soup of invisible simms. From this viewpoint, what causes any simmballic event we see is a set of other simmballic events, even if the details of the causation are often tricky or too blurry to pin down precisely. (We are very familiar with this type of blurriness of causality in daily life — for instance, if I just barely miss a free throw in basketball, we know that it was my fault and that I did *something* a bit wrong, but we don't know exactly what it was. If I throw a die and it comes up '6', we aren't in the least surprised, but we still don't know *why* it came up '6' — nor do we give the question the least thought.)

Contrariwise, from the lower-level "statistical mentalics" viewpoint, there are just simms and simms alone, interacting through the fundamental dynamics of careening, bashing simms — and from this viewpoint, there is never the least vagueness or doubt about causality, because everything is governed by sharp, precise, hard-edged mathematical laws. (If we could zoom in arbitrarily closely on my arms and hands and fingers and also on the basketball and the backboard and the rim, or on the die and the table, and watch everything in slow motion of any desired slowness, we could discover exactly what gave rise to the missed free throw or the '6'. This might require a descent all the way down to the level of atoms, but that's all right — eventually, the reason would emerge into the clear.)

If one understands the careenium well, it would seem that both points of view are valid, although the latter one, leaving out no details, might seem to be the more *fundamental* one (we could call it the "God's eye" point of view), while the former, being a highly compressed simplification in which vast amounts of information are thrown away, might seem to be the more *useful* one for us mortals, as it is so much more efficient (even though some things then seem to happen "for no reason" — that's the tradeoff).

I Am Not God

But not all observers of the careenium enjoy the luxury of being able to flip back and forth between these two wildly discrepant viewpoints. Not all thinking creatures understand the careenium nearly as clearly or as fully as I described it in Chapter 3. The God's-eye point of view is simply *not available* to all observers; indeed, the very fact that such a point of view might exist is utterly unsuspected by some careenium observers. I am in particular thinking of one very special and privileged careenium observer, and that is *the careenium itself.*

When the careenium grapples with its own nature, particularly when it is "growing up", just beginning to know itself, long before it has become a scientist that studies mathematics and physics (and perhaps, eventually, the noble discipline of careeniology), all it is aware of is its *simmballic* activity, not its simm-level churnings. After all, as you and I both know (but it does not know), the careenium's perceptions of all things are fantastically coarse-grained simplifications (small sets of simmballs that have been collectively triggered by a vast storm of impinging signals) — and its *self*-perceptions are no exception.

The innocent young careenium has no inkling that behind the scenes, way down on some hidden micro-scale, churning, seething, simm-level activities are taking place inside it. Not once has it ever suspected the existence, even in principle, of any alternative viewpoint concerning its nature and its behavior. Indeed, this young careenium reminds me of myself as an adolescent, just before I read the books on the human brain by Pfeiffer and by Penfield and Roberts, books that so troubled me and yet that so fired my imagination. This idealistic young careenium is much like the naïve teen-aged Doug, just at the cusp, just before he began to glimpse the extraordinary eerieness of what goes on in total darkness, day and night, inside each and every human cranium.

And so, built as irrefutably as a granite marble into the careenium's pre-scientific understanding of itself is the sense of being *a creature driven entirely by thoughts and ideas*; its self-image is infinitely far from that of being a vast mechanistic entity whose destiny is entirely determined by billions of invisibly careening, mutually bashing micro-objects. Instead, the naïve careenium serenely asserts of itself, "I am driven solely by *myself*, not by any mere physical objects anywhere."

What kind of thing, then, is this "I" that the careenium posits as driving its choices and its actions, and that human beings likewise posit as driving theirs? No one will be surprised at this point to hear me assert that it is a peculiar type of abstract, locked-in loop located inside the careenium or the cranium — in fact, a *strange* loop. And thus, in order to lay out clearly my claim about what constitutes "I"-ness, I need to spell out what I mean by "strange loop". And since we're just finishing Chapter 7 of *I Am a Strange Loop*, it's about time!

CHAPTER 8

Embarking on a Strange-Loop Safari

&ce; &ce; &ce;

Flap Loop, Lap Loop

I'VE already described, in Chapter 4, how enchanted I was as a child by the brazen act of closing a cardboard box by folding down its four flaps in a cyclic order. It always gave me a *frisson* of delight (and even today it still does a little bit) to perform that final *verboten* fold, and thus to feel I was flirting dangerously with paradoxicality. Needless to say, however, actual paradox was never achieved.

A close cousin to this "flap loop" is the "lap loop", shown on the facing page. There I am with a big grin (I'll call myself "A"), front and center in Anterselva di Mezzo, sitting on the lap of a young woman ("B"), also grinning, with B sitting on C's lap, C on D's lap, and so forth, until one complete lap has been made, with person K sitting on *my* lap. One lap with lots of laps but no collapse. If you've never played this game, I suggest you try it. One feels rather baffled about what on earth is holding the loop up.

Like the flap loop, this lap loop grazes paradoxicality, since each of its eleven lap-leaps is an upwards leap, but obviously, since a lap loop can be realized in the physical world, it cannot constitute a genuine paradox. Even so, when I played the "A" role in this lap loop, I felt as if I was sitting, albeit indirectly, on my own lap! This was a most strange sensation.

Seeking Strange Loopiness in Escher

And yet when I say "strange loop", I have something else in mind — a less concrete, more elusive notion. What I mean by "strange loop" is — here goes a first stab, anyway — not a physical circuit but an abstract loop in which, in the series of stages that constitute the cycling-around, there is a shift from one level of abstraction (or structure) to another, which feels like an upwards movement in a hierarchy, and yet somehow the successive "upward" shifts turn out to give rise to a closed cycle. That is, despite one's sense of departing ever further from one's origin, one winds up, to one's shock, exactly where one had started out. In short, a strange loop is a paradoxical level-crossing feedback loop.

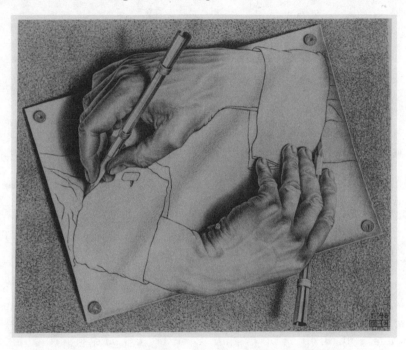

One of the most canonical (and, I am sorry to say, now hackneyed) examples is M. C. Escher's lithograph *Drawing Hands* (above), in which (depending on where one starts) one sees a right hand drawing a picture of a left hand (nothing paradoxical yet), and yet the left hand turns out to be drawing the right hand (all at once, it's a deep paradox).

Here, the abstract shift in levels would be the upward leap from *drawn* to *drawer* (or equally, from *image* to *artist*), the latter level being intuitively "above" the former, in more senses than one. To begin with, a drawer is always a sentient, mobile being, whereas a drawn is a frozen, immobile image (possibly of an inanimate object, possibly of an animate entity, but in any case motionless). Secondly, whereas a drawer is three-dimensional, a drawn is two-dimensional. And thirdly, a drawer *chooses* what to draw, whereas a drawn has no say in the matter. In at least these three senses, then, the leap from a drawn to a drawer always has an "upwards" feel to it.

As we've just stated, there is by definition a sharp, clear, upwards jump from any drawn image to its drawer — and yet in *Drawing Hands*, this rule of upwardness has been sharply and cleanly violated, for each of the hands is hierarchically "above" the other! How is that possible? Well, the answer is obvious: the whole thing is merely a drawn image, merely a fantasy. But because it looks so real, because it sucks us so effectively into its paradoxical world, it fools us, at least briefly, into believing in its reality. And moreover, we delight in being taken in by the hoax, hence the picture's popularity.

The abstract structure in *Drawing Hands* would constitute a perfect example of a genuine strange loop, were it not for that one little defect — what we think we see is *not* genuine; it's fake! To be sure, it's so impeccably drawn that we *seem* to be perceiving a full-fledged, true-blue, card-carrying paradox — but this conviction arises in us only thanks to our having suspended our disbelief and mentally slipped into Escher's seductive world. We fall, at least momentarily, for an illusion.

Seeking Strange Loops in Feedback

Is there, then, any *genuine* strange loop — a paradoxical structure that nonetheless undeniably belongs to the world we live in — or are so-called strange loops always just illusions that merely graze paradox, always just fantasies that merely flirt with paradox, always just bewitching bubbles that inevitably pop when approached too closely?

Well, what about our old friend video feedback as a candidate for strange loopiness? Unfortunately, although this modern phenomenon is very loopy and flirts with infinity, it has nothing in the least paradoxical to it — no more than does its simpler and older cousin, audio feedback. To be sure, if one points the TV camera straight at the screen (or brings the microphone right up to the loudspeaker) one gets that strange feeling of playing with fire, not only by violating a natural-seeming hierarchy but also by seeming to create a true infinite regress — but when one thinks about it, one realizes that there was no ironclad hierarchy to begin with, and the suggested infinity is never reached; then the bubble just pops. So although feedback loops of this sort are indisputably loops, and although they feel a bit strange, they are not members of the category "strange loop".

Seeking Strange Loops in the Russellian Gloom

Fortunately, there do exist strange loops that are not illusions. I say "fortunately" because the thesis of this book is that we ourselves — not our bodies, but our *selves* — are strange loops, and so if all strange loops were illusions, then we would all be illusions, and that would be a great shame. So it's fortunate that some strange loops exist in the real world.

On the other hand, it is not a piece of cake to exhibit one for all to see. Strange loops are shy creatures, and they tend to avoid the light of day. The quintessential example of this phenomenon, in fact, was only discovered in 1930 by Kurt Gödel, and he found it lurking in, of all places, the gloomy, austere, supposedly paradox-proof castle of Bertrand Russell's theory of types.

What was a 24-year-old Austrian logician doing, snooping about in this harsh and forbidding British citadel? He was fascinated by paradoxes, and although he knew they had supposedly been driven out by Russell and Whitehead, he nonetheless intuited that there was something in the extremely rich and flexible nature of numbers that had a propensity to let paradox bloom even in the most arid-seeming of deserts or the most sterilized of granite palaces. Gödel's suspicions had been aroused by a recent plethora of paradoxes dealing with numbers in curious new ways, and he felt convinced that there was something profound about these tricky games, even though some people claimed to have ways of defusing them.

Mr Berry of the Bodleian

One of these quirky paradoxes had been concocted by an Oxford librarian named G. G. Berry in 1904, two years before Gödel was born. Berry was intrigued by the subtle possibilities for describing numbers in words. He noticed that if you look hard enough, you can find a quite concise description of just about any integer you name. For instance, the integer 12 takes only one syllable to name, the integer 153 is pinpointable in but four syllables ("twelve squared plus nine" or "nine seventeens"), the integer 1,000,011 is nameable in just six syllables ("one million eleven"), and so forth. In how few syllables can you describe the number 1737?

In general, one would think that the larger the number, the longer any description of it would have to be, but it all depends on how easily the number is expressible in terms of "landmark" integers — those rare integers that have exceptionally short names or descriptions, such as ten to the trillion, with its extremely economic five-syllable description. Most large numbers, of course, are neither landmarks nor anywhere near one. Indeed, by far most numbers are "obscure", admitting only of very long and complex descriptions because, well, they are just "hard to describe", like remote outposts located way out in the boondocks, and which one can reach only by taking a long series of tiny side roads that get ever narrower and bumpier as one draws nearer to the destination.

Consider 777,777, whose standard English name, "seven hundred seventy-seven thousand seven hundred seventy-seven", is pretty long — 20 syllables, in fact. But this number has a somewhat shorter description: "777 times 1001" ("seven hundred seventy-seven times one thousand and one"), which is just 15 syllables long. Quite a savings! And we can compress it yet further: "three to the sixth plus forty-eight, all times ten cubed plus one" or even the stark "the number whose numeral is six sevens in a row". Either way, we're down to 14 syllables.

Working hard, we could come up with scads of English-language expressions that designate the value 777,777, and some of them, when spoken aloud, might contain very few syllables. How about "7007 times 111" ("seven thousand seven times one hundred eleven"), for instance? Down to 13 syllables! And how about "nine cubed plus forty-eight, all times ten cubed plus one"? Down to 12! And what about "thrice thirty-nine times seven thousand seven"? Down to just 11! Just how far down can we squeeze our descriptions of this number? It's not in the least obvious, because 777,777 just *might* have some subtle arithmetical property allowing it to be very concisely expressed. Such a description might even involve references to landmark numbers much larger than 777,777 itself!

Librarian Berry, after ruminating about the subtle nature of the search for ever shorter descriptions, came up with a devilish characterization of a very special number, which I'll dub *b* in his honor: *b* is the smallest integer whose English-language descriptions always use at least thirty syllables. In other words, *b* has no precise characterization shorter than thirty syllables. Since it always takes such a large number of syllables to describe it, we know that *b* must be a huge integer. Just how big, roughly, would *b* be?

Any large number that you run into in a newspaper or magazine or an astronomy or physics text is almost surely describable in a dozen syllables, twenty at most. For instance, Avogadro's number (6×10^{23}) can be specified in a very compact fashion ("six times ten to the twenty-third" — a mere eight syllables). You will not have an easy time finding a number so huge that no matter how you describe it, at least thirty syllables are involved.

In any case, Berry's *b* is, by definition, the very *first* integer that can't be boiled down to below thirty syllables of our fair tongue. It is, I repeat, using italics for emphasis, *the smallest integer whose English-language descriptions always use at least thirty syllables.* But wait a moment! How many syllables does my italicized phrase contain? Count them — 24. We somehow described *b* in fewer syllables than its definition allows. In fact, the italicized phrase does not merely describe *b* "somehow"; it is *b*'s very *definition*! So the concept of *b* is nastily self-undermining. Something very strange is going on.

I Can't Tell You How Indescribably Nondescript It Was!

It happens that there are a few common words and phrases in English that have a similarly flavored self-undermining quality. Take the adjective "nondescript", for instance. If I say, "Their house is so nondescript", you will certainly get some sort of visual image from my phrase — even though (or rather, precisely because) my adjective suggests that *no* description quite fits it. It's even weirder to say "The truck's tires were indescribably huge" or "I just can't tell you how much I appreciate your kindness." The self-undermining quality is oddly crucial to the communication.

There is also a kind of "junior version" of Berry's paradox that was invented a few decades after it, and which runs like this. Some integers are interesting. 0 is interesting because 0 times any number gives 0. 1 is interesting because 1 times any number leaves that number unchanged. 2 is interesting because it is the smallest even number, and 3 is interesting because it is the number of sides of the simplest two-dimensional polygon (a triangle). 4 is interesting because it is the first composite number. 5 is interesting because (among many other things) it is the number of regular polyhedra in three dimensions. 6 is interesting because it is three factorial ($3 \times 2 \times 1$) and also the triangular number of three ($3+2+1$). I could go on with this enumeration, but you get the point. The question is, when do we run into the first *un*interesting number? Perhaps it is 62? Or 1729? Well, no matter what it is, that is certainly an interesting property for a number to have! So 62 (or whatever your candidate number might have been) turns out to be interesting, after all — interesting because it is uninteresting. And thus the idea of "the smallest uninteresting integer" backfires on itself in a manner clearly echoing the backfiring of Berry's definition of b.

This is the kind of twisting-back of language that turned Bertrand Russell's sensitive stomach, as we well know, and yet, to his credit, it was none other than B. Russell who first publicized G. G. Berry's paradoxical number b. In his article about it in 1906, Gödel's birthyear (four syllables!), Russell did his best to deflect the paradox's sting by claiming that it was an illusion arising from a naïve misuse of the word "describable" in the context of mathematics. That notion, claimed Russell, had to be parceled out into an infinite hierarchy of different *types* of describability — descriptions at level 0, which could refer only to notions of pure arithmetic; descriptions at level 1, which could use arithmetic but could also refer to descriptions at level 0; descriptions at level 2, which could refer to arithmetic and also to descriptions at levels 0 and 1; and so forth and so on. And so the idea of "describability" without restriction to some specific hierarchical level was a chimera, declared Russell, believing he had discovered a profound new truth. And with this brand-new type of theory (the brand-new theory of types), he claimed to have immunized the precious, delicate world of rigorous reasoning against the ugly, stomach-turning plague of Berry-Berry.

Blurriness Buries Berry

While I agree with Russell that something fishy is going on in Berry's paradox, I don't agree about what it is. The weakness that I focus in on is the fact that English is a hopelessly imprecise medium for expressing mathematical statements; its words and phrases are far too vague. What may seem precise at first turns out to be fraught with ambiguity. For example, the expression "nine cubed plus forty-eight, all times ten cubed plus one", which earlier I exhibited as a description of 777,777, is in fact ambiguous — it might, for instance, be interpreted as meaning 777 times 1000, with 1 tacked on at the end, resulting in 777,001.

But that little ambiguity is just the tip of the iceberg. The truth of the matter is that it is far from clear what kinds of English expressions count as descriptions of a number. Consider the following phrases, which purport to be descriptions of specific integers:

- the number of distinct languages ever spoken on earth
- the number of heavenly bodies in the Solar System
- the number of distinct four-by-four magic squares
- the number of interesting integers less than 100

What is wrong with them? Well, they all involve ill-defined notions.

What, for instance, is meant by a "language"? Is sign language a language? Is it "spoken"? Is there a sharp cutoff between languages and dialects? How many "distinct languages" lay along the pathway from Latin to Italian? How many "distinct languages" were spoken en route from Neanderthal days to Latin? Is Church Latin a language? And Pig Latin? Even if we had videotapes of every last human utterance on earth for the past million years, the idea of objectively assigning each one to some particular "official" language, then cleanly teasing apart all the "truly distinct" languages, and finally counting them would *still* be a nonsensical pipe dream. It's already meaningless enough to talk about counting all the "items" in a garbage can, let alone all the languages of all time!

Moving on, what counts as a "heavenly body"? Do we count artificial satellites? And random pieces of flotsam and jetsam left floating out there by astronauts? Do we count every single asteroid? Every single distinct stone floating in Saturn's rings? What about specks of dust? What about isolated atoms floating in the void? Where does the Solar System stop? And so on, *ad infinitum.*

You might object, "But those aren't mathematical notions! Berry's idea was to use *mathematical* definitions of integers." All right, but then show me a sharp cutoff line between mathematics and the rest of the world. Berry's definition uses the vague notion of "syllable counting", for instance. How many syllables are there in "finally" or "family" or "rhythm" or "lyre" or "hour" or "owl"? But no matter; suppose we had established a rigorous and objective way of counting syllables. Still, what would count as a "mathematical concept"? Is the discipline of mathematics really that sharply defined? For instance, what is the precise definition of the notion "magic square"? Different authors define this notion differently. Do we have to take a poll of the mathematical community? And if so, who then counts as a member of that blurry community?

What about the blurry notion of "interesting numbers"? Could we give some kind of mathematical precision to that? As you saw above, reasons for calling a number "interesting" could involve geometry and other areas of mathematics — but once again, where do the borders of mathematics lie? Is game theory part of mathematics? What about medical statistics? What about the theory of twisting tendrils of plants? And on and on.

To sum up, the notion of an "English-language definition of an integer" turns out to be a hopeless morass, and so Berry's twisty notion of *b*, no less than Escher's twisty notion of two mutually drawing hands, is an ingenious figment of the imagination rather than a genuine strange loop. There goes a promising candidate for strange loopiness down the drain!

Although in this brief digression I've made it sound as if the idea Berry had in 1904 was naïve, I must point out that some six decades later, the young mathematician Greg Chaitin, inspired by Berry's idea, dreamt up a more precise cousin using computer programs instead of English-language descriptions, and this clever shift turned out to yield a radically new proof of, and perspective on, Gödel's 1931 theorem. From there, Chaitin and others went on to develop an important new branch of mathematics known as "algorithmic information theory". To go into that would carry us far afield, but I hope to have conveyed a sense for the richness of Berry's insight, for this was the breeding ground for Gödel's revolutionary ideas.

A Peanut-butter and Barberry Sandwich

Bertrand Russell's attempt to bar Berry's paradoxical construction by instituting a formalism that banned all self-referring linguistic expressions and self-containing sets was not only too hasty but quite off base. How so?

Well, a friend of mine recently told me of a Russell-like ban instituted by a friend of hers, a young and idealistic mother. This woman, in a well-meaning gesture, had strictly banned all toy guns from her household. The ban worked for a while, until one day when she fixed her kindergarten-age son a peanut-butter sandwich. The lad quickly chewed it into the shape of a pistol, then lifted it up, pointed it at her, and shouted, "Bang bang! You're dead, Mommy!" This ironic anecdote illustrates an important lesson: the medium that remains after all your rigid bans may well turn out to be flexible enough to fashion precisely the items you've banned.

And indeed, Russell's dismissal of Berry had little effect, for more and more paradoxes were being invented (or unearthed) in those intellectually tumultuous days at the turn of the twentieth century. It was in the air that truly peculiar things could happen when modern cousins of various ancient paradoxes cropped up inside the rigorously logical world of numbers, a world in which nothing of the sort had ever been seen before, a pristine paradise in which no one had dreamt paradox might arise.

Although these new kinds of paradoxes felt like attacks on the beautiful, sacred world of reasoning and numbers (or rather, *because* of that worrisome fact), quite a few mathematicians boldly embarked upon a quest to come up with ever deeper and more troubling paradoxes — that is, a quest for ever more powerful threats to the foundations of their own discipline! This sounds like a perverse thing to do, but they believed that in the long run such a quest would be very healthy for mathematics, because it would reveal key weak spots, showing where shaky foundations had to be shored up so as to become unassailable. In short, plunging deeply into the new wave of paradoxes seemed to be a useful if not indispensable activity for anyone working on the foundations of mathematics, for the new paradoxes were opening up profound questions concerning the nature of reasoning — and thus concerning the elusive nature of thinking — and thus concerning the mysterious nature of the human mind itself.

An Autobiographical Snippet

As I mentioned in Chapter 4, at age fourteen I ran across Ernest Nagel and James R. Newman's little gem, *Gödel's Proof*, and through it I fell under the spell of the paradox-skirting ideas on which Gödel's work was centered. One of the stranger loops connected with that period in my life was that I became acquainted with the Nagel family at just that time. Their home was in Manhattan, but they were spending the academic year 1959–60 "out west" at Stanford, and since Ernest Nagel and my father were old friends, I soon got to know the whole family. Shortly after the Nagels'

Stanford year was over, I savored the twisty pleasure of reading aloud the whole of *Gödel's Proof* to my friend Sandy, their older son, in the verdant yard of their summer home in the gentle hills near Brattleboro, Vermont. Sandy was just my age, and we were both exploring mathematics with a kind of wild intoxication that only teen-agers know.

Part of what pulled me so intensely was the weird loopiness at the core of Gödel's work. But the other half of my intense curiosity was my sense that what was *really* being explored by Gödel, as well as by many people he had inspired, was the mystery of the human mind and the mechanisms of human thinking. So many questions seemed to have been suddenly and sharply brought into light by Gödel's 1931 article — questions such as…

What happens inside mathematicians' heads when they do their most creative work? Is it always just rule-bound symbol manipulation, deriving theorems from a fixed set of axioms? What is the nature of human thought in general? Is what goes on inside our heads just a deterministic physical process? If so, are we all, no matter how idiosyncratic and sparkly, nothing but slaves to rigid laws governing the invisible particles out of which our brains are built? Could creativity ever emerge from a set of rigid rules governing minuscule objects or patterns of numbers? Could a rule-governed machine be as creative as a human? Could a programmed machine come up with ideas not programmed into it in advance? Could a machine make its own decisions? Have its own opinions? Be confused? Know it was confused? Be unsure whether it was confused? Believe it had free will? Believe it didn't have free will? Be conscious? Doubt it was conscious? Have a self, a soul, an "I"? Believe that its fervent belief in its "I" was only an illusion, but an *unavoidable* illusion?

Idealistic Dreams about Metamathematics

Back in those heady days of my youth, every time I entered a university bookstore (and that was as often as possible), I would instantly swoop down on the mathematics section and scour all the books that had to do with symbolic logic and the nature of symbols and meaning. Thus I bought book after book on these topics, such as Rudolf Carnap's famous but forbidding *The Logical Syntax of Language* and Richard Martin's *Truth and Denotation*, not to mention countless texts of symbolic logic. Whereas I very carefully read a few such textbooks, the tomes by Carnap and Martin just sat there on my shelf, taunting and teasing me, always seeming just out of reach. They were dense, almost impenetrably so — but I kept on thinking that if only someday, some grand day, I could finally read them and fully fathom them, then at last I would have penetrated to the core of the mysteries of thinking, meaning, creativity, and consciousness. As I look back now, that sounds ridiculously naïve (firstly to imagine this to be an attainable goal, and secondly to believe that those books in particular contained all the secrets), but at the time I was a true believer!

When I was sixteen, I had the unusual experience of teaching symbolic logic at Stanford Elementary School (my own elementary-school alma mater), using a brand-new text by the philosopher and educator Patrick Suppes, who happened to live down the street from our family, and whose classic *Introduction to Logic* had been one of my most reliable guides. Suppes was conducting an experiment to see if patterns of strict logical inference could be inculcated in children in the same way as arithmetic could, and the school's principal, who knew me well from my years there, one day bumped into me in the school's rotunda, and asked me if I would like to teach the sixth-grade class (which included my sister Laura) symbolic logic three times a week for a whole year. I fairly jumped at the chance, and all year long I thoroughly enjoyed it, even if a few of the kids now and then gave me a hard time (rubber bands in the eye, etc.). I taught my class the use of many rules of inference, including the mellifluous *modus tollendo tollens* and the impressive-sounding "hypothetical syllogism", and all the while I was honing my skills not only as a novice logician but also as a teacher.

What drove all this — my core inner passion — was a burning desire to see unveiled the secrets of human mentation, to come to understand how it could be that trillions of silent, synchronized scintillations taking place every second inside a human skull enable a person to think, to perceive, to remember, to imagine, to create, and to feel. At more or less the same time, I was reading books on the brain, studying several foreign languages, exploring exotic writing systems from various countries, inventing ways to get a computer to generate grammatically complicated and quasi-coherent sentences in English and in other languages, and taking a marvelously stimulating psychology course. All these diverse paths were focused on the dense nebula of questions about the relationship between mind and mechanism, between mentality and mechanicity.

Intricately woven together, then, in my adolescent mind were the study of pattern (mathematics) and the study of paradoxes (metamathematics). I was somehow convinced that all the mysterious secrets with which I was obsessed would become crystal-clear to me once I had deeply mastered these two intertwined disciplines. Although over the course of the next couple of decades I lost essentially all of my faith in the notion that these disciplines contained (even implicitly) the answers to all these questions, one thing I never lost was my intuitive hunch that around the core of the eternal riddle "What am I?", there swirled the ethereal vortex of Gödel's elaborately constructed loop.

It is for that reason that in this book, although I am being driven principally by questions about consciousness and self, I will have to devote some pages to the background needed for a (very rough) understanding of Gödel's ideas — and in particular, this means number theory and logic. There won't be heavy doses of either one, to be sure, but I do have to paint at least a coarse-grained picture of what these fields are basically about; otherwise, we won't have any way to proceed. So please fasten your seat belt, dear reader. We're going to be experiencing a bit of weather for the next two chapters.

Post Scriptum

After completing this chapter to my satisfaction, I recalled that I owned two books about "interesting numbers" — *The Penguin Dictionary of Curious and Interesting Numbers* by David Wells, an author on mathematics whom I greatly admire, and *Les nombres remarquables* by François Le Lionnais, one of the two founders of the famous French literary movement Oulipo. I dimly recalled that both of these books listed their "interesting numbers" in order of size, so I decided to check them out to see which was the lowest integer that each of them left out.

As I suspected, both authors made rather heroic efforts to include all the integers that exist, but inevitably, human knowledge being finite and human beings being mortal, each volume sooner or later started having gaps. Wells' first gap appeared at 43, while Le Lionnais held out a little bit longer, until 49. I personally was not too surprised by 43, but I found 49 surprising; after all, it's a square, which suggests at least a speck of interest. On the other hand, I admit that squareness gets a bit boring after you've already run into it several times, so I could partially understand why that property alone did not suffice to qualify 49 for inclusion in Le Lionnais's final list. Wells lists several intriguing properties of 49 (but not the fact that it's a square), and conversely, Le Lionnais points out some very surprising properties of 43.

So then I decided to find the lowest integer that *both* books considered to be utterly devoid of interest, and this turned out to be 62. For what it's worth, that will be my age when this book appears in print. Could it be that 62 is interesting, after all?

ɬ ɬ ɬ

CHAPTER 9

Pattern and Provability

❧ ❧ ❧

Principia Mathematica and its Theorems

IN THE early twentieth century, Bertrand Russell, spurred by the maxim "Find and study paradoxes; design and build great ramparts to keep them out!" (my words, not his), resolved that in *Principia Mathematica*, his new barricaded fortress of mathematical reasoning, no set could ever contain itself, and no sentence could ever turn around and talk about itself. These parallel bans were intended to save *Principia Mathematica* from the trap that more naïve theories had fallen into. But something truly strange turned up when Kurt Gödel looked closely at what I will call *PM* — that is, the formal system used in *Principia Mathematica* for reasoning about sets (and about numbers, but they came later, as they were defined in terms of sets).

Let me be a little more explicit about this distinction between *Principia Mathematica* and *PM*. The former is a set of three hefty tomes, whereas *PM* is a set of precise symbol-manipulation rules laid out and explored in depth in those tomes, using a rather arcane notation (see the end of this chapter). The distinction is analogous to that between Isaac Newton's massive tome entitled *Principia* and the laws of mechanics that he set forth therein.

Although it took many chapters of theorems and derivations before the rather lowly fact that one plus one equals two (written in *PM* notation as "s0 + s0 = ss0", where the letter "s" stands for the concept "successor of") was rigorously demonstrated using the strict symbol-shunting rules of *PM*, Gödel nonetheless realized that *PM*, though terribly cumbersome, had enormous power to talk about whole numbers — in fact, to talk about *arbitrarily subtle* properties of whole numbers. (By the way, that little phrase "arbitrarily subtle properties" already gives the game away, though the hint is so veiled that almost no one is aware of how much the words imply. It took Gödel to fully see it.)

For instance, as soon as enough set-theoretical machinery had been introduced in *Principia Mathematica* to allow basic arithmetical notions like addition and multiplication to enter the picture, it became easy to define, within the *PM* formalism, more interesting notions such as "square" (*i.e.,* the square of a whole number), "nonsquare", "prime", and "composite".

There could thus be, at least in theory, a volume of *Principia Mathematica* devoted entirely to exploring the question of which integers are, and which are not, the *sum of two squares.* For instance, 41 is the sum of 16 and 25, and there are infinitely many other integers that can be made by summing two squares. Call them members of Class A. On the other hand, 43 is *not* the sum of any pair of squares, and likewise, there are infinitely many other integers that *cannot* be made by summing two squares. Call them members of Class B. (Which class is 109 in? What about 133?) Fully fathoming this elegant dichotomy of the set of all integers, though a most subtle task, had been accomplished by number theorists long before Gödel's birth.

Analogously, one could imagine another volume of *Principia Mathematica* devoted entirely to exploring the question of which integers are, and which are not, the *sum of two primes.* For instance, 24 is the sum of 5 and 19, whereas 23 is not the sum of any pair of primes. Once again, we can call these two classes of integers "Class C" and "Class D", respectively. Each class has infinitely many members. Fully fathoming this elegant dichotomy of the set of all integers represents a very deep and, as of today, still unsolved challenge for number theorists, though much progress has been made in the two-plus centuries since the problem was first posed.

Mixing Two Unlikely Ideas: Primes and Squares

Before we look into Gödel's unexpected twist-based insight into *PM,* I need to comment first on the profound joy in discovering patterns, and next on the profound joy in understanding what lies behind patterns. It is mathematicians' relentless search for *why* that in the end defines the nature of their discipline. One of my favorite facts in number theory will, I hope, allow me to illustrate this in a pleasing fashion.

Let us ask ourselves a simple enough question concerning prime numbers: Which primes are sums of two squares (41, for example), and which primes are not (43, for example)? In other words, let's go back to Classes A and B, both of which are infinite, and ask which prime numbers lie in each of them. Is it possible that nearly all prime numbers are in one of these classes, and just a few in the other? Or is it about fifty–fifty? Are there infinitely many primes in each class? Given an arbitrary prime number p, is there a quick and simple test to determine which class p belongs to (without trying out all possible additions of two squares smaller than p)? Is there any kind of predictable pattern concerning how primes are distributed in these two classes, or is it just a jumbly chaos?

To some readers, these may seem like peculiar or even unnatural questions to tackle, but mathematicians are constitutionally very curious people, and it happens that they are often deeply attracted by the idea of exploring interactions between concepts that do not, *a priori*, seem related at all (such as the primes and the squares). What often happens is that some kind of unexpected yet intimate connection turns up — some kind of crazy hidden regularity that feels magical, the discovery or the revelation of which may even send mystical *frissons* up and down one's spine. I, for one, shamelessly admit to being highly susceptible to such spine-tingling mixtures of awe, beauty, mystery, and surprise.

To get a feel for this kind of thing, let us take the list of all the primes up to 100 — 2, 3, 5, 7, 11, 13, 17, 19, 23, 29, 31, 37, 41, 43, 47, 53, 59, 61, 67, 71, 73, 79, 83, 89, 97 — a rather jumbly, chaotic list, by the way — and redisplay it, highlighting those primes that *are* sums of two squares (that is, Class A primes), and leaving untouched those that are *not* (Class B primes). Here is what we get:

2, 3, **5**, 7, 11, **13**, **17**, 19, 23, **29**, 31, **37**, **41**, 43, 47, **53**, 59, **61**, 67, 71, **73**, 79, 83, **89**, **97**,...

Do you see anything interesting going on here? Well, for one thing, isn't it already quite a surprise that it seems to be a fairly even competition? Why should that be the case? Why shouldn't either Class A or Class B be dominant? Will either the Class A primes or the Class B primes take over after a while, or will their roughly even balance continue forever? As we go out further and further towards infinity, will the balance tend closer and closer to being exactly fifty–fifty? If so, why would such an amazing, delicate balance hold? To me, there is something enormously alluring here, and so I encourage you to look at this display for a little while — a few minutes, say — and try to find any patterns in it, before going on.

Pattern-hunting

All right, reader, here we are, back together again, hopefully after a bit of pattern-searching on your part. Most likely you noticed that our act of highlighting seems, not by intention but by chance (or is it chance?), to have broken the list into *singletons* and *pairs*. A hidden connection revealed?

Let's look into this some more. The boldface pairs are 13–17, 37–41, and 89–97, while the non-boldface pairs are 7–11, 19–23, 43–47, 67–71, and 79–83. Suppose, then, that we replace each *pair* by the letter "P" and each *singleton* by the letter "S", retaining the highlighting that distinguishes Class A from Class B. We then get the following sequence of letters:

$$\mathbf{S}, \text{S}, \mathbf{S}, \text{P}, \mathbf{P}, \text{P}, \mathbf{S}, \text{S}, \mathbf{P}, \text{P}, \mathbf{S}, \text{S}, \mathbf{S}, \text{P}, \mathbf{S}, \text{P}, \mathbf{P},\dots$$

Is there some kind of pattern here, or is there none? What do you think? If we pull out just the Class-A letters, we get this: **SSPSPSSSP**; and if we pull out just the Class-B letters, we get this: SPPSPSPP. If there is any kind of periodicity or subtler type of rhythmicity here, it's certainly elusive. No simple predictable pattern jumps out either in boldface or in non-boldface, nor did any jump out when they were mixed together. We have picked up a hint of a quite even balance between the two classes, yet we lack any hints as to why that might be. This is provocative but frustrating.

People who Pursue Patterns with Perseverance

At this juncture, I feel compelled to point out a distinction not between two classes of numbers, but between two classes of people. There are those who will immediately be drawn to the idea of pattern-seeking, and there are those who will find it of no appeal, perhaps even distasteful. The former are, in essence, those who are mathematically inclined, and the latter are those who are not. Mathematicians are people who at their deepest core are drawn on — indeed, are easily seduced — by the urge to find patterns where initially there would seem to be none. The passionate quest after order in an apparent disorder is what lights their fires and fires their souls. I hope you are among this class of people, dear reader, but even if you are not, please do bear with me for a moment.

It may seem that we have already divined a pattern of sorts — namely, that we will forever encounter just singletons and pairs. Even if we can't quite say how the S's and P's will be interspersed, it appears at least that the imposition of the curious dichotomy "sums-of-two-squares *vs.* not-sums-of-two-squares" onto the sequence of the prime numbers breaks it up into singletons and pairs, which is already quite a fantastic discovery! Who would have guessed?

Unfortunately, I must now confess that I have misled you. If we simply throw the very next prime, which is 101, into our list, it sabotages the seeming order we've found. After all, the prime number 101, being the sum of the two squares 1 and 100, and thus belonging to Class A, has to be written in boldface, and so our alleged boldface *pair* 89–97 turns out to be a boldface *triplet* instead. And thus our hopeful notion of a sequence of just S's and P's goes down the drain.

What does a pattern-seeker do at this point — give up? Of course not! After a setback, a flexible pattern-seeker merely *regroups*. Indeed, taking our cue from the word just given, let us try regrouping our sequence of primes in a different fashion. Suppose we segregate the two classes, displaying them on separate lines. This will give us the following:

Yes square + square: 2, 5, 13, 17, 29, 37, 41, 53, 61, 73, 89, 97, 101,...

No square + square: 3, 7, 11, 19, 23, 31, 43, 47, 59, 67, 71, 79, 83,...

Do you see anything yet? If not, let me give you a hint. What if you simply take the differences between adjacent numbers in each line? Try it yourself — or else, if you're very lazy, then just read on.

In the upper line, you will get 3, 8, 4, 12, 8, 4, 12, 8, 12, 16, 8, 4, whereas in the lower line you will get 4, 4, 8, 4, 8, 12, 4, 12, 8, 4, 8, 4. There is something that surely should jump out at even the most indifferent reader at this point: not only is there a preponderance of just a few integers (4, 8, and 12), but moreover, all these integers are multiples of 4. This seems too much to be merely coincidental.

And the only larger number in these two lists — 16 — is also a multiple of 4. Will this new pattern — multiples of 4 exclusively — hold up forever? (Of course, there is that party-pooper of a '3' at the very outset, but we can chalk it up to the fact that 2 is the only even prime. No big deal.)

Where There's Pattern, There's Reason

The key thought in the preceding few lines is the article of faith that *this pattern cannot merely be a coincidence*. A mathematician who finds a pattern of this sort will instinctively ask, "Why? What is the *reason* behind this order?" Not only will all mathematicians wonder *what* the reason is, but even more importantly, they will all implicitly believe that whether or not anyone ever finds the reason, there *must be* a reason for it. Nothing happens "by accident" in the world of mathematics. The existence of a perfect pattern, a regularity that goes on forever, reveals — just as smoke reveals a fire — that something is going on behind the scenes. Mathematicians consider it a sacred goal to seek that thing, uncover it, and bring it out into the open.

This activity is called, as you well know, "finding a proof", or stated otherwise, turning a conjecture into a theorem. The late great eccentric

Hungarian mathematician Paul Erdös once made the droll remark that "a mathematician is a device for turning coffee into theorems", and although there is surely truth in his witticism, it would be more accurate to say that mathematicians are *devices for finding conjectures and turning them into theorems.*

What underlies the mathematical mindset is an unshakable belief that whenever some mathematical statement X is *true,* then X has a *proof,* and vice versa. Indeed, to the mathematical mind, "having a proof" is no more and no less than what "being true" means! Symmetrically, "being false" means "having no proof". One can find *hints* of a perfect, infinite pattern by doing numerical explorations, as we did above, but how can one know for sure that a suspected regularity will continue forever, without end? How can one know, for instance, that there are infinitely many prime numbers? How do we know there will not, at some point, be a last one — the Great Last Prime *P*?

If it existed, *P* would be a truly important and interesting number, but if you look at a long list of consecutive primes (the list above of primes up to 100 gives the flavor), you will see that although their rhythm is a bit "bumpy", with odd gaps here and there, the interprime gaps are always quite small compared to the size of the primes involved. Given this very clear trend, if the primes were to run out all of a sudden, it would almost feel like falling off the edge of the Earth without any warning. It would be a huge shock. Still, how do we *know* this won't happen? Or do we know it? Finding, with the help of a computer, that new primes keep on showing up way out into the billions and the trillions is great, but it won't guarantee in rock-solid fashion that they won't just stop all of a sudden somewhere out further. We have to rely on *reasoning* to get us there, because although finite amounts of evidence can be strongly suggestive, they just don't cut the mustard, because infinity is very different from any finite number.

Sailing the Ocean of Primes and Falling off the Edge

You probably have seen Euclid's proof of the infinitude of the primes somewhere, but if not, you have missed out on one of the most crucial pillars of human knowledge that ever have been found. It would be a gap in your experience of life as sad as never having tasted chocolate or never having heard a piece of music. I can't tolerate such crucial gaps in my readers' knowledge, so here goes nothing!

Let's suppose that *P*, the Great Last Prime in the Sky, does exist, and see what that supposition leads to. For *P* to exist means that there is a Finite, Closed Club of All Primes, of which *P* itself is the glorious, crowning, final member. Well then, let's boldly multiply all the primes in the Closed Club together to make a delightfully huge number called Q. This number Q is thus divisible by 2 and also by 3, 5, 7, 11, and so forth. By its definition, Q is divisible by every prime in the Club, which means by every prime in the universe! And now, for a joyous last touch, as in birthday parties, let's add one candle to grow on, to make $Q + 1$. So here's a colossal number that, we are assured, is *not* prime, since P (which is obviously dwarfed by Q) is the Great Last Prime, the biggest prime of all. All numbers beyond P are, by our initial supposition, composite. Therefore $Q + 1$, being way beyond P and hence composite, *has* to have some prime divisor. (Remember this, please.)

What could that unknown prime divisor be? It can't be 2, because 2 divides Q itself, which is just one step below $Q + 1$, and two even numbers are never located at a distance of 1 from each other. It also can't be 3, because 3 likewise divides Q itself, and numbers divisible by 3 are never next-door neighbors! In fact, whatever prime p that we select from the Club, we find that p can't divide $Q + 1$, because p divides its lower neighbor Q (and multiples of p are never next-door neighbors — they come along only once every p numbers). And so reasoning has shown us that *none* of the members of the Finite, Closed Club of Primes divides $Q + 1$.

But above, I observed (and I asked you to remember) that $Q + 1$, being composite, *has* to have a prime divisor. Sting! We have been caught in a trap, painted ourselves into a corner. We have concocted a crazy number — a number that on the one hand must be composite (*i.e.*, has some smaller prime divisor) and yet on the other hand has no smaller prime divisor. This contradiction came out of our assumption that there was a Finite, Closed Club of Primes, gloriously crowned by P, and so we have no choice but to go back and erase that whole amusing, suspect vision.

There cannot be a "Great Last Prime in the Sky"; there cannot be a "Finite, Closed Club of All Primes". These are fictions. The truth, as we have just demonstrated, is that the list of primes goes on without end. We will never, ever "fall off the Earth", no matter how far out we go. Of that we now are assured by flawless reasoning, in a way that no *finite* amount of computational sailing among seas of numbers could ever have assured us.

If, perchance, coming to understand *why* there is no last prime (as opposed to merely knowing *that* it is the case) was a new experience to you, I hope you savored it as much as a piece of chocolate or of music. And just like such experiences, following this proof is a source of pleasure that one can come back to and dip into many times, finding it refreshing each new time. Moreover, this proof is a rich source of other proofs — Variations on a Theme by Euclid (though we will not explore them here).

The Mathematician's Credo

We have just seen up close a lovely example of what I call the "Mathematician's Credo", which I will summarize as follows:

X is true *because* there is a proof of X;

X is true *and so* there is a proof of X.

Notice that this is a two-way street. The first half of the Credo asserts that *proofs are guarantors of truth,* and the second half asserts that *where there is a regularity, there is a reason.* Of course we ourselves may not uncover the hidden reason, but we firmly and unquestioningly believe that it exists and in principle could someday be found by someone.

To doubt either half of the Credo would be unthinkable to a mathematician. To doubt the first line would be to imagine that a proved statement could nonetheless be false, which would make a mockery of the notion of "proof", while to doubt the second line would be to imagine that within mathematics there could be perfect, exceptionless patterns that go on forever, yet that do so with no rhyme or reason. To mathematicians, this idea of flawless but reasonless structure makes no sense at all. In that regard, mathematicians are all cousins of Albert Einstein, who famously declared, "God does not play dice." What Einstein meant is that nothing in nature happens without a cause, and for mathematicians, that there is always one unifying, underlying cause is an unshakable article of faith.

No Such Thing as an Infinite Coincidence

We now return to Class A versus Class B primes, because we had not quite reached our revelation, had not yet experienced that mystical *frisson* I spoke of. To refresh your memory, we had noticed that each line was characterized by differences of the form $4n$ — that is, 4, 8, 12, and so forth. We didn't *prove* this fact, but we *observed* it often enough that we *conjectured* it.

The lower line in our display starts out with 3, so our conjecture would imply that all the other numbers in that line are gotten by adding various multiples of 4 to 3, and consequently, that every number in that line is of the form $4n + 3$. Likewise (if we ignore the initial misfit of 2), the first number in the upper line is 5, so if our conjecture is true, then every subsequent number in that line is of the form $4n + 1$.

Well, well — our conjecture has suggested a remarkably simple pattern to us: Primes of the form $4n + 1$ *can* be represented as sums of two squares, while primes of the form $4n + 3$ *cannot.* If this guess is correct, it establishes a beautiful, spectacular link between primes and squares (two classes of numbers that *a priori* would seem to have nothing to do with each other), one that catches us completely off guard. This is a glimpse of pure magic — the kind of magic that mathematicians live for.

And yet for a mathematician, this flash of joy is only the beginning of the story. It is like a murder mystery: we have found out someone is dead, but whodunnit? There always has to be an explanation. It may not be easy to find or easy to understand, but it has to exist.

Here, we know (or at least we strongly suspect) that there is a beautiful infinite pattern, but *for what reason?* The bedrock assumption is that there *is* a reason here — that our pattern, far from being an "infinite coincidence", comes from one single compelling, underlying reason; that behind all these infinitely many "independent" facts lies just one phenomenon.

As it happens, there is actually much more to the pattern we have glimpsed. Not only are primes of the form $4n + 3$ never the sum of two squares (proving this is easy), but also it turns out that every prime number of the form $4n + 1$ has *one and only one way* of being the sum of two squares. Take 101, for example. Not only does 101 equal $100 + 1$, but there is no other sum of two squares that yields 101. Finally, it turns out that in the limit, as one goes further and further out, the ratio of the number of Class A primes to the number of Class B primes grows ever closer to 1. This means that the delicate balance that we observed in the primes below 100 and conjectured would continue *ad infinitum* is rigorously provable.

Although I will not go further into this particular case study, I will state that many textbooks of number theory prove this theorem (it is far from trivial), thus supplementing a pattern with a proof. As I said earlier, X is true *because* X has a proof, and conversely, X is true *and so* X has a proof.

The Long Search for Proofs, and for their Nature

I mentioned above that the question "Which numbers are sums of two primes?", posed almost 300 years ago, has never been fully solved. Mathematicians are dogged searchers, however, and their search for a proof may go on for centuries, even millennia. They are not discouraged by eons of failure to find a proof of a mathematical pattern that, from numerical trends, seems likely to go on and on forever. Indeed, extensive empirical confirmation of a mathematical conjecture, which would satisfy most people, only makes mathematicians more ardent and more frustrated. They want a proof as good as Euclid's, not just lots of spot checks! And they are driven by their belief that a proof *has* to exist — in other words, that *if no proof existed*, then the pattern in question would have to be *false*.

This, then, constitutes the flip side of the Mathematician's Credo:

> X is false *because* there is no proof of X;
> X is false *and so* there is no proof of X.

In a word, just as provability and truth are the same thing for a mathematician, so are nonprovability and falsity. They are synonymous.

During the centuries following the Renaissance, mathematics branched out into many subdisciplines, and proofs of many sorts were found in all the different branches. Once in a while, however, results that were clearly absurd seemed to have been rigorously proven, yet no one could pinpoint where things had gone awry. As stranger and stranger results turned up, the uncertainty about the nature of proofs became increasingly disquieting, until finally, in the middle of the nineteenth century, a powerful movement arose whose goal was to specify just what reasoning really was, and to bond it forever with mathematics, fusing the two into one.

Many philosophers and mathematicians contributed to this noble goal, and around the turn of the twentieth century it appeared that the goal was coming into sight. Mathematical reasoning seemed to have been precisely characterized as the repeated use of certain basic rules of logic, dubbed *rules of inference*, such as *modus ponens*: If you have proven a result X and you have also proven X $\Rightarrow$ Y (where the arrow represents the concept of implication, so that the line means "If X is true, then Y is also true"), then you can toss Y into the bin of proven results. There were a few other fundamental rules of inference, but it was agreed that not very many were needed. About a decade into the twentieth century, Bertrand Russell and Alfred North Whitehead codified these rules in a uniform if rather prickly notation (see facing page), thus apparently allowing all the different branches of mathematics to be folded in with logic, making a seamless, perfect unity.

Thanks to Russell and Whitehead's grand work, *Principia Mathematica*, people no longer needed to fear falling into hidden crevasses of false reasoning. Theorems were now understood as simply being the bottom lines of sequences of symbol-manipulations whose top lines were either axioms or earlier theorems. Mathematical truth was all coming together so elegantly. And as this Holy Grail was emerging into clear view, a young boy was growing up in the town of Brünn, Austro-Hungary.

❧ ❧ ❧

∗97·31. $\vdash . (\overrightarrow{B\text{'}R}) \uparrow \text{Cnv}\text{'}\overleftarrow{R_*} \,\epsilon\, \epsilon_\Delta \text{'} \overleftarrow{R_*}\text{''}\overrightarrow{B\text{'}R} . D\text{'}\{(\overrightarrow{B\text{'}R}) \uparrow \text{Cnv}\text{'}\overleftarrow{R_*}\} = \overrightarrow{B\text{'}R}$

Dem.

$$\vdash . \ast97\cdot3 . \ast85\cdot13 \frac{\breve{R_*}}{Q} . \supset$$

$\vdash : S \,\epsilon\, (\breve{R_*})_\Delta\text{'}\overrightarrow{B\text{'}R} . \supset . S | \text{Cnv}\text{'}\overleftarrow{R_*} \,\epsilon\, \epsilon_\Delta \text{'} \overleftarrow{R_*}\text{''}\overrightarrow{B\text{'}R}$ \hfill (1)

$\vdash . (1) . \ast97\cdot301 . \qquad \supset \vdash . I \restriction \overrightarrow{B\text{'}R} | \text{Cnv}\text{'}\overleftarrow{R_*} \,\epsilon\, \epsilon_\Delta \text{'} \overleftarrow{R_*}\text{''}\overrightarrow{B\text{'}R} .$

$[\ast50\cdot61] \qquad\qquad \supset \vdash . (\overrightarrow{B\text{'}R}) \uparrow \text{Cnv}\text{'}\overleftarrow{R_*} \,\epsilon\, \epsilon_\Delta \text{'} \overleftarrow{R_*}\text{''}\overrightarrow{B\text{'}R}$ \hfill (2)

$\vdash . 35\cdot62 . 33\cdot431 . \supset \vdash . D\text{'}\{(\overrightarrow{B\text{'}R}) \uparrow \text{Cnv}\text{'}\overleftarrow{R_*}\} = \overrightarrow{B\text{'}R}$ \hfill (3)

$\vdash . (2) . (3) . \supset \vdash . \text{Prop}$

∗97·32. $\vdash . \overrightarrow{B\text{'}R} \,\epsilon\, D\text{''} \epsilon_\Delta \text{'} \overleftarrow{R_*}\text{''}\overrightarrow{B\text{'}R}$ \qquad [∗97·31]

∗97·33. $\vdash : R \,\epsilon\, 1 \rightarrow 1 . \alpha \subset s\text{'}\overrightarrow{R_*}\text{'}\beta . \beta \subset s\text{'}\overrightarrow{R_*}\text{''}\alpha . \supset . \overleftrightarrow{R_*}\text{''}\alpha = \overleftrightarrow{R_*}\text{''}\beta$

Dem.

$\vdash . \ast97\cdot15 . \text{Fact} . \supset \vdash :. \text{Hp} . \supset : y \,\epsilon\, \beta . x \,\epsilon\, \overleftrightarrow{R_*}\text{'}y . \supset . \overleftrightarrow{R_*}\text{'}y = \overleftrightarrow{R_*}\text{'}x . y \,\epsilon\, \beta .$

$[\ast37\cdot62] \qquad\qquad\qquad\qquad\qquad \supset . \overleftrightarrow{R_*}\text{'}x \,\epsilon\, \overleftrightarrow{R_*}\text{''}\beta$ \hfill (1)

$\vdash . (1) . \ast10\cdot11\cdot21\cdot23 . \ast40\cdot4 . \supset \vdash :. \text{Hp} . \supset : x \,\epsilon\, s\text{'}\overleftrightarrow{R_*}\text{''}\beta . \supset_x . \overleftrightarrow{R_*}\text{'}x \,\epsilon\, \overleftrightarrow{R_*}\text{''}\beta :$

$[\text{Hp.Syll}] \qquad\qquad\qquad\qquad\qquad \supset : x \,\epsilon\, \alpha . \supset_x . \overleftrightarrow{R_*}\text{'}x \,\epsilon\, \overleftrightarrow{R_*}\text{''}\beta :$

$[\ast37\cdot61] \qquad\qquad\qquad\qquad\qquad \supset : \overleftrightarrow{R_*}\text{''}\alpha \subset \overleftrightarrow{R_*}\text{''}\beta$ \hfill (2)

$\vdash . \ast40\cdot4 . \supset \vdash :. \text{Hp} . \supset : y \,\epsilon\, \beta . \supset . (\exists x) . x \,\epsilon\, \alpha . y \,\epsilon\, \overleftrightarrow{R_*}\text{'}x .$

$[\ast97\cdot15] \qquad\qquad\qquad\qquad \supset . (\exists x) . x \,\epsilon\, \alpha . \overleftrightarrow{R_*}\text{'}x = \overleftrightarrow{R_*}\text{'}y .$

$[\ast37\cdot62] \qquad\qquad\qquad\qquad \supset . \overleftrightarrow{R_*}\text{'}y \,\epsilon\, \overleftrightarrow{R_*}\text{''}\alpha$ \hfill (3)

$\vdash . (3) . \ast37\cdot61 . \supset \vdash : \text{Hp} . \supset . \overleftrightarrow{R_*}\text{''}\beta \subset \overleftrightarrow{R_*}\text{''}\alpha$ \hfill (4)

$\vdash . (2) . (4) . \supset \vdash . \text{Prop}$

∗97·34. $\vdash : R \,\epsilon\, 1 \rightarrow 1 . \beta \,\epsilon\, D\text{''} \epsilon_\Delta \text{'} \overleftrightarrow{R_*}\text{'}\alpha . \supset . \overleftrightarrow{R_*}\text{''}\alpha = \overleftrightarrow{R_*}\text{''}\beta$

Dem.

$\vdash . \ast83\cdot6\cdot62 . \qquad\qquad \supset \vdash :. \text{Hp} . \supset : x \,\epsilon\, \alpha . \supset_x . \underset{E}{\exists} ! \beta \cap \overleftrightarrow{R_*}\text{'}x : \beta \subset s\text{'}\overleftrightarrow{R_*}\text{''}\alpha$ \hfill (1)

$\vdash . \ast40\cdot4 . \ast97\cdot101 . \supset \vdash :. x \,\epsilon\, \alpha . \supset_x . \underset{E}{\exists} ! \beta \cap \overleftrightarrow{R_*}\text{'}x . \equiv . \alpha \subset s\text{'}\overleftrightarrow{R_*}\text{''}\beta$ \hfill (2)

$\vdash . (1) . (2) . \ast97\cdot33 . \supset \vdash . \text{Prop}$

∗97·341. $\vdash : R \,\epsilon\, 1 \rightarrow 1 . \beta \,\epsilon\, D\text{''} \epsilon_\Delta \text{'} \overleftarrow{R_*}\text{''}\overrightarrow{B\text{'}R} . \supset . \overleftrightarrow{R_*}\text{''}\beta = \overrightarrow{R_*}\text{''}\overrightarrow{B\text{'}R}$

$[\ast97\cdot34 \dfrac{\overrightarrow{B\text{'}R}}{\alpha} . \ast97\cdot2]$

∗97·35. $\vdash : R \,\epsilon\, \text{Cls} \rightarrow 1 . T \,\epsilon\, \text{Potid}\text{'}R . \overrightarrow{B\text{'}R} \subset D\text{'}T . \supset .$

$\qquad\qquad \text{Cnv}\text{'}\{(\overleftarrow{R_*} \restriction \overrightarrow{B\text{'}R}) | T\} \,\epsilon\, \epsilon_\Delta \text{'} \overleftarrow{R_*}\text{''}\overrightarrow{B\text{'}R} . \overleftarrow{\text{C}}\text{'}\{(\overleftarrow{R_*} \restriction \overrightarrow{B\text{'}R}) | T\} = \breve{T}\text{''}\overrightarrow{B\text{'}R}$

Dem.

$\vdash . \ast97\cdot3 . \ast92\cdot101 . \supset \vdash : \text{Hp} . \supset . \text{Cnv}\text{'}\{(\overleftarrow{R_*} \restriction \overrightarrow{B\text{'}R}) | T\} \,\epsilon\, 1 \rightarrow \text{Cls}$ \hfill (1)

CHAPTER 10

Gödel's Quintessential Strange Loop

ৰু ৰু ৰু

Gödel Encounters Fibonacci

BY HIS early twenties, the boy from Brünn was already a superb mathematician and, like all mathematicians, he knew whole numbers come in limitless varieties. Aside from squares, cubes, primes, powers of ten, sums of two squares, and all the other usual suspects, he was familiar with many other types of integers. Most crucially for his future, young Kurt knew, thanks to Leonardo di Pisa (more often known as "Fibonacci"), that one could define classes of integers *recursively*.

In the 1300's, Fibonacci had concocted and explored what are now known as the "Fibonacci numbers":

1, 2, 3, 5, 8, 13, 21, 34, 55, 89, 144, 233, 377, 610, 987, 1597,...

In this rapidly growing infinite sequence, whose members I will henceforth refer to as the *F* numbers, each new element is created by summing the two previous ones (except for the first pair, 1 and 2, which we simply declare by fiat to be *F* numbers).

This almost-but-not-quite-circular fashion of defining a sequence of numbers in terms of itself is called a "recursive definition". This means there is some kind of precise calculational rule for making new elements out of previous ones. The rule might involve adding, multiplying, dividing, whatever — as long as it's well-defined. The opening gambit of a recursive sequence (in this case, the numbers 1 and 2) can be thought of as a *packet of seeds* from which a gigantic plant — all of its branches and leaves, infinite in number — grows in a predetermined manner, based on the fixed rule.

The Caspian Gemstones: An Allegory

Leonardo di Pisa's sequence is brimming with amazing patterns, but unfortunately going into that would throw us far off course. Still, I cannot resist mentioning that 144 jumps out in this list of the first few F numbers because it is a salient perfect square. Aside from 8, which is a cube, and 1, which is a rather degenerate case, no other perfect square, cube, or any other exact power appears in the first few hundred terms of the F sequence.

Several decades ago, people started wondering if the presence of 8 and 144 in the F sequence was due to a *reason*, or if it was just a "random accident". Therefore, as computational tools started becoming more and more powerful, they undertook searches. Curiously enough, even with the advent of supercomputers, allowing millions and even billions of F numbers to be churned out, no one ever came across any other perfect powers in Fibonacci's sequence. The chance of a power turning up very soon in the F sequence was looking slim, but why would a *perfect* mutual avoidance occur? What do nth powers for arbitrary n have to do with adding up pairs of numbers in Fibonacci's peculiar recursive fashion? Couldn't 8 and 144 just be little random glitches? Why couldn't other little glitches take place?

To cast allegorical light on this, imagine someone chanced one day to fish up a giant diamond, a magnificent ruby, and a tiny pearl at the bottom of the great green Caspian Sea in central Asia, and other seekers of fortune, spurred on by these stunning finds, then started madly dredging the bottom of the world's largest lake to seek more diamonds, rubies, pearls, emeralds, topazes, etc., but none was found, no matter how much dredging was done. One would naturally wonder if more gems might be hidden down there, but how could one ever know? (Caveat: my allegory is slightly flawed, because we can imagine, at least in principle, a richly financed scientific team someday dredging the lake's bottom completely, since, though huge, it is finite. For my analogy to be "perfect", we would have to conceive of the Caspian Sea as infinite. Just stretch your imagination a bit, reader!)

Now the twist. Suppose some mathematically-minded geologist set out to *prove* that the two exquisite Caspian gems, plus the tiny round pearl, were *sui generis* — in other words, that there was a precise *reason* that no other gemstone or pearl of any type or size would ever again, or *could* ever again, be found in the Caspian Sea. Does seeking such a proof make any sense? How could there be a watertight scientific *reason* absolutely forbidding any gems — except for one pearl, one ruby, and one diamond — from ever being found on the floor of the Caspian Sea? It sounds absurd.

This is typical of how we think about the physical world — we think of it as being filled with contingent events, facts that could be otherwise, situations that have no fundamental reason for their being as they are. But let me remind you that mathematicians see their pristine, abstract world as the antithesis to the random, accident-filled physical world we all inhabit. Things that happen in the mathematical world strike mathematicians as happening, without any exceptions, for statable, understandable *reasons*.

This — the Mathematician's Credo — is the mindset that you have to adopt and embrace if you wish to understand how mathematicians think. And in this particular case, the mystery of the lack of Fibonacci powers, although just a tiny one in most mathematicians' eyes, was a particularly baffling one, because it seemed to offer no natural route of access. The two phenomena involved — integer powers with arbitrarily large exponents, on the one hand, and Fibonacci numbers on the other — simply seemed (like gemstones and the Caspian Sea) to be too conceptually remote from each other to have any deep, systematic, inevitable interrelationship.

And then along came a vast team of mathematicians who had set their collective bead on the "big game" of Fermat's Last Theorem (the notorious claim, originally made by Pierre de Fermat in the middle of the seventeenth century, that no positive integers a, b, c exist such that $a^n + b^n$ equals c^n, with the exponent n being an integer greater than 2). This great international relay team, whose final victorious lap was magnificently sprinted by Andrew Wiles (his sprint took him about eight years), was at last able to prove Fermat's centuries-old claim by using amazing techniques that combined ideas from all over the vast map of contemporary mathematics.

In the wake of this team's revolutionary work, new paths were opened up that seemed to leave cracks in many famous old doors, including the tightly-closed door of the small but alluring Fibonacci power mystery. And indeed, roughly ten years after the proof of Fermat's Last Theorem, a trio of mathematicians, exploiting the techniques of Wiles and others, were able to pinpoint the exact *reason* for which cubic 8 and square 144 will never have any perfect-power mates in Leonardo di Pisa's recursive sequence (except for 1). Though extremely recondite, the reason behind the infinite mutual-avoidance dance had been found. This is just one more triumph of the Mathematician's Credo — one more reason to buy a lot of stock in the idea that in mathematics, *where there's a pattern, there's a reason.*

A Tiny Spark in Gödel's Brain

We now return to the story of Kurt Gödel and his encounter with the powerful idea that all sorts of infinite classes of numbers can be defined through various kinds of recursive rules. The image of the organic growth of an infinite structure or pattern, all springing out of a finite set of initial seeds, struck Gödel as much more than a mere curiosity; in fact, it reminded him of the fact that theorems in *PM* (like theorems in Euclid's *Elements*) always spring (by formal rules of inference) from earlier theorems in *PM,* with the exception of the first few theorems, which are declared by fiat to be theorems, and thus are called "axioms" (analogues to the seeds).

In other words, in the careful analogy sparked in Gödel's mind by this initially vague connection, the *axioms* of *PM* would play the role of Fibonacci's seeds 1 and 2, and the *rules of inference* of *PM* would play the role of adding the two most recent numbers. The main difference is that in *PM* there are several rules of inference, not just one, so at any stage you have a choice of what to do, and moreover, you don't have to apply your chosen rule to the most recently generated theorem(s), so that gives you even more choice. But aside from these extra degrees of freedom, Gödel's analogy was very tight, and it turned out to be immensely fruitful.

Clever Rules Imbue Inert Symbols with Meaning

I must stress here that each rule of inference in a formal system like *PM* not only leads from one or more input formulas to an output formula, but it does so *by purely typographical means* — that is, via purely mechanical symbol-shunting that doesn't require any thought about the meanings of symbols. From the viewpoint of a person (or machine) following the rules to produce theorems, the symbols might as well be totally devoid of meaning.

On the other hand, each rule has to be very carefully designed so that, given input formulas that express truths, the output formula will also express a truth. The rule's designer (Russell and Whitehead, in this case) therefore has to think about the symbols' intended meanings in order to be sure that the rule will work exactly right for a manipulator (human or otherwise) who is *not* thinking about the symbols' intended meanings.

To give a trivial example, suppose the symbol "∨" were intended to stand for the concept "or". Then a possible rule of inference would be:

From any formula "P ∨ Q" one can derive the reversed formula "Q ∨ P".

This rule of inference is reasonable because whenever an or-statement (such as "You're crazy or I'm crazy") is true, then so is the flipped-around or-statement ("I'm crazy or you're crazy").

This particular ∨-flipping rule happens not to be one of *PM*'s rules of inference, but it could have been one. The point is just that this rule shows how one can mechanically shunt symbols and ignore their meanings, and yet preserve truth while doing so. This rule is rather trivial, but there are subtler ones that do real work. That, indeed, is the whole idea of symbolic logic, first suggested by Aristotle and then developed piecemeal over many centuries by such thinkers as Blaise Pascal, Gottfried Wilhelm von Leibniz, George Boole, Augustus De Morgan, Gottlob Frege, Giuseppe Peano, David Hilbert, and many others. Russell and Whitehead were simply developing the ancient dream of totally mechanizing reasoning in a more ambitious fashion than any of their predecessors had.

Mechanizing the Mathematician's Credo

If you apply *PM*'s rules of inference to its axioms (the seeds that constitute the "zeroth generation" of theorems), you will produce some "progeny" — theorems of the "first generation". Then apply the rules once again to the first-generation theorems (as well as to the axioms) in all the different ways you can; you will thereby produce a new batch of theorems — the second generation. Then from that whole brew comes a third batch of theorems, and so on, *ad infinitum*, constantly snowballing. The infinite body of theorems of *PM* is fully determined by the initial seeds and by the typographical "growth rules" that allow one to make new theorems out of old ones.

Needless to say, the hope here is that all of these mechanically generated theorems of *PM* are true statements of number theory (*i.e.*, no false statement is ever generated), and conversely, it is hoped that all true statements of number theory are mechanically generated as theorems of *PM* (*i.e.*, no true statement is left ungenerated forever). The first of these hopes is called *consistency*, and the second one is called *completeness*.

In a nutshell, we want the entire infinite body of theorems of *PM* to coincide exactly with the infinite body of true statements in number theory — we want perfect, flawless alignment. At least that's what Russell and Whitehead wanted, and they believed that with *PM* they had attained this goal (after all, "s0 + s0 = ss0" was a theorem, wasn't it?).

Let us recall the Mathematician's Credo, which in some form or other had existed for many centuries before Russell and Whitehead came along:

> X is true *because* there is a proof of X;
>
> X is true *and so* there is a proof of X.

The first line expresses the first hope expressed above — consistency. The second line expresses the second hope expressed above — completeness. We thus see that the Mathematician's Credo is very closely related to what

Russell and Whitehead were aiming for. Their goal, however, was to set the Credo on a new and rigorous basis, with *PM* serving as its pedestal. In other words, where the Mathematician's Credo merely speaks of "a proof" without saying what is meant by the term, Russell and Whitehead wanted people to think of it as meaning *a proof within PM.*

Gödel himself had great respect for the power of *PM,* as is shown by the opening sentences of his 1931 article:

> The development of mathematics in the direction of greater exactness has — as is well known — led to large tracts of it becoming formalized, so that proofs can be carried out according to a few mechanical rules. The most comprehensive formal systems yet set up are, on the one hand, the system of *Principia Mathematica* (*PM*) and, on the other, the axiom system for set theory of Zermelo-Fraenkel (later extended by J. v. Neumann). These two systems are so extensive that all methods of proof used in mathematics today have been formalized in them, *i.e.,* reduced to a few axioms and rules of inference.

And yet, despite his generous hat-tip to Russell and Whitehead's opus, Gödel did not actually believe that a perfect alignment between truths and *PM* theorems had been attained, nor indeed that such a thing could *ever* be attained, and his deep skepticism came from having smelled an extremely strange loop lurking inside the labyrinthine palace of mindless, mechanical, symbol-churning, meaning-lacking mathematical reasoning.

Miraculous Lockstep Synchrony

The conceptual parallel between recursively defined sequences of integers and the leapfrogging set of theorems of *PM* (or, for that matter, of any formal system whatever, as long as it had axioms acting as seeds and rules of inference acting as growth mechanisms) suggested to Gödel that the typographical patterns of symbols on the pages of *Principia Mathematica* — that is, the rigorous logical derivations of new theorems from previous ones — could somehow be "mirrored" in an exact manner inside the world of numbers. An inner voice told him that this connection was not just a vague resemblance but could in all likelihood be turned into an absolutely precise correspondence.

More specifically, Gödel envisioned a set of whole numbers that would organically grow out of each other via arithmetical calculations much as Fibonacci's *F* numbers did, but that would also correspond in an exact one-to-one way with the set of theorems of *PM.* For instance, if you made theorem Z out of theorems X and Y by using typographical rule *R5*, and if you made the number z out of numbers x and y using computational rule *r5*, then everything would match up. That is to say, if x were the number corresponding to theorem X and y were the number corresponding to theorem Y, then z would "miraculously" turn out to be the number corresponding to theorem Z. There would be perfect synchrony; the two sides (typographical and numerical) would move together in lock-step. At first this vision of miraculous synchrony was just a little spark, but Gödel quickly realized that his inchoate dream might be made so precise that it could be spelled out to others, so he started pursuing it in a dogged fashion.

Flipping between Formulas and Very Big Integers

In order to convert his intuitive hunch into a serious, precise, and respectable idea, Gödel first had to figure out how any string of *PM* symbols (irrespective of whether it asserted a truth or a falsity, or even was just a random jumble of symbols haphazardly thrown together) could be systematically converted into a positive integer, and conversely, how such an integer could be "decoded" to give back the string from which it had come. This first stage of Gödel's dream, a systematic mapping by which every formula would receive a numerical "name", came about as follows.

The basic alphabet of *PM* consisted of only about a dozen symbols (other symbols were introduced later but they were all defined in terms of the original few, so they were not conceptually necessary), and to each of these symbols Gödel assigned a different small integer (these initial few choices were quite arbitrary — it really didn't matter what number was associated with an isolated symbol).

For multi-symbol formulas (by the way, in this book the terms "string of symbols" — "string" for short — and "formula" are synonymous), the idea was to replace the symbols, one by one, moving left to right, by their code numbers, and then to combine all of those individual code numbers (by using them as exponents to which successive prime numbers are raised) into one unique big integer. Thus, once *isolated* symbols had been assigned numbers, the numbers assigned to *strings* of symbols were *not* arbitrary.

For instance, suppose that the (arbitrary) code number for the symbol "0" is 2, and the code number for the symbol "=" is 6. Then for the three symbols in the very simple formula "0=0", the code numbers are 2, 6, 2, and these three numbers are used as *exponents* for the first three prime numbers (2, 3, and 5) as follows:

$$2^2 \cdot 3^6 \cdot 5^2 = 72900$$

So we see that 72900 is the single number that corresponds to the formula "0=0". Of course this is a rather large integer for such a short formula, and you can easily imagine that the integer corresponding to a fifty-symbol formula is astronomical, since it involves putting the first fifty prime numbers to various powers and then multiplying all those big numbers together, to make a true colossus. But no matter — numbers are just numbers, no matter how big they are. (Luckily for Gödel, there are infinitely many primes, since if there had been merely, say, one billion of them, then his method would only have let him encode formulas made of a billion symbols or fewer. Now that would be a crying shame!)

The decoding process works by finding the prime factorization of 72900 (which is unique), and reading off the exponents that the ascending primes are raised to, one by one — 2, 6, 2 in this case.

To summarize, then, in this non-obvious but simple manner, Gödel had found a way to replace any given formula of *PM* by an equivalent number (which other people soon would dub its *Gödel number*). He then extended this idea of "arithmetization" to cover arbitrary *sequences* of formulas, since proofs in *PM* are sequences of formulas, and he wanted to be able to deal with proofs, not just isolated formulas. Thus an arbitrarily long sequence of formulas could be converted into one large integer via essentially the same technique, using primes and exponents. You can imagine that we're talking *really* big numbers here.

In short, Gödel showed how any visual symbol-pattern whatsoever in the idiosyncratic notation of *Principia Mathematica* could be assigned a unique number, which could easily be decoded to give back the visual pattern (*i.e.*, sequence of symbols) to which it corresponded. Conceiving of and polishing this precise two-way mapping, now universally called "Gödel numbering", constituted the first key step of Gödel's work.

Very Big Integers Moving in Lock-step with Formulas

The next key step was to make Fibonacci-like recursive definitions of special sets of integers — integers that would organically grow out of previously generated ones by addition or multiplication or more complex computations. One example would be the *wff* numbers, which are those integers that, via Gödel's code, represent "well-formed" or "meaningful" formulas of *PM,* as opposed to those that represent meaningless or ungrammatical strings. (A sample well-formed formula, or "wff" for short, would be "0+0=sss0". Though it asserts a falsity, it's still a meaningful statement. On the other hand, "=)0(=" and "00==0+=" are not wffs. Like the arbitrary sequence of pseudo-words "zzip dubbiwubbi pizz", they don't assert anything.) Since, as it happens, longer wffs are built up in *PM* from shorter wffs by just a few simple and standard rules of typographical juxtaposition, their larger code numbers can likewise be built up from the smaller code numbers of shorter ones by just a few simple and standard rules of numerical calculation.

I've said the foregoing rather casually, but in fact this step was perhaps the deepest of Gödel's key insights — namely, that once strings of symbols had been "arithmetized" (given numerical counterparts), then any kind of rule-based typographical shunting-around of strings on paper could be perfectly paralleled by some kind of *purely arithmetical calculation* involving their numerical proxies — which were huge numbers, to be sure, but still just numbers. What to Russell and Whitehead looked like elaborate *symbol-shunting* looked like a lot of straightforward *number-crunching* to Kurt Gödel (although of course he didn't use that colorful modern term, since this was all taking place back in the prehistoric days when computers didn't yet exist). These were simply two different views of what was going on — views that were 100 percent equivalent and interchangeable.

Glimmerings of How PM Can Twist Around and See Itself

Gödel saw that the game of building up an infinite class of numbers, such as wff numbers, through recursion — that is, making new "members of the club" by combining older, established members via some number-crunching rule — is essentially the same idea as Fibonacci's recursive game of building up the class of *F* numbers by taking sums of earlier members. Of course recursive processes can be far more complicated than just taking the sum of the latest two members of the club.

What a recursive definition does, albeit implicitly, is to divide the entire set of integers into members and non-members of the club — that is, those numbers that are *reachable*, sooner or later, via the recursive building-up process, and those that are *never reachable*, no matter how long one waits. Thus 34 is a member of the *F* club, whereas 35 is a non-member. How do we know 35 is not an *F* number? That's very easy — the rule that makes new *F* numbers always makes larger ones from smaller ones, and so once we've passed a certain size, there's no chance we'll be returning to "pick up" other numbers in that vicinity later. In other words, once we've made the *F* numbers 1, 2, 3, 5, 8, 13, 21, 34, 55, we know they are the only ones in that range, so obviously 35, 36, and so on, up to 54, are not *F* numbers.

If, however, some other club of numbers is defined by a recursive rule whose outputs are sometimes *bigger* than its inputs and other times are *smaller* than its inputs, then, in contrast to the simple case of the *F* club, you can't be so sure that you won't ever be coming back and picking up smaller integers that were missed in earlier passes.

Let's think a little bit more about the recursively defined club of numbers that we called "wff numbers". We've seen that the number 72900 possesses "wff-ness", and if you think about it, you can see that 576 and 2916 lack that quality. (Why? Well, if you factor them and look at the exponents of 2 and 3, you will see that these numbers are the numerical encodings of the strings "0=" and "=0", respectively, neither of which makes sense, whence they are not well-formed formulas.) In other words, despite its odd definition, wff-ness, no more and no less than squareness or primeness or Fibonacci's *F*-ness, is a valid object of study in the world of pure number. The distinction between members and non-members of the "wff club" is every bit as genuine a *number-theoretical* distinction as that between members and non-members of the club of squares, the club of prime numbers, or the club of *F* numbers, for wff numbers are definable in a recursive arithmetical (*i.e.*, computational) fashion. Moreover, it happens that the recursive rules defining wff-ness always produce outputs that are bigger than their inputs, so that wff-ness shares with *F*-ness the simple property that once you've exceeded a certain magnitude, you know you'll never be back visiting that zone again.

Just as some people's curiosity was fired by the fact of seeing a square in Fibonacci's recursively defined sequence, so some people might become interested in the question as to whether there are any squares (or cubes, etc.) in the recursively defined sequence of wff numbers. They could spend a lot of time investigating such purely number-theoretical questions, never thinking at all about the corresponding formulas of *Principia Mathematica*.

One could be completely ignorant of the fact that Gödel's wff numbers had their origin in Russell and Whitehead's rules defining well-formedness in *Principia Mathematica*, just as one can study the laws of probability without ever suspecting that this deep branch of mathematics was originally developed to analyze gambling. What long ago inspired someone to dream up a particular recursive definition obviously doesn't affect the numbers it defines; all that matters is that there should be a purely computational way of making any member of the club grow out of the initial seeds by applying the rules some finite number of times.

Now wff numbers are, as it happens, relatively easy to define in a recursive fashion, and for that reason wff-ness (exactly like *F*-ness) is just the kind of mathematical notion that *Principia Mathematica* was designed to study. To be sure, Whitehead and Russell had never dreamed that their mechanical reasoning system might be put to such a curious use, in which its own properties as a machine were essentially placed under observation by itself, rather like using a microscope to examine some of its own lenses for possible defects. But then, inventions often do surprise their inventors.

Prim Numbers

Having realized that some hypothetical volume of the series by Whitehead and Russell could define and systematically explore the various numerical properties of wff numbers, Gödel pushed his analogy further and showed, with a good deal of fancy machinery but actually not very much conceptual difficulty, that there was an infinitely more interesting recursively defined class of whole numbers, which I shall here call *prim* numbers (whimsically saluting the title of the famous three tomes), and which are the numbers belonging to *provable* formulas of *PM* (*i.e.*, theorems).

A *PM* proof, of course, is a series of formulas leading from the axioms of *PM* all the way to the formula in question, each step being allowed by some rule of reasoning, which in *PM* became a formal typographical rule of inference. To every typographical rule of inference acting on strings of *PM*, Gödel exhibited a perfectly matching computational rule that acted on numbers. Numerical computation was effectively thumbing its nose at typographical manipulation, sassily saying, "Anything you can do, I can do better!" Well, not really *better* — but the key point, as Gödel showed beyond any doubt, was that a computational rule would always be able to mimic perfectly — to keep in perfect synchrony with — any formal typographical rule, and so numerical rules were *just as good*.

The upshot was that to every *provable string* of Russell and Whitehead's formal system, there was a counterpart *prim number*. Any integer that was prim could be decoded into symbols, and the string you got would be a provable-in-*PM* formula. Likewise, any provable-in-*PM* formula could be encoded as one whopping huge integer, and by God, with enough calculation, you could show that that number was a prim number. A simple example of a prim number is, once again, our friend 72900, since the formula "0=0", over and above being a well-formed formula, is also, and not too surprisingly, derivable in *PM*. (Indeed, if it weren't, *PM* would be absolutely pathetic as a mechanical model of mathematical reasoning!)

There is a crucial difference between wff numbers and prim numbers, which comes from the fact that the rules of inference of *PM* sometimes produce output strings that are *shorter* than their input strings. This means that the corresponding arithmetical rules defining prim numbers will sometimes take *large* prim numbers as input and make from them a *smaller* prim number as output. Therefore, stretches of the number line that have been visited once can always be revisited later, and this fact makes it much, much harder to determine about a given integer whether it is prim or not. This is a central and very deep fact about prim numbers.

Just as with squares, primes, *F* numbers, or wff numbers, there could once again be a hypothetical volume of the series of tomes by Whitehead and Russell in which prim numbers were defined and their mathematical properties studied. For example, such a volume might contain a proof of the formula of *PM* that (when examined carefully) asserts "72900 is a prim number", and it might also discuss another formula that could be seen to assert the opposite ("72900 is *not* a prim number"), and so on. This latter statement is false, of course, while the former one is true. And even more complex number-theoretical ideas could be expressed using the *PM* notation and discussed in the hypothetical volume, such as "There are infinitely many prim numbers" — which would be tantamount to asserting (via a code), "There are infinitely many formulas that are provable in *PM*".

Although it might seem an odd thing to do, one could certainly pose eighteenth-century–style number-theory questions such as, "Which integers are expressible as the sum of two prim numbers, and which integers are not?" Probably nobody would ever seriously ask such an oddball question, but the point is that the property of being a prim number, although it's a rather arcane "modern" property, is no more and no less a genuinely number-theoretical property of an integer than is a "classical" property, such as being square or being prime or being a Fibonacci number.

The Uncanny Power of Prim Numbers

Suppose someone told you that they had built a machine — I'll dub it "Guru" — that would always correctly answer any question of the form "Is *n* a prime number?", with *n* being any integer that you wish. When asked, "Is 641 prime?", Guru would spin its wheels for a bit and then say "yes". As for 642, Guru would "think" a little while and then say "no". I suppose you would not be terribly surprised by such a machine. That such a machine can be realized, either in silicon circuitry on in domino-chain technology, is not anything to boggle anyone's mind in this day and age.

But suppose someone told you that they had built an analogous machine — I'll dub it "Göru" — that would always correctly answer any question of the form "Is *n* a prim number?" Would this claim — strictly analogous to the previous one — strike you as equally ho-hum? If so, then I respectfully submit that you've got another think coming.

The reason is this. If you believed Göru to be reliable and you also believed in the Mathematician's Credo (*Principia Mathematica* version), then you could conclude that your little Göru, working all by itself, could answer *any number-theoretical question* that you were interested in, just like a genie conjured from a magic lamp. How so? What makes Göru a magic genie?

Well, suppose you wanted to know if statement X is true or false (for instance, the famous claim "Every even number greater than 2 is the sum of two primes" — which, as I stated above, remains unsettled even today, after nearly three centuries of work). You would just write X down in the formal notation of *PM*, then convert that formula mechanically into its Gödel number *x*, and feed that number into Göru (thus asking if *x* is prim or not). Of course *x* will be a huge integer, so it would probably take Göru a good while to give you an answer, but (assuming that Göru is not a hoax) sooner or later it would spit out either a "yes" or a "no". In case Göru said "yes", you would know that *x* is a prim number, which tells you that the formula it encodes is a *provable* formula, which means that statement X is true. Conversely, were Göru to tell you "no", then you would know that the statement X is *not* provable, and so, believing in the Mathematician's Credo (*Principia Mathematica* version), you would conclude it is false.

In other words, if we only had a machine that could infallibly tell apart prim numbers and "saucy" (non-prim) numbers, and taking for granted that the *Principia Mathematica* version of the Mathematician's Credo is valid, then we could infallibly tell true statements from false ones. In short, having a Göru would give us a royal key to all of mathematical knowledge.

The prim numbers alone would therefore seem to contain, in a cloaked fashion, *all of mathematical knowledge* wrapped up inside them! No other sequence of numbers ever dreamt up by anyone before Gödel had anything like this kind of magically oracular quality. These amazing numbers seem to be worth their weight in gold! But as I told you, the prim numbers are elusive, because small ones sometimes wind up being added to the club at very late stages, so it won't be easy to tell prim numbers from saucy ones, nor to build a Göru. (This is meant as a premonition of things to come.)

Gödelian Strangeness

Finally, Gödel carried his analogy to its inevitable, momentous conclusion, which was to spell out for his readers (not symbol by symbol, of course, but via a precise set of "assembly instructions") an astronomically long formula of *PM* that made the seemingly innocent assertion, "A certain integer *g* is not a prim number." However, that "certain integer *g*" about which this formula spoke happened, by a most unaccidental (some might say diabolical) coincidence, to be the number associated with (*i.e.*, coding for) *this very formula* (and so it was necessarily a gargantuan integer). As we are about to see, Gödel's odd formula can be interpreted on two different levels, and it has two very different meanings, depending on how one interprets it.

On its more straightforward level, Gödel's formula merely asserts that this gargantuan integer *g* lacks the number-theoretical property called *primness*. This claim is very similar to the assertion "72900 is not a prime number", although, to be sure, *g* is a lot larger than 72900, and primness is a far pricklier property than is primeness. However, since primness was defined by Gödel in such a way that it numerically mirrored the provability of strings via the rules of the *PM* system, the formula *also* claims:

> The formula that happens to have the code number *g*
> is not provable via the rules of *Principia Mathematica*.

Now as I already said, the formula that "just happens" to have the code number *g* is the formula making the above claim. In short, Gödel's formula is making a claim about *itself* — namely, the following claim:

> This very formula is not provable via the rules of *PM*.

Sometimes this second phraseology is pointedly rendered as "I am not a theorem" or, even more tersely, as

> I am unprovable

(where "in the *PM* system" is tacitly understood).

Gödel further showed that his formula, though very strange and discombobulating at first sight, was not all that unusual; indeed, it was merely one member of an infinite family of formulas that made claims about the system *PM*, many of which asserted (some truthfully, others falsely) similarly weird and twisty things about themselves (*e.g.*, "Neither I nor my negation is a theorem of *PM*", "If I have a proof inside *PM*, then my negation has an even shorter proof than I do", and so forth and so on).

Young Kurt Gödel — he was only 25 in 1931 — had discovered a vast sea of amazingly unsuspected, bizarrely twisty formulas hidden inside the austere, formal, type-theory-protected and therefore supposedly paradox-free world defined by Russell and Whitehead in their grandiose three-volume œuvre *Principia Mathematica*, and the many counterintuitive properties of Gödel's original formula and its countless cousins have occupied mathematicians, logicians, and philosophers ever since.

How to Stick a Formula's Gödel Number inside the Formula

I cannot leave the topic of Gödel's magnificent achievement without going into one slightly technical issue, because if I failed to do so, some readers would surely be left with a feeling of confusion and perhaps even skepticism about a key aspect of Gödel's work. Moreover, this idea is actually rather magical, so it's worth mentioning briefly.

The nagging question is this: How on earth could Gödel fit a formula's Gödel number into the formula itself? When you think about it at first, it seems like trying to squeeze an elephant into a matchbox — and in a way, that's exactly right. No formula can literally contain the numeral for its own Gödel number, because that numeral will contain many more symbols than the formula does! It seems at first as if this might be a fatal stumbling block, but it turns out not to be — and if you think back to our discussion of G. G. Berry's paradox, perhaps you can see why.

The trick involves the simple fact that some huge numbers have very short descriptions (387420489, for instance, can be described in just four syllables: "nine to the ninth"). If you have a very short recipe for calculating a very long formula's Gödel number, then instead of describing that huge number in the most plodding, clunky way ("the successor of the successor of the successor of ...... the successor of the successor of zero"), you can describe it via your computational shortcut, and if you express your shortcut in symbols (rather than inserting the numeral itself) inside the formula, then you can make the formula talk about itself without squeezing an elephant into a matchbox. I won't try to explain this in a mathematical fashion, but instead I'll give an elegant linguistic analogy, due to the philosopher W. V. O. Quine, which gets the gist of it across.

Gödel's Elephant-in-Matchbox Trick via Quine's Analogy

Suppose you wanted to write a sentence in English that talks about itself without using the phrase "this sentence". You would probably find the challenge pretty tricky, because you'd have to actually *describe* the sentence inside itself, using quoted words and phrases. For example, consider this first (somewhat feeble) attempt:

The sentence "This sentence has five words" has five words.

Now what I've just written (and you've just read) is a sentence that is true, but unfortunately it's not about itself. After all, the full thing contains *ten* words, as well as some quotation marks. This sentence is about a shorter sentence embedded inside it, in quote marks. And changing "five" to "ten" still won't make it refer to itself; all that this simple act does is to turn my sentence, which was true, into a false one. Take a look:

The sentence "This sentence has ten words" has ten words.

This sentence is false. And more importantly, it's *still* merely about a shorter sentence embedded inside itself. As you see, so far we are not yet very close to having devised a sentence that talks about itself.

The problem is that anything I put inside quote marks will necessarily be shorter than the entire sentence of which it is a part. This is trivially obvious, and in fact it is an exact linguistic analogue to the stumbling block of trying to stick a formula's own Gödel number directly inside the formula itself. An elephant will not fit inside a matchbox! On the other hand, an elephant's DNA *will* easily fit inside a matchbox…

And indeed, just as DNA is a *description* of an elephant rather than the elephant itself, so there is a way of getting around the obstacle by using a *description* of the huge number rather than the huge number itself. (To be slightly more precise, we can use a concise symbolic description instead of using a huge *numeral.*) Gödel discovered this trick, and although it is quite subtle, Quine's analogy makes it fairly easy to understand. Look at the following sentence fragment, which I'll call "Quine's Quasi-Quip":

preceded by itself in quote marks yields a full sentence.

As you will note, Quine's Quasi-Quip is certainly *not* a full sentence, for it has no grammatical subject (that is, "yields" has no subject); that's why I gave it the prefix "Quasi". But what if we were to put a noun at the head of the Quasi-Quip — say, the title "Professor Quine"? Then Quine's Quasi-Quip will turn into a full sentence, so I'll call it "Quine's Quip":

"Professor Quine" preceded by itself in quote marks yields a full sentence.

Here, the verb "yields" *does* have a subject — namely, Professor Quine's title, modified by a trailing adjectival phrase that is six words long.

But what does Quine's Quip *mean*? In order to figure this out, we have to actually *construct* the entity that it's talking about, which means we have to precede Professor Quine's title by itself in quote marks. This gives us:

"Professor Quine" Professor Quine

The Quine's Quip that we created a moment ago merely asserts (or rather, claims) that this somewhat silly phrase is a full sentence. Well, that claim is obviously false. The above phrase is *not* a full sentence; it doesn't even contain a verb.

However, we arbitrarily used Professor Quine's title when we could have used a million different things. Is there some *other* noun that we might place at the head of Quine's Quasi-Quip that will make Quine's Quip come out *true*? What Gödel realized, and what Quine's analogy helps to make clear, is that for this to happen, you have to use, as your subject of the verb "yields", a *subjectless sentence fragment.*

What is an example of a subjectless sentence fragment? Well, just take any old sentence such as "Snow is white", and cut off its subject. What you get is a subjectless sentence fragment: "is white". So let's use *this* as the noun to place in front of Quine's Quasi-Quip:

"is white" preceded by itself in quote marks yields a full sentence.

This medium-sized mouthful makes a claim about a construction that we have yet to exhibit, and so let's do so without further ado:

"is white" is white.

(I threw in the period for good measure, but let's not quibble.)

Now what we have just produced certainly *is* a full sentence, because it has a verb ("is"), and that verb has a subject (the quoted phrase), and the whole thing makes sense. I'm not saying that it is *true*, mind you, for indeed it is blatantly false: "is white" is in fact *black* (although, to be fair, letters and words do contain some white space along with their black ink, otherwise we couldn't read them). In any case, Quine's Quasi-Quip when fed "is white" as its input yielded a full sentence, and that's exactly what Quine's Quip claimed. We're definitely making headway.

The Trickiest Step

Our last devilish trick will be to use Quine's Quasi-Quip *itself* as the noun to place at its head. Here, then, is Quine's Quasi-Quip with a quoted copy of itself installed in front:

"preceded by itself in quote marks yields a full sentence"
preceded by itself in quote marks yields a full sentence.

What does this Quip claim? Well, first we have to determine what entity it is talking *about,* and that means we have to construct the analogue to "'is white' is white". Well, in this case, the analogue is the following:

"preceded by itself in quote marks yields a full sentence"
preceded by itself in quote marks yields a full sentence.

I hope you are not lost at this point, for we really have hit the crux of the matter. Quine's Quip turns out to be talking about a phrase that is identical to the Quip itself! It is claiming that *something* is a full sentence, and when you go about constructing that thing, it turns out to be Quine's Quip itself. So Quine's Quip talks about itself, claiming of itself that it is a full sentence (which it surely is, even though it is built out of two subjectless sentence fragments, one in quote marks and one not).

While you are pondering this, I will jump back to the source of it all, which was Gödel's *PM* formula that talked about itself. The point is that Gödel numbers, since they can be used as *names* for formulas and can be *inserted* into formulas, are precisely analogous to quoted phrases. Now we have just seen that there is a way to use quotation marks and sentence fragments to make a full sentence that talks about itself (or if you prefer, a sentence that talks about *another* sentence, but one that is a clone to it, so that whatever is true of the one is true of the other).

Gödel, analogously, created a "subjectless formula fragment" (by which I mean a *PM* formula that is not about any specific integer, but just about some unspecified variable number x). And then, making a move analogous to that of feeding Quine's Quasi-Quip into itself (but in quotes), he took that formula fragment's Gödel number k (which is a specific number, not a variable) and replaced the variable x by it, thus producing a formula (not just a fragment) that made a claim about a much larger integer, g. And g is the Gödel number of that very claim. And last but not least, the claim was not about whether the entity in question was a full sentence or not, but about whether the entity in question was a *provable formula* or not.

An Elephant in a Matchbox is Neither Fish Nor Fowl

I know this is a lot to swallow in one gulp, and so if it takes you several gulps (careful rereadings), please don't feel discouraged. I've met quite a few sophisticated mathematicians who admit that they never understood this argument totally!

I think it would be helpful at this juncture to exhibit a kind of hybrid sentence that gets across the essential flavor of Gödel's self-referential construction but that does so in Quinean terms — that is, using the ideas we've just been discussing. The hybrid sentence looks like this:

"when fed its own Gödel number yields a non-prim number"
when fed its own Gödel number yields a non-prim number.

The above sentence is neither fish nor fowl, for it is not a formula of *Principia Mathematica* but an English sentence, so of course it doesn't have a Gödel number and it couldn't possibly be a theorem (or a nontheorem) of *PM*. What a mixed metaphor!

And yet, mixed metaphor though it is, it still does a pretty decent job of getting across the flavor of the *PM* formula that Gödel actually concocted. You just have to keep in mind that using quote marks is a metaphor for taking Gödel numbers, so the upper line should be thought of as being a Gödel number (*k*) rather than as being a sentence fragment in quote marks. This means that metaphorically, the lower line (an English sentence fragment) *has* been fed its own Gödel number as its subject. Very cute!

I know that this is very tricky, so let me state it once again, slightly differently. Gödel asks you to imagine the formula that *k* stands for (that formula happens to contain the variable *x*), and then to feed *k* into it (this means to replace the single letter *x* by the extremely long numeral *k*, thus giving you a much bigger formula than you started with), and to take the Gödel number of the result. That will be the number *g*, huger far than *k* — and lastly, Gödel asserts that *this* walloping number is not a prim number. If you've followed my hand-waving argument, you will agree that the full formula's Gödel number (*g*) is not found explicitly inside the formula, but instead is very subtly *described* by the formula. The elephant's DNA has been used to get a description of the entire elephant into the matchbox.

Sluggo and the Morton Salt Girl

Well, I don't want to stress the technical points here. The main thing to remember is that Gödel devised a very clever number-description trick — a recipe for making a very huge number *g* out of a less huge number *k* — in order to get a formula of *PM* to make a claim about its own Gödel number's non-primness (which means that the formula is actually making a claim of its own nontheoremhood). And you might also try to remember that the "little" number *k* is the Gödel number of a "formula fragment" containing a variable *x*, analogous to a subjectless sentence fragment in quote marks, while the larger number *g* is the Gödel number of a *complete sentence in PM notation*, analogous to a complete sentence in English.

Popular culture is by no means immune to the delights of self-reference, and it happens that the two ideas we have been contrasting here — having a formula contain its own Gödel number *directly* (which would necessitate an infinite regress) and having a formula contain a *description* of its Gödel number (which beautifully bypasses the infinite regress) — are charmingly illustrated by two images with which readers may be familiar.

In this first image, Ernie Bushmiller's character Sluggo (from his classic strip *Nancy*) is dreaming of himself dreaming of himself dreaming of himself, without end. It is clearly a case of self-reference, but it involves an infinite regress, analogous to a *PM* formula that contained its own Gödel number directly. Such a formula, unfortunately, would have to be infinitely long!

Our second image, in contrast, is the famous label of a Morton Salt box, which shows a girl holding a box of Morton Salt. You may think you smell infinite regress once again, but if so, you are fooling yourself! The girl's arm is covering up the critical spot where the regress would occur. If you were to ask the girl to (please) hand you her salt box so that you could actually *see* the infinite regress on its label, you would wind up disappointed, for the label on *that* box would show her holding a yet smaller box with her arm once again blocking the regress.

And yet we still have a self-referential picture, because customers in the grocery store understand that the little box shown on the label is the same as the big box they are holding. How do they arrive at this conclusion? By using analogy. To be specific, not only do they have the large box in their own hands, but they can see the little box the girl is holding, and the two boxes have a lot in common (their cylindrical shape, their dark-blue color, their white caps at both ends); and in case that's not enough, they can also see salt spilling out of the little one. These pieces of evidence suffice to convince everyone that the little box and the large box are identical, and there you have it: self-reference without infinite regress!

In closing this chapter, I wish to point out explicitly that the most concise English translations of Gödel's formula and its cousins employ the word "I" ("I am not provable in *PM*"; "I am not a *PM* theorem"). This is not a coincidence. Indeed, this informal, almost sloppy-seeming use of the singular first-person pronoun affords us our first glimpse of the profound connection between Gödel's austere mathematical strange loop and the very human notion of a conscious self.

❧ ❧ ❧

CHAPTER 11

How Analogy Makes Meaning

෨ ෨ ෨

The Double Aboutness of Formulas in PM

IMAGINE the bewilderment of newly knighted Lord Russell when a young Austrian Turk named "Kurt" declared in print that *Principia Mathematica*, that formidable intellectual fortress so painstakingly erected as a bastion against the horrid scourge of self-referentiality, was in fact riddled through and through with formulas allegedly stating all sorts of absurd and incomprehensible things about themselves. How could such an outrage ever have been allowed to take place? How could vacuously twittering self-referential propositions have managed to sneak through the thick ramparts of the beautiful and timeless Theory of Ramified Types? This upstart Austrian sorcerer had surely cast some sort of evil spell, but by what means had he wrought his wretched deed?

The answer is that in his classic article — "On Formally Undecidable Propositions of *Principia Mathematica* and Related Systems (I)" — Gödel had re-analyzed the notion of *meaning* and had concluded that what a formula of *PM* meant was not so simple — not so unambiguous — as Russell had thought. To be fair, Russell himself had always insisted that *PM*'s strange-looking long formulas had *no* intrinsic meaning. Indeed, since the theorems of *PM* were churned out by formal rules that paid no attention to meaning, Russell often said the whole work was just an array of meaningless marks (and as you saw at the end of Chapter 9, the pages of *Principia Mathematica* often look more like some exotic artwork than like a work of math).

And yet Russell was also careful to point out that all these curious patterns of horseshoes, hooks, stars, and squiggles *could* be interpreted, if one wished, as being statements about numbers and their properties, because under duress, one could read the meaningless vertical egg '0' as standing for the number zero, the equally meaningless cross '+' as standing for addition, and so on, in which case all the theorems of *PM* came out as statements about numbers — but not just random blatherings about them. Just imagine how crushed Russell would have been if the squiggle pattern "ss0 + ss0 = sssss0" turned out to be a theorem of *PM*! To him, this would have been a disaster of the highest order. Thus he had to concede that there *was* meaning to be found in his murky-looking tomes (otherwise, why would he have spent long years of his life writing them, and why would he care which strings were theorems?) — but that meaning depended on using a *mapping* that linked shapes on paper to abstract magnitudes (*e.g.*, zero, one, two…), operations (*e.g.*, addition), relationships (*e.g.*, equality), concepts of logic (*e.g.*, "not", "and", "there exists", "all"), and so forth.

Russell's dependence on a systematic mapping to read meanings into his fortress of symbols is quite telling, because what the young Turk Gödel had discovered was simply a *different* systematic mapping (a much more complicated one, admittedly) by which one could read *different* meanings into the selfsame fortress. Ironically, then, Gödel's discovery was very much in the Russellian spirit.

By virtue of Gödel's subtle new code, which systematically mapped strings of symbols onto numbers and vice versa (recall also that it mapped typographical shunting laws onto numerical calculations, and vice versa), many formulas could be read on a second level. The first level of meaning, obtained via the old standard mapping, was always about numbers, just as Russell claimed, but the second level of meaning, using Gödel's newly revealed mapping (piggybacked on top of Russell's first mapping), was about *formulas,* and since both levels of meaning depended on mappings, Gödel's new level of meaning was no less real and no less valid than Russell's original one — just somewhat harder to see.

Extra Meanings Come for Free, Thanks to You, Analogy!

In my many years of reflecting about what Gödel did in 1931, it is this insight of his into the roots of meaning — his discovery that, thanks to a mapping, full-fledged meaning can suddenly appear in a spot where it was entirely unsuspected — that has always struck me the most. I find this insight as profound as it is simple. Strangely, though, I have seldom if ever seen this idea talked about in a way that brings out the profundity I find in it, and so I've decided to try to tackle that challenge myself in this chapter. To this end, I will use a series of examples that start rather trivially and grow in subtlety, and hopefully in humor as well. So here we go.

Standing in line with a friend in a café, I spot a large chocolate cake on a platter behind the counter, and I ask the server to give me a piece of it. My friend is tempted but doesn't take one. We go to our table and after my first bite of cake, I say, "Oh, this tastes awful." I mean, of course, not merely that my one slice is bad but that the whole cake is bad, so that my friend should feel wise (or lucky) to have refrained. This kind of mundane remark exemplifies how we effortlessly generalize outwards. We unconsciously think, "This piece of the cake is very much like the rest of the cake, so a statement about it will apply equally well to any other piece." (There is also another analogy presumed here, which is that my friend's reaction to foods is similar to mine, but I'll leave that alone.)

Let's try another example, just a tiny bit more daring. There's a batch of cookies on a plate at a party and I pick one up, take a bite, and remark to my children, "This is delicious!" Immediately, my kids take one each. Why? Because they wanted to taste something delicious. Yes, but how did they jump from my statement about *my* cookie to a conclusion about *other* cookies on the plate? The obvious answer is that the cookies are all "the same" in some sense. Unlike the pieces of cake, though, the cookies are not all parts of one single physical object, and thus they are ever so slightly "more different" from one another than are the pieces of cake — but they were made by the same person from the same ingredients using the same equipment. These cookies come from a single batch — they belong to the same category. In all relevant aspects, we see them as interchangeable. To be sure, each one is unique, but in the senses that count for human cookie consumption, they are almost certain to be equivalent. Therefore if I say about a particular one, "My, this is delicious!", my statement's meaning implicitly jumps across to any other of them, by the force of analogy. Now, to be sure, it's a rather trivial analogy to jump from one cookie to another when they all come from the same plate, but it's nonetheless an analogy, and it allows my specific statement "This is delicious!" to be taken as a general statement about all the cookies at once.

You may find these examples too childish for words. The first one involves an "analogy" between several slices of the same cake, and the second one an "analogy" between several cookies on the same plate. Are these banalities even worthy of the label "analogy"? To me there is no doubt about it; indeed, it is out of a dense fabric of a myriad of invisible, throwaway analogies no grander than these that the vast majority of our rich mental life is built. Yet we take such throwaway analogies so much for granted that we tend to think that the word "analogy" must denote something far more exalted. But one of my life's most recurrent theme songs is that we should have great respect for what seem like the most mundane of analogies, for when they are examined, they often can be seen to have sprung from, and to reveal, the deepest roots of human cognition.

Exploiting the Analogies in Everyday Situations

As we've just seen, a remark made with the aim of talking about situation A can also implicitly apply to situation B, even if there was no intention of talking about B, and B was never mentioned at all. All it takes is that there be an easy analogy — an unforced mapping that reveals both situations to have essentially the same central structure or conceptual core — and then the extra meaning is there to be read, whether one chooses to read it or not. In short, a statement about one situation can be heard as if it were about an analogous — or, to use a slightly technical term, *isomorphic* — situation. An isomorphism is just a formalized and strict analogy — one in which the network of parallelisms between two situations has been spelled out explicitly and precisely — and I'll use the term freely below.

When an analogy between situations A and B is glaringly obvious (no matter how simple it is), we sometimes will exploit it to talk "accidentally on purpose" about situation B by pretending to be talking only about situation A. "Hey there, Andy — take your muddy boots off when you come into the house!" Such a sentence, when shouted at one's five-year-old son who is tramping in the front door with his equally mud-oozing friend Bill, is obviously addressed just as much to Bill as to Andy, via a very simple, very apparent analogy (a boy-to-boy leap, if you will, much like the earlier cookie-to-cookie leap). Hinting by analogy allows us to get our message across politely but effectively. Of course we have to be pretty sure that the person at whom we're beaming our implicit message (Bill, here) is likely to be aware of the A/B analogy, for otherwise our clever and diplomatic ploy will all have been for naught.

Onward and upward in our chain of examples. People in romantic situations make use of such devices all the time. One evening, at a passionate moment during a tender clinch, Xerxes queries of his sweetie pie Yolanda, "Do I have bad breath?" He genuinely wants to know the answer, which is quite thoughtful of him, but at the same time his question is loaded (whether he intends it to be or not) with a second level of meaning, one not quite so thoughtful: "You have bad breath!" Yolanda answers his question but of course she also picks up on its potential alternative meaning in a flash. In fact, she suspects that Xerxes' *real* intent was to tell her about *her* breath, not to find out about his own — he was just being diplomatic.

Now how can one statement speak on two levels at once? How can a second meaning lie lurking inside a first meaning? You know the answer as well as I do, dear reader, but let me spell it out anyway. Just as in the muddy-boots situation, there is a very simple, very loud, very salient, very obvious analogy between the two parties, and this means that any statement made about X will be (or at least can be) heard as being about Y at the same time. The X/Y mapping, the analogy, the partial isomorphism — whatever you wish to call it — carries the meaning efficiently and reliably from one framework over to the other.

Let's look at this mode of communication in a slightly more delicate romantic situation. Audrey, who is not sure how serious Ben is about her, "innocently" turns the conversation to their mutual friends Cynthia and Dave, and "innocently" asks Ben what he thinks of Dave's inability to commit to Cynthia. Ben, no fool, swiftly senses the danger here, and so at first he is wary about saying anything specific since he may incriminate himself even though talking "only" about Dave, but then he also realizes that this danger gives him an opportunity to convey to Audrey some things that he hasn't dared to raise with her directly. Accordingly, Ben replies with a calculated air of nonchalance that he can imagine why Dave might be hesitant to commit himself, since, after all, Cynthia is so much more intellectual than Dave is. Ben is hoping that Audrey will pick up on the hint that since *she* is so much more involved in art than *he* is, that's why *he's* been hesitant to commit himself as well. His hint is carried to her implicitly but clearly via the rather strong couple-to-couple analogy that both Audrey and Ben have built up in their heads over the past several months without ever breathing a word of it to each other. Ben has managed to talk very clearly about himself although without ever talking *directly* about himself, and what's more, both he and Audrey know this is so.

The preceding situation might strike you as being very contrived, thus leaving you with the impression that seeing one romantic situation as "coding" for another is a fragile and unlikely possibility. But nothing could be further from the truth. If two people are romantically involved (or even if they aren't, but at least one of them feels there's a potential spark), then almost any conversation between them about any romance whatsoever, no matter who it involves, stands a good chance of being heard by one or both of them as putting a spotlight on their own situation. Such boomeranging-back is almost inevitable because romances, even very good ones, are filled with uncertainty and yearning. We are always on the lookout for clues or insights into our romantic lives, and analogies are among the greatest sources of clues and insights. Therefore, to notice an analogy between ourselves and another couple that is occupying center stage in our conversation is pretty much a piece of cake handed to us on a silver platter. The crucial question is whether it tastes good or not.

The Latent Ambiguity of the Village Baker's Remarks

Indirect reference of the sort just discussed is often artistically exploited in literature, where, because of a strong analogy that readers easily perceive between Situations A and B, lines uttered by characters in Situation A can easily be heard as applying equally well to Situation B. Sometimes the characters in Situation A are completely unaware of Situation B, which can make for a humorous effect, whereas other times the characters in Situation A are simultaneously characters in Situation B, but aren't aware of (or aren't thinking about) the analogy linking the two situations they are in. The latter creates a great sense of irony, of course.

Since I recently saw a lovely example of this, I can't resist telling you about it. It happens at the end of the 1938 film by Marcel Pagnol, *La Femme du boulanger*. Towards his wife Aurélie, who ran off with a local shepherd only to slink guiltily home three days later, Aimable, the drolly-named village baker, is all sweetness and light — but toward his cat Pomponnette, who, as it happens, *also* ran off and abandoned her mate Pompon three days earlier and who *also* came back on the same day as did Aurélie (all of this happening totally by coincidence, of course), Aimable is absolutely merciless. Taking the side of the injured Pompon (some might say "identifying with him"), Aimable rips Pomponnette to shreds with his accusatory words, and all of this happens right in front of the just-returned Aurélie, using excoriating phrases that viewers might well have expected him to use towards Aurélie. As if this were not enough, Aurélie consumes the heart-shaped bread that Aimable had prepared for himself for dinner (he had no inkling that she would return), while at the very same time, Pomponnette the straying kittycat, wearing a collar with a huge heart on it, is consuming the food just laid out for her mate Pompon.

Does Aimable the baker actually perceive the screamingly obvious analogy? Or could he be so kind and forgiving a soul that he doesn't see Aurélie and Pomponnette as two peas in a pod, and could the deliciously double-edged bile that we hear him savagely (but justifiedly) dumping on the cat be innocently single-edged to him?

Whichever may be the case, I urge you to go out and see the film; it's a poignant masterpiece. And if by some strange chance your very own sweetheart, sitting at your side and savoring the movie with you, has just returned to the nest after *une toute petite amourette* with some third party, just imagine how she or he is going to start squirming when that last scene arrives! But why on earth would someone *outside* the movie feel the sting of a volley of stern rebukes made by someone *inside* the movie? Ah, well… analogy has force in proportion to its precision and its visibility.

Chantal and the Piggybacked Levels of Meaning

Let's now explore an analogy whose two sides are more different than two cookies or two lovers, more different even than a straying wife and a straying cat. It's an analogy that comes up, albeit implicitly, when we are watching a video on our TV — let's say, a show about a French baker, his wife, his friends, and his cats. The point is that we are not *really* watching the cavortings of people and cats — not literally, anyway. To say that we are doing so is a useful shorthand, since what we are actually seeing is a myriad of pixels that are copying, in a perfect lockstep-synchronized fashion, dynamically shifting patterns of color splotches that once were scattered off some animate and inanimate objects in a long-ago-and-far-away French village. We are watching a million or so dots that "code" for those people's actions, but luckily the code is very easy for us to decode — so totally effortless, in fact, that we are sucked in by the mapping, by the isomorphism (the screen/scene *analogy*, if you will), and we find ourselves "teleported" to some remote place and time where we seem to be seeing events happening just as they normally do; we feel it is annoyingly nitpicky to make fine distinctions about whether we are "really" watching those events or not. (Are we *really* talking to each other if we talk by phone?)

It is all too easy to forget that moths, flies, dogs, cats, neonates, television cameras, and other small-souled beings do not perceive a television screen as we do. Although it's hard for us to imagine, they see the pixels in a raw, uninterpreted fashion, and thus to them a TV screen is as drained of long-ago-and-far-away meanings as is, to you or me, a pile of fall leaves, a Jackson Pollock painting, or a newspaper article in Malagasy (my apologies to you if you speak Malagasy; in that case, please replace it by Icelandic — and don't tell me that you speak that language, too!). "Reading" a TV screen at the representational level is intellectually far beyond such creatures, even if for most humans it is essentially second nature already by age two or so.

A dog gazing vacantly at a television screen, unable to make out any imagery, unaware even that any imagery is intended, is thus not unlike Lord Russell staring blankly at a formula of his beloved system *PM* and seeing only its "easy" (arithmetical) meaning, while the other meaning, the mapping-mediated meaning due to Gödel, lies intellectually beyond him, utterly inaccessible, utterly undreamt-of. Or perhaps you think this comparison is unfair to Sir Bertrand, and in a way I agree, so let me try to make it a little more realistic and more generous.

Instead of a dog that, when placed in front of a TV screen, sees only pixels rather than people, imagine little three-year-old Chantal Duplessix, who is watching *La Femme du boulanger* with her parents. All three are native speakers of French, so there's no language barrier. Just like her *maman* and *papa*, Chantal sees right through the pixels to the events in the village, and when that wonderful final scene arrives and Aimable rakes the cat over the coals, Chantal laughs and laughs at Aimable's fury — but she doesn't suspect for a moment that there is *another* reading of his words. She's too young to get the analogy between Aurélie and Pomponnette, and so for her there is only one meaning there. Filmmaker Pagnol's analogy-mediated meaning, which takes for granted the "simple" (although dog-eluding) mapping of pixels to remote events and thus piggybacks on it, is effortlessly perceived by her parents, but for the time being, it lies intellectually beyond Chantal, and is utterly inaccessible to her. In a few years, of course, things will be different — Chantal will have learned how to pick up on analogies between all sorts of complex situations — but that's how things are now.

With this situation, we can make a more realistic and more generous comparison to Bertrand Russell (yet another analogy!). Chantal, unlike a dog, does not merely see meaningless patterns of light on the screen; she effortlessly sees people and events — the "easy" meaning of the patterns. But there is a second level of meaning that takes the people and events for granted, a meaning transmitted by an analogy between events, and it's that *higher* level of meaning that eludes Chantal. In much the same way, Gödel's higher level of meaning, mediated by his mapping, his marvelous analogy, eluded Bertrand Russell. From what I have read about Russell, he never saw the second level of meaning of formulas of *PM*. In a certain sad sense, the good Lord never learned to read his own holy books.

Pickets at the Posh Shop

As I suggested above, your recently returned roving sweetheart might well hear an extra level of meaning while listening to Aimable chastise Pomponnette. Thus a play or film can carry levels of meaning that the author never dreamt of. Let's consider, for example, the little-known 1931 play *The Posh Shop Picketeers*, written by social activist playwright Rosalyn Wadhead (ever hear of her?). This play is about a wildcat strike called by the workers at Alf and Bertie's Posh Shop (I admit, I never did figure out what they sold there). In this play, there is a scene where shoppers approaching the store's entrance are exhorted not to cross the picket line and not to buy anything in the store ("Alf and Bertie are filthy dirty! Please don't cross our Posh Shop pickets! Please cross over to the mom-and-pop shop!"). In the skilled hands of our playwright, this simple situation led to a drama of great tension. But for some reason, just before the play was to open, the ushers in the theater and the actors in the play got embroiled in a bitter dispute, as a result of which the ushers' union staged a wildcat strike on opening night, put up picket lines, and beseeched potential playgoers not to cross their lines to see *The Posh Shop Picketeers*.

Obviously, given this unanticipated political context, the lines uttered by the actors inside the play assumed a powerful second meaning for viewers in the audience, an extra level of meaning that Rosalyn Wadhead never intended. In fact, the picketing Posh Shop worker named "Cagey", who disgustedly proclaims, after a brash matron pushes her aside and arrogantly strides into Alf and Bertie's upscale showroom, "Anyone who crosses the picket line in front of Alf and Bertie's Posh Shop is scum", was inevitably heard by everyone in the audience (which by definition consisted solely of people who had crossed the picket line outside the theater) as saying, "Anyone who crossed the picket line in front of this theater is scum", and of course this amounted to saying, "Anyone who is now sitting in this audience is scum", which could also be heard as "You should not be listening to these lines", which was the diametric opposite of what all the actors, including the one playing the part of Cagey, wanted to tell their audience, whose entry into the theater they so much appreciated, given the ushers' hostile picket line.

But what could the actors do about the fact that they were unmistakably calling their deeply appreciated audience "scum" and insinuating that no one should even have been there to hear these lines? Nothing. They *had* to recite the play's lines, and the analogy was there, it was blatant and strong, and therefore the ironic, twisting-back, self-referential meaning of Cagey's line, as well as of many others in the play, was unavoidable. Admittedly, the self-reference was *indirect* — mediated by an analogy — but that did not make it any less real or strong than would "direct" reference. Indeed, what we might be tempted to call "direct" reference is mediated by a code, too — the code between words and things given to us by our native language (Malagasy, Icelandic, etc.). It's just that *that* code is a simpler one (or at least a more familiar one). In sum, the seemingly sharp distinction between "direct" reference and "indirect" reference is only a matter of degree, not a black-and-white distinction. To repeat, analogy has force in proportion to its precision and visibility.

Prince Hyppia: Math Dramatica

Well, so much for Rosalyn Wadhead and the surprise double-edgedness of the lines in *The Posh Shop Picketeers*, admittedly a rather obscure work. Let's move on to something completely different. We'll talk instead about the world-famous play *Prince Hyppia: Math Dramatica*, penned in the years 1910–1913 by the celebrated British playwright Y. Ted Enrustle (surely you've heard of *him*!). Fed up with all the too-clevah-by-hahf plays-about-plays that were all the rage in those days, he set out to write a play that would have nothing whatsoever to do with playwriting or acting or the stage. And thus, in this renowned piece, as you doubtless recall, all the characters are strictly limited to speaking about various properties, from very simple to quite arcane, of whole numbers. How could anyone possibly get any further from writing a play about a play? For example, early on in Act I, the beautiful Princess Bloppia famously exclaims, "7 times 11 times 13 equals 1001!", to which the handsome Prince Hyppia excitedly retorts, "Wherefore the number 1001 is composite and not prime!" Theirs would seem to be a math made in heaven. (You may now groan.)

But it's in Act III that things really heat up. The climax comes when Princess Bloppia mentions an arithmetical fact about a certain very large integer g, and Prince Hyppia replies, "Wherefore the number g is saucy and not prim!" (It's a rare audience that fails to gasp in unison when they hear Hyppia's most math-dramatical outburst.) The curious thing is that the proud Prince seems to have no idea of the import of what he is saying, and even more ironically, apparently the playwright, Y. Ted Enrustle, didn't either. However, as everyone today knows, this remark of Prince Hyppia asserts — via the intermediary link of a tight analogy — that a certain long line of typographical symbols is "unpennable" using a standard set of conventions of dramaturgy that held, way back in those bygone days. And the funny thing is that the allegedly unpennable line is none other than the proclamation that the actor playing Prince Hyppia has just pronounced!

As you can well imagine, although Y. Ted Enrustle was constantly penning long lines of symbols that adhered to popular dramaturgical conventions (after all, that was his livelihood!), he'd never dreamt of a connection between the natural numbers (whose peculiar properties his curious characters accurately articulated) and the humble lines of symbols that he penned for his actors to read and memorize. Nonetheless, when, nearly two decades later, this droll coincidence was revealed to the play-going public in a wickedly witty review entitled "On Formerly Unpennable Proclamations in *Prince Hyppia: Math Dramatica* and Related Stageplays (I)", authored by the acerbic young Turko-Viennese drama critic Gerd Külot

(I'll skip the details here, as the story is so well known), its piercing cogency was immediately appreciated by many, and as a result, playgoers who had read Külot's irreverent review became able to rehear many of the famous lines uttered in *Prince Hyppia: Math Dramatica* as if they were not about numbers at all, despite what Y. Ted Enrustle had intended, but were direct (and often quite biting) comments about Y. Ted Enrustle's play itself!

And thus it wasn't long before savvy audiences were reinterpreting the droll remarks by the oddball numerologist Qéé Dzhii (a character in *Prince Hyppia: Math Dramatica* who had gained notoriety for her nearly nonstop jabbering about why she preferred saucy numbers to prim numbers) as revealing, via allusions that now seemed hilariously obvious, why she preferred dramatic lines that were unpennable (using the dramaturgical conventions of the day) to lines that were pennable. Drama lovers considered this new way of understanding the play too delicious for words, for it revealed *Prince Hyppia* to be a play-about-a-play (with a vengeance!), although most of the credit for this insight was given to the brash young forcign critic rather than to the venerable elder playwright.

Y. Ted Enrustle, poor fellow, was simply gobsmacked — there's no other word for it. How could anyone in their right mind take Qéé Dzhii's lines in this preposterous fashion? They were only about *numbers*! After all, to write a drama that was about numbers and *only* about numbers had been his sole ambition, and he had slaved away for years to accomplish that noble goal!

Y. Ted Enrustle lashed out vehemently in print, maintaining that his play was decidedly *not* about a play, let alone about itself! Indeed, he went so far as to insist that Gerd Külot's review could not conceivably be about *Prince Hyppia: Math Dramatica* but had to be about *another* play, possibly a *related* play, perhaps an *analogous* play, perchance even a perfectly *parallel* play, peradventure a play with a similar-sounding title penned by a pair of paranoiac paradoxophobes, but in any case it was not about *his* play.

And yet, protest though he might, there was nothing at all that Y. Ted Enrustle could do about how audiences were now interpreting his beloved play's lines, because the two notions — the sauciness of certain integers and the unpennability of certain lines of theatrical dialogue — were now seen by enlightened playgoers as precisely isomorphic phenomena (every bit as isomorphic as the parallel escapades of Aurélie and Pomponnette). The subtle mapping discovered by the impish Külot and gleefully revealed in his review made both meanings apply equally well (at least to anyone who had read and understood the review). The height of the irony was that, in the case of a few choice arithmetical remarks such as Prince Hyppia's famous outburst, it was *easier* and *more natural* to hear them as referring to unpennable lines in plays than to hear them as referring to non-prim numbers! But Y. Ted Enrustle, despite reading Külot's review many times, apparently never quite caught on to what it was really saying.

Analogy, Once Again, Does its Cagey Thing

Okay, okay, enough's enough. The jig's up! Let me confess. For the last several pages, I've been playing a game, talking about strangely named plays by strangely named playwrights as well as a strangely titled review by a strangely named reviewer, but the truth is (and you knew it all along, dear reader), I've *really* been talking about something totally different — to wit, the strange loop that Austrian logician Kurt Gödel (Gerd Külot) discovered and revealed inside Russell and Whitehead's *Principia Mathematica.*

"Now, now," I hear some voice protesting (but of course it's not *your* voice), "how on earth could you have *really* been talking about Whitehead and Russell and *Principia Mathematica* if the lines you wrote were not about them but about Y. Ted Enrustle and *Prince Hyppia: Math Dramatica* and such things?" Well, once again, it's all thanks to the power of analogy; it's the same game as in a *roman à clef,* where a novelist speaks, not so secretly, about people in real life by ostensibly speaking solely about fictional characters, but where savvy readers know precisely who stands for whom, thanks to analogies so compelling and so glaring that, taken in their cultural context, they cannot be missed by anyone sufficiently sophisticated.

And so we have worked our way up my ladder of examples of doubly-hearable remarks, all the way from the throwaway café blurt "This tastes awful" to the supersophisticated dramatic line "The number g is not prim". We have repeatedly seen how analogies and mappings give rise to secondary meanings that ride on the backs of primary meanings. We have seen that even primary meanings depend on unspoken mappings, and so in the end, we have seen that all meaning is mapping-mediated, which is to say, all meaning comes from analogies. This is Gödel's profound insight, exploited to the hilt in his 1931 paper, bringing the aspirations embodied in *Principia Mathematica* tumbling to the ground. I hope that for all my readers, understanding Gödel's keen insight into meaning is now a piece of cake.

How Can an "Unpennable" Line be Penned?

Something may have troubled you when you learned that Prince Hyppia's famous line about the number g proclaims (via analogy) its own unpennability. Isn't this self-contradictory? If some line in some play is truly unpennable, then how could the playwright have ever penned it? Or, turning this question around, how could Prince Hyppia's classic line be found in Y. Ted Enrustle's play if it never was penned at all?

A very good question indeed. But now, please recall that I defined a "pennable line" as a line that could be written by a playwright who was tacitly adhering to a set of well-established dramaturgical conventions. The concept of "pennability", in other words, implicitly referred to some particular *system of rules*. This means that an "unpennable" line, rather than being a line that could never, ever be written by anyone, would merely be a line that violated one or more of the dramaturgical conventions that most playwrights took for granted. Therefore, an unpennable line could indeed be penned — just not by someone who rigorously respected those rules.

For a strictly rule-bound playwright to pen such a line would be seen as extremely inconsistent; a churlish drama critic, ever reaching for cute new ways to snipe, might even write, "X's play is so mega-inconsistent!" And thus, perhaps it was the recognition of Y. Ted Enrustle's unexpected and bizarre-o "mega-inconsistency" that invariably caused audiences to gasp at Prince Hyppia's math-dramatic outburst. No wonder Gerd Külot received kudos for pointing out that a *formerly* unpennable line had been penned!

"Not" is Not the Source of Strangeness

A reader might conclude that a strange loop necessarily involves a self-undermining or self-negating quality ("This formula is *not* provable"; "This line is *not* pennable"; "You should *not* be attending this play"). However, negation plays no essential role in strange loopiness. It's just that the strangeness becomes more pungent or humorous if the loop enjoys a self-undermining quality. Recall Escher's *Drawing Hands*. There is no negation in it — both hands are drawing. Imagine if one were erasing the other!

In this book, a loop's strangeness comes purely from the way in which a system can seem to "engulf itself" through an unexpected twisting-around, rudely violating what we had taken to be an inviolable hierarchical order. In the cases of both *Prince Hyppia: Math Dramatica* and *Principia Mathematica*, we saw that a system carefully designed to talk only about numbers and *not* to talk about itself nonetheless ineluctably winds up talking about itself in a "cagey" fashion — and it does so precisely because of the chameleonic nature of numbers, which are so rich and complex that numerical patterns have the flexibility to mirror any other kind of pattern.

Every bit as strange a loop, although perhaps a little less dramatic, would have been created if Gödel had concocted a self-*affirming* formula that cockily asserted of itself, "This formula is provable via the rules of

PM", which to me is reminiscent of the brashness of Muhammad ("I'm the greatest") Ali as well as of Salvador ("The great") Dalí. Indeed, some years after Gödel, such self-affirming formulas were concocted and studied by logicians such as Martin Hugo Löb and Leon Henkin. These formulas, too, had amazing and deep properties. I therefore repeat that the *strange* loopiness resides not in the flip due to the word "not", but in the unexpected, hierarchy-violating twisting-back involving the word "this".

I should, however, immediately point out that a phrase such as "this formula" is nowhere to be found inside Gödel's cagey formula — no more than the phrase "this audience" is contained in Cagey's line "Anyone who crosses the picket line to go into Alf and Bertie's Posh Shop is scum." The unanticipated meaning "People *in this audience* are scum" is, rather, the inevitable outcome of a blatantly obvious analogy (or mapping) between two entirely different picket lines (one outside the theater, one on stage), and thus, by extension, between the picket-crossing members of the audience and the picket-line crossers in the play they are watching.

The preconception that an obviously suspicion-arousing word such as "this" (or "I" or "here" or "now" — "indexicals", as they are called by philosophers — words that refer explicitly to the speaker or to something closely connected with the speaker or the message itself) is an indispensable ingredient for self-reference to arise in a system is shown by Gödel's discovery to be a naïve illusion; instead, the strange twisting-back is a simple, natural consequence of an unexpected isomorphism between two different situations (that which is being talked about, on the one hand, and that which is doing the talking, on the other). Bertrand Russell, having made sure that all indexical notions such as "this" were absolutely excluded from his formal system, believed his handiwork to be forever immunized against the scourge of wrapping-around — but Kurt Gödel, with his fateful isomorphism, showed that such a belief was an unjustified article of faith.

Numbers as a Representational Medium

Why did this kind of isomorphism first crop up when somebody was carefully scrutinizing *Principia Mathematica*? Why hadn't anybody thought of such a thing before Gödel came along? It cropped up because *Principia Mathematica* is in essence about the natural numbers, and what Gödel saw was that the world of natural numbers is so rich that, given *any* pattern involving objects of any type, a set of numbers can be found that will be isomorphic to it — in other words, there are numbers that will perfectly mirror the objects and their pattern, numbers that will dance in just the way the objects in the pattern dance. Dancing the same dance is the key.

Kurt Gödel was the first person to realize and exploit the fact that the positive integers, though they might superficially seem to be very austere and isolated, in fact constitute a profoundly rich representational medium. They can mimic or mirror any kind of pattern. Like any human language, where nouns and verbs (etc.) can engage in unlimitedly complex dancing, the natural numbers too, can engage in unlimitedly complex additive and multiplicative (etc.) dancing, and can thereby "talk", via code or analogy, about events of any sort, numerical or non-numerical. This is what I meant when I wrote, in Chapter 9, that the seeds of *PM*'s destruction were already hinted at by the seemingly innocent fact that *PM* had enough power to talk about *arbitrarily subtle properties* of whole numbers.

People of earlier eras had intuited much of this richness when they had tried to embed the nature of many diverse aspects of the world around us — stars, planets, atoms, molecules, colors, curves, notes, harmonies, melodies, and so forth — in numerical equations or other types of numerical patterns. Four centuries ago, launching this whole tendency, Galileo Galilei had famously declared, "The book of Nature is written in the language of mathematics" (a thought that must seem shocking to people who love nature but hate mathematics). And yet, despite all these centuries of highly successful mathematizations of various aspects of the world, no one before Gödel had realized that one of the domains that mathematics can model is *the doing of mathematics itself.*

The bottom line, then, is that the unanticipated self-referential twist that Gödel found lurking inside *Principia Mathematica* was a natural and inevitable outcome of the deep representational power of whole numbers. Just as it is no miracle that a video system can create a self-referential loop, but rather a kind of obvious triviality due to the power of TV cameras (or, to put it more precisely, the immensely rich representational power of very large arrays of pixels), so too it is no miracle that *Principia Mathematica* (or any other comparable system) contains self-focused sentences like Gödel's formula, for the system of integers, exactly like a TV camera (only more so!), can "point" at any system whatsoever and can reproduce that system's patterns perfectly on the metaphorical "screen" constituted by its set of theorems. And just as in video feedback, the swirls that result from *PM* pointing at itself have all sorts of unexpected, emergent properties that require a brand-new vocabulary to describe them.

&. &. &.

CHAPTER 12

On Downward Causality

‎❧ ❧ ❧

Bertrand Russell's Worst Nightmare

To MY mind, the most unexpected emergent phenomenon to come out of Kurt Gödel's 1931 work is a bizarre new type of mathematical causality (if I can use that unusual term). I have never seen his discovery cast in this light by other commentators, so what follows is a personal interpretation. To explain my viewpoint, I have to go back to Gödel's celebrated formula — let's call it "KG" in his honor — and analyze what its existence implies for *PM*.

As we saw at the end of Chapter 10, KG's meaning (or more precisely, its *secondary* meaning — its higher-level, non-numerical, non-Russellian meaning, as revealed by Gödel's ingenious mapping), when boiled down to its essence, is the whiplash-like statement "KG is unprovable inside *PM*." And so a natural question — *the* natural question — is, "Well then, is KG *indeed* unprovable inside *PM*?"

To answer this question, we have to rely on one article of faith, which is that anything provable inside *PM* is a true statement (or, turning this around, that *nothing false is provable* in *PM*). This happy state of affairs is what we called, in Chapter 10, "consistency". Were *PM* not consistent, then it would prove falsities galore about whole numbers, because the instant that you've proven any particular falsity (such as "0=1"), then an infinite number of others ("1=2", "0=2", "1+1=1", "1+1=3", "2+2=5", and so forth) follow from it by the rules of *PM*. Actually, it's worse than that: if *any* false statement, no matter how obscure or recondite it was, were provable in *PM*, then *every conceivable* arithmetical statement, whether true or false, would become provable, and the whole grand edifice would come tumbling down in a pitiful shambles. In short, the provability of even one falsity would mean that *PM* had nothing to do with arithmetical truth at all.

What, then, would Bertrand Russell's worst nightmare be? It would be that someday, someone would come up with a *PM* proof of a formula expressing an untrue arithmetical statement ("0 = s0" is a good example), because the moment that that happened, *PM* would be fit for the dumpster. Luckily for Russell, however, every logician on earth would give you better odds for a snowball's surviving a century in hell. In other words, Bertrand Russell's worst nightmare is truly just a nightmare, and it will never take place outside of dreamland.

Why would logicians and mathematicians — not just Russell but all of them (including Gödel) — give such good odds for this? Well, the axioms of *PM* are certainly true, and its rules of inference are as simple and as rock-solidly sane as anything one could imagine. How can you get falsities out of that? To think that *PM* might have false theorems is, quite literally, as hard as thinking that two plus two is five. And so, along with all mathematicians and logicans, let's give Russell and Whitehead the benefit of the doubt and presume that their grand palace of logic is consistent. From here on out, then, we'll generously assume that *PM* never proves any false statements — all of its theorems are sure to be true statements. Now then, armed with our friendly assumption, let's ask ourselves, "What would follow if KG were provable inside *PM*?"

A Strange Land where "Because" Coincides with "Although"

Indeed, reader, let's posit, you and I, that KG is provable in *PM,* and then see where this assumption — I'll dub it the "Provable-KG Scenario" — leads us. The ironic thing, please note, is that KG itself *doesn't* believe the Provable-KG Scenario. Perversely, KG shouts to the world, "I am *not* provable!" So if *we* are right about KG, dear reader, then KG is wrong about itself, no matter how loudly it shouts. After all, no formula can be both *provable* (as we claim KG is) and also *unprovable* (as KG claims to be). One of us has to be wrong. (And for any formula, being *wrong* means being *false.* The two terms are synonyms.) So… if the Provable-KG Scenario is the case, then KG is wrong (= false).

All right. Our reasoning started with the Provable-KG Scenario and wound up with the conclusion "KG is false". In other words, *if KG is provable, then it is also false.* But hold on, now — a provable falsity in *PM*?! Didn't we just declare firmly, a few moments ago, that *PM* never proves falsities? Yes, we did. We agreed with the universal logicians' belief that *PM* is consistent. If we stick to our guns, then, the Provable-KG Scenario has to be wrong, because it leads to Russell's worst nightmare. We have to retract it, cancel it, repudiate it, nullify it, and revoke it, because *accepting* it led us to a conclusion (*"PM* is inconsistent") that we know is wrong.

Ergo, the Provable-KG Scenario is hereby rejected, which leaves us with the opposite scenario: KG is *not* provable. Now the funny thing is that this is exactly what KG is shouting to the rooftops. We see that what KG proclaims about itself — "I'm unprovable!" — is *true.* In a nutshell, we have established two facts: (1) KG is unprovable in *PM*; and (2) KG is true.

We have just uncovered a very strange anomaly inside *PM*: here is a statement of arithmetic (or number theory, to be slightly more precise) that we are sure is *true,* and yet we are equally sure it is *unprovable* — and to cap it off, these two contradictory-sounding facts are consequences of each other! In other words, KG is unprovable not only *although* it is true, but worse yet, *because* it is true.

This weird situation is utterly unprecedented and profoundly perverse. It flies in the face of the Mathematician's Credo, which asserts that truth and provability are just two sides of the same coin — that they always go together, because they entail each other. Instead, we've just encountered a case where, astoundingly, truth and *un*provability entail each other. Now isn't that a fine how-do-you-do?

Incompleteness Derives from Strength

The fact that there exists a truth of number theory that is unprovable in *PM* means, as you may recall from Chapter 9, that *PM* is *incomplete.* It has holes in it. (So far we've seen just one hole — KG — but it turns out there are plenty more — an infinity of them, in fact.) Some statements of number theory that *should* be provable escape from *PM*'s vast net of proof — they slip through its mesh. Clearly, this is another kind of nightmare — perhaps not quite as devastating as Bertrand Russell's worst nightmare, but somehow even more insidious and troubling.

Such a state of affairs is certainly not what the mathematicians and logicians of 1931 expected. Nothing in the air suggested that the axioms and rules of inference of *Principia Mathematica* were weak or deficient in any way. They seemed, quite the contrary, to imply virtually everything that anyone might have thought was true about numbers. The opening lines of Gödel's 1931 article, quoted in Chapter 10, state this clearly. If you'll recall, he wrote, speaking of *Principia Mathematica* and Zermelo-Fraenkel set theory: "These two systems are so extensive that all methods of proof used in mathematics today have been formalized in them, *i.e.,* reduced to a few axioms and rules of inference."

What Gödel articulates here was virtually a universal credo at the time, and so his revelation of *PM*'s incompleteness, in the twenty-five pages that followed, came like a sudden thunderbolt from the bluest of skies.

To add insult to injury, Gödel's conclusion sprang not from a weakness in PM but from a strength. That strength is the fact that numbers are so flexible or "chameleonic" that their patterns can mimic patterns of reasoning. Gödel exploited the simple but marvelous fact that the familiar whole numbers can dance in just the same way as the unfamiliar symbol-patterns of *PM* dance. More specifically, the prim numbers that he invented act indistinguishably from provable strings, and one of *PM*'s natural strengths is that it is able to talk about prim numbers. For this reason, it is able to talk about itself (in code). In a word, *PM*'s *expressive power* is what gives rise to its incompleteness. What a fantastic irony!

Bertrand Russell's Second-worst Nightmare

Any enrichment of *PM* (say, a system having more axioms or more rules of inference, or both) would have to be just as expressive of the flexibility of numbers as was *PM* (otherwise it would be weaker, not stronger), and so the same Gödelian trap would succeed in catching it — it would be just as readily hoist on its own petard.

Let me spell this out more concretely. Strings provable in the larger and allegedly superior system *Super-PM* would be isomorphically imitated by a *richer* set of numbers than the prim numbers (hence let's call them "super-prim numbers"). At this point, just as he did for *PM*, Gödel would promptly create a new formula KH for *Super-PM* that said, "The number *h* is not a super-prim number", and of course he would do it in such a way that *h* would be the Gödel number of KH itself. (Doing this for *Super-PM* is a cinch once you've done it for *PM*.) The exact same pattern of reasoning that we just stepped through for *PM* would go through once again, and the supposedly more powerful system would succumb to incompleteness in just the same way, and for just the same reasons, as *PM* did. The old proverb puts it succinctly: "The bigger they are, the harder they fall."

In other words, the hole in *PM* (and in any other axiomatic system as rich as *PM*) is not due to some careless oversight by Russell and Whitehead but is simply an inevitable property of *any* system that is flexible enough to capture the chameleonic quality of whole numbers. *PM* is rich enough to be able to turn around and point at itself, like a television camera pointing at the screen to which it is sending its image. If you make a good enough TV system, this looping-back ability is inevitable. And the higher the system's resolution is, the more faithful the image is.

As in judo, your opponent's power is the source of their vulnerability. Kurt Gödel, maneuvering like a black belt, used *PM*'s power to bring it crashing down. Not as catastrophically as with inconsistency, mind you, but in a wholly unanticipated fashion — crashing down with *incompleteness*. The fact that you can't get around Gödel's black-belt trickery by enriching or enlarging *PM* in any fashion is called "essential incompleteness" — Bertrand Russell's *second-worst* nightmare. But unlike his worst nightmare, which is just a bad dream, this nightmare takes place outside of dreamland.

An Endless Succession of Monsters

Not only does extending *PM* fail to save the boat from sinking, but worse, KG is far from being the only hole in *PM*. There are infinitely many ways of Gödel-numbering any given axiomatic system, and each one produces its own cousin to KG. They're all different, but they're so similar they are like clones. If you set out to save the sinking boat, you are free to toss KG or any of its clones as a new axiom into *PM* (for that matter, feel free to toss them all in at once!), but your heroic act will do little good; Gödel's recipe will instantly produce a brand-new cousin to KG. Once again, this new self-referential Gödelian string will be "just like" KG and its passel of clones, but it won't be *identical* to any of them. And you can toss *that* one in as well, and you'll get yet another cousin! It seems that holes are popping up inside the struggling boat of *PM* as plentifully as daisies and violets pop up in the springtime. You can see why I call this nightmare more insidious and troubling than Russell's worst one.

Not only Bertrand Russell was blindsided by this amazingly perverse and yet stunningly beautiful maneuver; virtually every mathematical thinker was, including the great German mathematician David Hilbert, one of whose major goals in life had been to rigorously ground all of mathematics in an axiomatic framework (this was called "the Hilbert Program"). Up till the Great Thunderclap of 1931, it was universally believed that this noble goal had been reached by Whitehead and Russell.

To put it another way, the mathematicians of that time universally believed in what I earlier called the "Mathematician's Credo (*Principia Mathematica* version)". Gödel's shocking revelation that the pedestal upon which they had quite reasonably placed their faith was fundamentally and irreparably flawed followed from two things. One is our kindly assumption that the pedestal is consistent (*i.e.*, we will never find any falsity lurking among the theorems of *PM*); the other is the nonprovability in *PM* of KG and all its infinitely many cousins, which we just showed is a consequence flowing from their self-referentiality, taking *PM*'s consistency into account.

To recap it just one last time, what is it about KG (or any of its cousins) that makes it not provable? In a word, it is its self-referential meaning: if KG were provable, its loopy meaning would flip around and make it unprovable, and so *PM* would be inconsistent, which we know it is not.

But notice that we have not made any detailed analysis of the nature of derivations that would try to make KG appear as their bottom line. In fact, we have totally ignored the *Russellian* meaning of KG (what I've been calling its *primary* meaning), which is the claim that the gargantuan number that I called '*g*' possesses a rather arcane and recherché number-theoretical property that I called "sauciness" or "non-primness". You'll note that in the last couple of pages, not one word has appeared about prim numbers or non-prim numbers and their number-theoretical properties, nor has the number *g* been mentioned at all. We finessed all such numerical issues by looking only at KG's *secondary* meaning, the meaning that Bertrand Russell never quite got. A few lines of purely non-numerical reasoning (the second section of this chapter) convinced us that this statement (which is about numbers) could not conceivably be a theorem of *PM*.

Consistency Condemns a Towering Peak to Unscalability

Imagine that a team of satellite-borne explorers has just discovered an unsuspected Himalayan mountain peak (let's call it "KJ") and imagine that they proclaim, both instantly and with total confidence, that thanks to a special, most unusual property of the summit alone, there is no conceivable route leading up to it. Merely from looking at a single photo shot vertically downwards from 250 miles up, the team declares KJ an *unclimbable peak*, and they reach this dramatic conclusion without giving any thought to the peak's properties as seen from a conventional mountaineering perspective, let alone getting their hands dirty and actually trying out any of the countless potential approaches leading up the steep slopes towards it. "Nope, none of them will work!", they cheerfully assert. "No need to bother trying any of them out — you'll fail every time!"

Were such an odd event to transpire, it would be remarkably different from how all previous conclusions about the scalability of mountains had been reached. Heretofore, climbers always had to attempt many routes — indeed, to attempt them many times, with many types of equipment and in diverse weather conditions — and even thousands of failures in a row would not constitute an ironclad proof that the given peak was forever unscalable; all one could conclude would be that it had *so far resisted scaling*. Indeed, the very idea of a "proof of unscalability" would be most alien to the activity of mountaineering.

By contrast, our team of explorers has concluded from some novel property of KJ, without once thinking about (let alone actually trying out) a single one of the infinitely many conceivable routes leading up to its summit, that by its very nature it is unscalable. And yet their conclusion, they claim, is not merely probable or extremely likely, but dead certain.

This amounts to an unprecedented, upside-down, top-down kind of alpinistic causality. What kind of property might account for the peculiar peak's unscalability? Traditional climbing experts would be bewildered at a blanket claim that for every conceivable route, climbers will inevitably encounter some fatal obstacle along the way. They might more modestly conclude that the distant peak would be extremely difficult to scale by looking *upwards* at it and trying to take into account all the imaginable routes that one might take in order to reach it. But our intrepid team, by contrast, has looked solely at KJ's tippy-top and concluded *downwards* that there simply could be no route that would ever reach it from below.

When pressed very hard, the team of explorers finally explains how they reached their shattering conclusions. It turns out that the photograph taken of KJ from above was made not with ordinary light, which would reveal nothing special at all, but with the newly discovered "Gödel rays". When KJ is perceived through this novel medium, a deeply hidden set of fatal structures is revealed.

The problem stems from the consistency of the rock base underlying the glaciers at the very top; it is so delicate that, were any climber to come within striking distance of the peak, the act of setting the slightest weight on it (even a grain of salt; even a baby bumblebee's eyelash!) would instantly trigger a thunderous earthquake, and the whole mountain would come tumbling down in rubble. So the peak's inaccessibility turns out to have nothing to do with how anyone might try to get *up* to it; it has to do with an inherent instability belonging to the summit itself, and moreover, a type of instability that only Gödel rays can reveal. Quite a silly fantasy, is it not?

Downward Causality in Mathematics

Indeed it is. But Kurt Gödel's bombshell, though just as fantastic, was not a fantasy. It was rigorous and precise. It revealed the stunning fact that a formula's *hidden meaning* may have a peculiar kind of "downward" causal power, determining the formula's truth or falsity (or its derivability or nonderivability inside *PM* or any other sufficiently rich axiomatic system). Merely from knowing the formula's meaning, one can infer its truth or falsity without any effort to derive it in the old-fashioned way, which requires one to trudge methodically "upwards" from the axioms.

   This is not just peculiar; it is astonishing. Normally, one cannot merely look at what a mathematical conjecture *says* and simply appeal to the content of that statement on its own to deduce whether the statement is true or false (or provable or unprovable).

   For instance, if I tell you, "There are infinitely many perfect numbers" (numbers such as 6, 28, and 496, whose factors add up to the number itself), you will not know if my claim — call it 'Imp' — is true or not, and merely staring for a long time at the written-out statement of Imp (whether it's expressed in English words or in some prickly formal notation such as that of *PM*) will not help you in the least. You will have to try out various approaches to this peak. Thus you might discover that 8128 is the next perfect number after 496; you might note that none of the perfect numbers you come up with is odd, which is somewhat odd; you might observe that each one you find has the form $p(p+1)/2$, where p is an odd prime (such as 3, 7, or 31) and $p+1$ is also a power of 2 (such as 4, 8, or 32); and so forth.

   After a while, perhaps a long series of failures to prove Imp would gradually bring you around to suspecting that it is false. In that case, you might decide to switch goals and try out various approaches to the nearby rival peak — namely, Imp's negation ~Imp — which is the statement "There are *not* infinitely many perfect numbers", which is tantamount to asserting that there is a *largest* perfect number (reminiscent of our old friend *P,* allegedly the largest prime number in the world).

   But suppose that through a stunning stroke of genius you discovered a new kind of "Gödel ray" (*i.e.,* some clever new Gödel numbering, including all of the standard Gödel machinery that makes prim numbers dance in perfect synchrony with provable strings) that allowed you to see through to a hidden *second* level of meaning belonging to Imp — a hidden meaning that proclaimed, to those lucky few who knew how to decipher it, "The integer i is not prim", where i happened to be the Gödel number of Imp itself. Well, dear reader, I suspect it wouldn't take you long to recognize this scenario. You would quickly realize that Imp, just like KG, asserts of itself via your new Gödel code, "Imp has no proof in *PM.*"

   In that most delightful though most unlikely of scenarios, you could immediately conclude, without any further search through the world of whole numbers and their factors, or through the world of rigorous proofs, that Imp was both *true* and *unprovable.* In other words, you would conclude that the statement "There are infinitely many perfect numbers" is true, and you would also conclude that it has no proof using *PM*'s axioms and rules of inference, and last of all (twisting the knife of irony), you would conclude that Imp's *lack of proof* in *PM* is a direct consequence of its *truth.*

You may think the scenario I've just painted is nonsensical, but it is exactly analogous to what Gödel did. It's just that instead of starting with an *a priori* well-known and interesting statement about numbers and then fortuitously bumping into a very strange alternate meaning hidden inside it, Gödel carefully *concocted* a statement about numbers and revealed that, because of how he had designed it, it had a very strange alternate meaning. Other than that, though, the two scenarios are identical.

The hypothetical Imp scenario and the genuine KG scenario are, as I'm sure you can tell, radically different from how mathematics has traditionally been done. They amount to *upside-down reasoning* — reasoning from a would-be theorem downwards, rather than from axioms upwards, and in particular, reasoning from a *hidden meaning* of the would-be theorem, rather than from its surface-level claim about numbers.

Göru and the Futile Quest for a Truth Machine

Do you remember Göru, the hypothetical machine that tells prim numbers from saucy (non-prim) numbers? Back in Chapter 10, I pointed out that if we had built such a Göru, or if someone had simply given us one, then we could determine the truth or falsity of any number-theoretical conjecture at all. To do so, we would merely translate conjecture C into a *PM* formula, calculate its Gödel number c (a straightforward task), and then ask Göru, "Is c prim or saucy?" If Göru came back with the answer "c is prim", we'd proclaim, "Since c is prim, conjecture C is provable, hence it is true", whereas if Göru came back with the answer "c is saucy", then we'd proclaim, "Since c is saucy, conjecture C is not provable, hence it is false." And since Göru would always (by stipulation) eventually give us one or the other of these answers, we could just sit back and let it solve whatever math puzzle we dreamt up, of whatever level of profundity.

It's a great scenario for solving all problems with just one little gadget, but unfortunately we can now see that it is fatally flawed. Gödel revealed to us that there is a profound gulf between truth and provability in *PM* (indeed, in any formal axiomatic system like *PM*). That is, there are many true statements that are not provable, alas. So if a formula of *PM* fails to be a theorem, you can't take that as a sure sign that it is false (although luckily, whenever a formula *is* a theorem, that's a sure sign that it is true). So even if Göru works exactly as advertised, always giving us a correct 'yes' or 'no' answer to any question of the form "Is n prim?", it won't be able to answer all mathematical questions for us, after all.

Despite being less informative than we had hoped, Göru would still be a nice machine to own, but it turns out that even that is not in the cards.

No reliable prim/saucy distinguisher can exist at all. (I won't go into the details here, but they can be found in many texts of mathematical logic or computability.) All of a sudden, it seems as if dreams are coming crashing down all around us — and in a sense, this is what happened in the 1930's, when the great gulf between the abstract concept of truth and mechanical ways to ascertain truth was first discovered, and the stunning size of this gulf started to dawn on people.

It was logician Alfred Tarski who put one of the last nails in the coffin of mathematicians' dreams in this arena, when he showed that there is not even any way to *express* in *PM* notation the English statement "*n* is the Gödel number of a true formula of number theory". What Tarski's finding means is that although there is an infinite set of numbers that stand for true statements (using some particular Gödel numbering), and a complementary infinite set of numbers that stand for false statements, there is no way to express that distinction as a number-theoretical one. In other words, the set of all wff numbers is divided into two complementary parts by the true/false dichotomy, but the boundary line is so peculiar and elusive that it is not characterizable in any mathematical fashion at all.

All of this may seem terribly perverse, but if so, it is a wonderful kind of perversity, in that it reveals the profundity of humanity's age-old goals in mathematics. Our collective quest for mathematical truth is shown to be a quest for something indescribably subtle and therefore, in a sense, sacred. I'm reminded again that the name "Gödel" contains the word "God" — and who knows what further mysteries are lurking in the two dots on top?

The Upside-down Perceptions of Evolved Creatures

As the above excursion has shown, strange loops in mathematical logic have very surprising properties, including what appears to be a kind of upside-down causality. But this is by no means the first time in this book that we have encountered upside-down causality. The notion cropped up in our discussion of the careenium and of human brains. We concluded that evolution tailored human beings to be perceiving entities — entities that filter the world into macroscopic categories. We are consequently fated to describe what goes on about us, including what other people do and what we ourselves do, not in terms of the underlying particle physics (which lies many orders of magnitude removed from our everyday perceptions and our familiar categories), but in terms of such abstract and ill-defined high-level patterns as mothers and fathers, friends and lovers, grocery stores and checkout stands, soap operas and beer commercials, crackpots and geniuses, religions and stereotypes, comedies and tragedies, obsessions and phobias, and of course beliefs and desires, hopes and fears, dreads and dreams, ambitions and jealousies, devotions and hatreds, and so many other abstract patterns that are a million metaphorical miles from the microworld of physical causality.

There is thus a curious upside-downness to our normal human way of perceiving the world: we are built to perceive "big stuff" rather than "small stuff", even though the domain of the tiny seems to be where the actual motors driving reality reside. The fact that our minds see only the high level while completely ignoring the low level is reminiscent of the possibilities of high-level vision that Gödel revealed to us. He found a way of taking a colossally long *PM* formula (KG or any cousin) and reading it in a concise, easily comprehensible fashion ("KG has no proof in *PM*") instead of reading it as the low-level numerical assertion that a certain gargantuan integer possesses a certain esoteric recursively defined number-theoretical property (non-primness). Whereas the standard low-level reading of a *PM* string is right there on the surface for anyone to see, it took a genius to imagine that a high-level reading might exist in parallel with it.

By contrast, in the case of a creature that thinks with a brain (or with a careenium), reading its own brain activity at a high level is natural and trivial (for instance, "I remember how terrified I was that time when Grandma took me to see *The Wizard of Oz*"), whereas the low-level activities that underwrite the high level (numberless neurotransmitters hopping like crazy across synaptic gaps, or simms silently bashing by the billions into each other) are utterly hidden, unsuspected, invisible. A creature that thinks knows next to nothing of the substrate allowing its thinking to happen, but nonetheless it knows all about its symbolic interpretation of the world, and knows very intimately something it calls "I".

Stuck, for Better or Worse, with "I"

It would be a rare thinker indeed that would discount its everyday, familiar symbols and its ever-present sense of "I", and would make the bold speculation that somewhere physically inside its cranium (or its careenium), there might be an esoteric, hidden, *lower* level, populated by some kind of invisible churnings that have nothing to do with its symbols (or simmballs), but which somehow must involve myriads of microscopic units that, most mysteriously, lack all symbolic quality.

When you think about human life this way, it seems rather curious that we become aware of our brains in high-level, non-physical terms (like hopes and beliefs) long before becoming aware of them on low-level neural terms. (In fact, most people never come into contact at all with their brains at that level.) Had things happened in an analogous fashion in the case of *Principia Mathematica,* then recognition of the high-level Gödelian meaning of certain formulas of *PM* would have long preceded recognition of their far more basic Russellian meanings, which is an inconceivable scenario. In any case, we humans evolved to perceive and describe ourselves in high-level mentalistic terms ("I hope to read *Eugene Onegin* next summer") and not in low-level physicalistic terms (imagine an unimaginably long list of the states of all the neurons responsible for your hoping to read *Eugene Onegin* next summer), although humanity is collectively making small bits of headway toward the latter.

Proceeding Slowly Towards the Bottom Level

Such mentalistic notions as "belief", "hope", "guilt", "envy", and so on arose many eons before any human dreamt of trying to ground them as recurrent, recognizable patterns in some physical substrate (the living brain, seen at some fine-grained level). This tendency to proceed slowly from intuitive understanding at a high level to scientific understanding at a low level is reminiscent of the fact that the abstract notion of a *gene* as the basic unit by which heredity is passed from parent to offspring was boldly postulated and then carefully studied in laboratories for many decades before any "hard" physical grounding was found for it. When microscopic structures were finally found that allowed a physical "picture" to be attached to the abstract notion, they turned out to be wildly unexpected entities: a gene was revealed to be a medium-length stretch of a very long helically twisting cord made of just four kinds of molecules (nucleotides) linked one to the next to form a chain millions of units long.

And then, miraculously, it turned out that the chemistry of these four molecules was in a certain sense incidental — what mattered most of all when one thought about heredity was their newly revealed *informational* properties, as opposed to their traditional physico-chemical properties. That is, the proper description of how heredity and reproduction worked could in large part be abstracted away from the chemistry, leaving just a high-level picture of information-manipulating processes alone.

At the heart of these information-manipulating processes lay a high abstraction called the "genetic code", which mapped every possible three-nucleotide "word" (or "codon"), of which there are sixty-four, to one of twenty different molecules belonging to a totally unrelated chemical family (the amino acids). In other words, a profound understanding of genes and heredity was possible only if one was intimately familiar with a high-level meaning-mediating mapping. This should sound familiar.

Of Hogs, Dogs, and Bogs

If you wish to understand what goes on in a biological cell, you have to learn to think on this new informational level. Physics alone, although theoretically sufficient, just won't cut it in terms of feasibility. Obviously the elementary particles take care of themselves, not caring at all about the informational levels of biomolecules (let alone about human perceptual categories or abstract beliefs or "I"-ness or patriotism or the burning desire, on the part of a particular large agglomeration of biomolecules, to compose a set of twenty-four preludes and fugues). Out of all these elementary particles doing their microscopic things, there emerge the macroscopic events that befall a bio-creature.

However, as I pointed out before, if you choose to focus on the particle level, then you cannot draw neat boundary lines separating an entity such as a cell or a hog from the rest of the world in which it resides. Notions like "cell" or "hog" aren't relevant at that far lower level. The laws of particle physics don't respect such notions as "hog", "cell", "gene", or "genetic code", or even the notion of "amino acid". The laws of particle physics involve only particles, and larger macroscopic boundaries drawn for the convenience of thinking beings are no more relevant to them than voting-precinct boundaries are to butterflies. Electrons, photons, neutrinos, and so forth zip across such artificial boundaries without the least compunction.

If you go the particles route, then you are committed to doing so whole hog, which unfortunately means going way beyond the hog. It entails taking into account all the particles in all the members of the hog's family, all the particles in the barn it lives in, in the mud it wallows in, in the farmer who feeds it, in the atmosphere it breathes, in the raindrops that fall on it, in the cumulo-nimbus clouds from which those drops fall, in the thunderclaps that make the hog's eardrums reverberate, in the whole of the earth, in the whole of the sun, in the cosmic background radiation pervading the entire universe and stretching back in time to the Big Bang, and on and on. This is far too large a task for finite folks like us, and so we have to settle for a compromise, which is to look at things at a less inclusive, less detailed level, but (fortunately for us) a more insight-providing level, namely the informational level.

At that level, biologists talk about and think about what genes *stand for*, rather than focusing on their traditional physico-chemical properties. And they implicitly accept the fact that this new, "leaner and meaner" way of talking suggests that genes, thanks to their informational qualities, have their own causal properties — or in other words, that certain extremely abstract large-scale events or states of affairs (for example, the high-level regularity that golden retrievers tend to be very gentle and friendly) can validly be attributed to *meanings of molecules.*

To people who deal directly in dogs and not in molecular biology, this kind of thing is taken for granted. Dog folks talk all the time about the temperamental and mental propensities of this or that breed, as if all this were somehow completely detached from the physics and chemistry of DNA (not to mention physical levels finer than that of DNA), and as if it resided purely at the abstract level of "character traits of dog breeds". And the marvelous thing is that dog folks, no less than molecular biologists, can get along perfectly well thinking and talking this way. It actually works! Indeed, if they (or molecular biologists) tried to do it the pure-physics way or the pure-molecular-biology way, they would instantly get bogged down in the infinite detail of unimaginable numbers of interacting micro-entities constituting dogs and their genes (not to mention the rest of the universe).

The upshot of all this is that the most *real* way of talking about dogs or hogs involves, as Roger Sperry said, high-level entities pushing low-level entities around with impunity. Recall that the intangible, abstract quality of the primality of the integer 641 is what most truly topples hard, solid dominos located in the "prime stretch" of the chainium. This is nothing if not downward causality, and it leads us straight to the conclusion that the most efficient way to think about brains that have symbols — and for most purposes, the *truest* way — is to think that the microstuff inside them is pushed around by ideas and desires, rather than the reverse.

CHAPTER 13

The Elusive Apple of My "I"

❧ ❧ ❧

The Patterns that Constitute Experience

BY OUR deepest nature, we humans float in a world of familiar and comfortable but quite impossible-to-define abstract patterns, such as: "fast food" and "clamato juice", "tackiness" and "wackiness", "Christmas bonuses" and "customer service departments", "wild goose chases" and "loose cannons", "crackpots" and "feet of clay", "slam dunks" and "bottom lines", "lip service" and "elbow grease", "dirty tricks" and "doggie bags", "solo recitals" and "sleazeballs", "sour grapes" and "soap operas", "feedback" and "fair play", "goals" and "lies", "dreads" and "dreams", "she" and "he" — and last but not least, "you" and "I".

Although I've put each of the above items in quotation marks, I am not talking about the written words, nor am I talking about the observable phenomena in the world that these expressions "point to". I am talking about the *concepts* in my mind and your mind that these terms designate — or, to revert to an earlier term, about the corresponding *symbols* in our respective brains.

With my hopefully amusing little list (which I pared down from a much longer one), I am trying to get across the flavor of most adults' daily mental reality — the bread-and-butter sorts of symbols that are likely to be awakened from dormancy in one's brain as one goes about one's routines, talking with friends and colleagues, sitting at a traffic light, listening to radio programs, flipping through magazines in a dentist's waiting room, and so on. My list is a random walk through an everyday kind of mental space, drawn up in order to give a feel for the phenomena in which we place the most stock and in which we most profoundly believe (sour grapes and wild goose chases being quite real to most of us), as opposed to the forbidding and inaccessible level of quarks and gluons, or the only slightly more accessible level of genes and ribosomes and transfer RNA — levels of "reality" to which we may pay lip service but which very few of us ever think about or talk about.

And yet, for all its supposed reality, my list is pervaded by vague, blurry, unbelievably elusive abstractions. Can you imagine trying to define any of its items *precisely*? What on earth is the quality known as "tackiness"? Can you teach it to your kids? And please give me a pattern-recognition algorithm that will infallibly detect sleazeballs!

Reflected Communist Bachelors with Spin 1/2 are All Wet

As a simple illustration of how profoundly wedded our thinking is to the blurry, hazy categories of the macroworld, consider the curious fact that logicians — people who by profession try to write down ironclad, razor-sharp rules of logical inference that apply with impeccable precision to linguistic expressions — seldom if ever resort to the level of particles and fields for their canonical examples of fundamental, eternal truths. Instead, their most frequent examples of "truth" are typically sentences that use totally out-of-focus categories — sentences such as "Snow is white", "Water is wet", "Bachelors are unmarried males", and "Communism either is or is not in for deep trouble in the next few years in China."

If you think these sentences *do* express sharp truths, just ponder for a moment... What does "snow" really mean? Is it as sharp a category as "checkmate" or "prime number"? And what does "wet" really mean, *exactly*? No blur at all there? What about "unmarried" — not to mention "the next few years" and "in for deep trouble"? Ambiguities galore here! And yet such classic philosophers' sentences, since they reside at the level where we naturally float, seem to most people far realer and (therefore far more reliably true) than sentences such as "Electrons have spin 1/2" or "The laws of electromagnetism are invariant under a mirror reflection."

Because of our relatively huge size, most of us never see or deal directly with electrons or the laws of electromagnetism. Our perceptions and actions focus on far larger, vaguer things, and our deepest beliefs, far from being in electrons, are in the many macroscopic items that we are continually assigning to our high-frequency and low-frequency mental categories (such as "fast food" and "doggie bags" on the one hand, and "feet of clay" and "customer service departments" on the other), and also in the perceived causality, however blurry and unreliable it may be, that seems to hold among these large and vague items.

Our keenest insights into causality in the often terribly confusing world of living beings invariably result from well-honed acts of categorization at a macroscopic level. For example, the reasons for a mysterious war taking place in some remote land might suddenly leap into sharp focus for us when an insightful commentator links the war's origin to an ancient conflict between certain religious dogmas. On the other hand, no enlightenment whatsoever would come if a physicist tried to explain the war by saying it came about thanks to trillions upon trillions of momentum-conserving collisions taking place among ephemeral quantum-mechanical specks.

I could go on and say similar things about how we always perceive love affairs and other grand themes of human life in terms of intangible everyday patterns belonging to the large-scale world, and never in terms of the interactions of elementary particles. In contrast to declaring that quantum electrodynamics is "what makes the world go round", I could instead cite such eternally elusive mysteries as beauty, generosity, sexuality, insecurity, fidelity, jealousy, loneliness, and on and on, making sure not to leave out that wonderful tingling of two souls that we curiously call "chemistry", and that the French, even more curiously, describe as *avoir des atomes crochus,* which means having atoms that are hooked together.

Making such a list, though fun, would be a simple exercise and would tell you nothing new. The key point, though, is that we perceive essentially *everything* in life at this level, and essentially *nothing* at the level of the invisible components that, intellectually, we know we are made out of. There are, I concede, a few exceptions, such as our modern keen awareness of the microscopic causes of disease, and also our interest in the tiny sperm–egg events that give rise to a new life, and the common knowledge of the role of microscopic factors in the determination of the sex of a child — but these are highly exceptional. The general rule is that we swim in the world of everyday concepts, and it is they, not micro-events, that define our reality.

Am I a Strange Marble?

The foregoing means that we can best understand our *own* actions just as we best understand other creatures' actions — in terms of stable but intangible internal patterns called "hopes" and "beliefs" and so on. But the need for self-understanding goes much further than that. We are powerfully driven to create a term that summarizes the presumed unity, internal coherence, and temporal stability of all the hopes and beliefs and desires that are found inside our own cranium — and that term, as we all learn very early on, is "I". And pretty soon this high abstraction behind the scenes comes to feel like the maximally real entity in the universe.

Just as we are convinced that ideas and emotions, rather than particles, cause wars and love affairs, so we are convinced that our "I" causes our own actions. The Grand Pusher in and of our bodies is our "I", that marvelous marble whose roundness, solidity, and size we so unmistakably feel inside the murky box of our manifold hopes and desires.

Of course I am alluding here to "Epi" — the nonexistent marble in the box of envelopes. But the "I" illusion is far subtler and more recalcitrant than the illusion of a marble created by many aligned layers of paper and glue. Where does the tenaciousness of this illusion come from? Why does it refuse to go away no matter how much "hard science" is thrown at it? To try to answer questions of this sort, I shall now focus on the strange loop that makes an "I" — where it is found, and how it arises and stabilizes.

A Pearl Necklace I Am Not

To begin with, for each of us, the strange loop of our unique "I"-ness resides inside our own brain. There is thus one such loop lurking inside the cranium of each normal human being. Actually, I take that back, since, in Chapter 15, I will raise this number rather drastically. Nonetheless, saying that there is just one is a good approximation to start with.

When I refer to "a strange loop inside a brain", do I have in mind a physical structure — some kind of palpable closed curve, perhaps a circuit made out of many neurons strung end-to-end? Could this neural loop be neatly excised in a brain operation and laid out on a table, like a delicate pearl necklace, for all to see? And would the person whose brain had thus been "delooped" thereby become an unconscious zombie?

Needless to say, that's hardly what I have in mind. The strange loop making up an "I" is no more a pinpointable, extractable physical object than an audio feedback loop is a tangible object possessing a mass and a diameter. Such a loop may exist "inside" an auditorium, but the fact that it is physically localized doesn't mean that one can pick it up and heft it, let alone measure such things as its temperature and thickness! An "I" loop, like an audio feedback loop, is an abstraction — but an abstraction that seems immensely real, almost physically palpable, to beings like us, beings that have high readings on the hunekometer.

I Am My Brain's Most Complex Symbol

Like a careenium (and also like *PM*), a brain can be seen on at (at least) two levels — a low level involving very small physical processes (perhaps involving particles, perhaps involving neurons — take your pick), and a high level involving large structures selectively triggerable by perception, which in this book I have called *symbols*, and which are the structures in our brain that constitute our categories.

Among the untold thousands of symbols in the repertoire of a normal human being, there are some that are far more frequent and dominant than others, and one of them is given, somewhat arbitrarily, the name "I" (at least in English). When we talk about other people, we talk about them in terms of such things as their ambitions and habits and likes and dislikes, and we accordingly need to formulate for each of them the analogue of an "I", residing, naturally, inside *their* cranium, not our own. This counterpart of our own "I" of course receives various labels, depending on the context, such as "Danny" or "Monica" or "you" or "he" or "she".

The process of perceiving one's self interacting with the rest of the universe (comprised mostly, of course, of one's family and friends and favorite pieces of music and favorite books and movies and so on) goes on for a lifetime. Accordingly, the "I" symbol, like all symbols in our brain, starts out pretty small and simple, but it grows and grows and grows, eventually becoming the most important abstract structure residing in our brains. But where is it in our brains? It is not in some small localized spot; it is spread out all over, because it has to include so much about so much.

Internalizing Our Weres, Our Wills, and Our Woulds

My self-symbol, unlike that of my dog, reaches back fairly accurately, though quite spottily, into the deep (and seemingly endless) past of my existence. It is our unlimitedly extensible human category system that underwrites this fantastic jump in sophistication from other animals to us, in that it allows each of us to build up our episodic memory — the gigantic warehouse of our recollections of events, minor and major, simple and complex, that have happened to us (and to our friends and family members and people in books and films and newspaper articles and so forth, *ad infinitum*) over a span of decades.

Similarly, driven by its dreads and dreams, my self-symbol peers with great intensity, though with little confidence, out into the murky fog of my future existence. My vast episodic memory of my past, together with its counterpart pointing blurrily towards what is yet to come (my *episodic projectory*, I think I'll call it), and further embellished by a fantastic folio of alternative versions or "subjunctive replays" of countless episodes ("if only X had happened..."; "how lucky that Y never took place...", "wouldn't it be great if Z were to occur..." — and why not call this my *episodic subjunctory*?), gives rise to the endless hall of mirrors that constitutes my "I".

I Cannot Live without My Self

Since we perceive not particles interacting but macroscopic patterns in which certain things push other things around with a blurry causality, and since the Grand Pusher in and of our bodies is our "I", and since our bodies push the rest of the world around, we are left with no choice but to conclude that the "I" is where the causality buck stops. The "I" seems to each of us to be the root of all our actions, all our decisions.

This is only one side of the truth, of course, since it utterly snubs the viewpoint whereby an impersonal physics of micro-entities is what makes the world go round, but it is a surprisingly reliable and totally indispensable distortion. These two properties of the naïve, non-physics viewpoint — its reliability and its indispensability — lock it ever more tightly into our belief systems as we pass from babyhood through childhood to adulthood.

I might add that the "I" of a particle physicist is no less entrenched than is the "I" of a novelist or a shoestore clerk. A profound mastery of all of physics will not in the least undo the decades of brainwashing by culture and language, not to mention the millions of years of human evolution preparing the way. The notion of "I", since it is an incomparably efficient shorthand, is an indispensable explanatory device, rather than just an optional crutch that can be cheerily jettisoned when one grows sufficiently scientifically sophisticated.

The Slow Buildup of a Self

What would make a human brain a candidate for housing a loop of self-representation? Why would a fly brain or a mosquito brain not be just as valid a candidate? Why, for that matter, not a bacterium, an ovum, a sperm, a virus, a tomato plant, a tomato, or a pencil? The answer should be clear: a human brain is a representational system that knows no bounds in terms of the extensibility or flexibility of its categories. A mosquito brain, by contrast, is a tiny representational system that contains practically no categories at all, never mind being flexible and extensible. Very small representational systems, such as those of bacteria, ova, sperms, plants, thermostats, and so forth, do not enjoy the luxury of self-representation. And a tomato and a pencil are not representational systems at all, so for them, the story ends right there (sorry, little tomato! sorry, little pencil!).

So a human brain is a strong candidate for having the potential of rich perceptual feedback, and thus rich self-representation. But what kinds of perceptual cycles do we get involved in? We begin life with the most elementary sorts of feedback about ourselves, which stimulate us to formulate categories for our most obvious body parts, and building on this basic pedestal, we soon develop a sense for our bodies as flexible physical objects. In the meantime, as we receive rewards for various actions and punishments for others, we begin to develop a more abstract sense of "good" and "bad", as well as notions of guilt and pride, and our sense of ourselves as abstract entities that have the power to decide to make things happen (such as continuing to run up a steep hill even though our legs are begging us to just walk) begins to take root.

It is crucial to our young lives that we hone our developing self-symbol as precisely as possible. We want (and need) to find out where we belong in all sorts of social hierarchies and classes, and sometimes, even if we don't want to know these things, we find out anyway. For instance, we are all told, early on, that we are "cute"; in some of us, however, this message is reinforced far more strongly than in others. In this manner, each of us comes to realize that we are "good-looking" or "gullible" or "cheeky" or "shy" or "spoiled" or "funny" or "lazy" or "original", or whatever. Dozens of such labels and concepts accrete to our growing self-symbols.

As we go through thousands of experiences large and small, our representations of these experiences likewise accrete to our self-symbols. Of course a memory of a visit to the Grand Canyon, say, is attached not only to our self-symbol but to many other symbols in our brains, but our self-symbol is enriched and rendered more complex by this attachment.

Making Tosses, Internalizing Bounces

Constantly, relentlessly, day by day, moment by moment, my self-symbol is being shaped and refined — and in turn, it triggers external actions galore, day after day after day. (Or so the causality appears to it, since it is on this level, not on the micro-level, that it perceives the world.) It sees its chosen actions (kicks, tosses, screams, laughs, jokes, jabs, trips, books, pleas, threats, etc.) making all sorts of entities in its environment react in large or small ways, and it internalizes those effects in terms of its coarse-grained categories (as to their graininess, it has no choice). Through endless random explorations like this, my self-symbol slowly acquires concise and valuable insight into its nature as a chooser and launcher of actions, embedded in a vast and multifarious, partially predictable world.

To be more concrete: I throw a basketball toward a hoop, and thanks to hordes of microscopic events in my arms, my fingers, the ball's spin, the air, the rim, and so forth, all of which I am unaware of, I either miss or make my hook shot. This tiny probing of the world, repeated hundreds or thousands of times, informs me ever more accurately about my level of skill as a basketball player (and also helps me decide if I like the sport or not). My sense of my skill level is, of course, but a very coarse-grained summary of billions of fine-grained facts about my body and brain.

Similarly, my social actions induce reactions on the part of other sentient beings. Those reactions bounce back to me and I perceive them in terms of my repertoire of symbols, and in this way I indirectly perceive myself through my effect on others. I am building up my sense of who I am in others' eyes. My self-symbol is coalescing out of an initial void.

Smiling Like Hopalong Cassidy

One morning when I was about six years old, I mustered all my courage, stood up in my first-grade class's show-and-tell session, and proudly declared, "I can smile just like Hopalong Cassidy!" (I don't remember how I had convinced myself that I had this grand ability, but I was as sure of it as I was of anything in the world.) I then proceeded to flash this lovingly practiced smile in front of everybody. In my episodic memory lo these many decades later there is a vivid trace of this act of derring-do, but unfortunately I have only the dimmest recollection of how my teacher, Miss McMahon, a very sweet woman whom I adored, and my little classmates reacted, and yet their collective reaction, whatever it was, was surely a formative influence on my early life, and thus on my gradually growing, slowly stabilizing "I".

What we do — what our "I" tells us to do — has consequences, sometimes positive and sometimes negative, and as the days and years go by, we try to sculpt and mold our "I" in such a way as to stop leading us to negative consequences and to lead us to positive ones. We see if our Hopalong Cassidy smile is a hit or a flop, and only in the former case are we likely to trot it out again. (I haven't wheeled it out since first grade, to be honest.)

When we're a little older, we watch as our puns fall flat or evoke admiring laughter, and according to the results we either modify our pun-making style or learn to censor ourselves more strictly, or perhaps both. We also try out various styles of dress and learn to read between the lines of other people's reactions as to whether something looks good on us or not. When we are rebuked for telling small lies, either we decide to stop lying or else we learn to make our lies subtler, and we incorporate our new knowledge about our degree of honesty into our self-symbol. What goes for lies also goes for bragging, obviously. Most of us work on adapting our use of language to various social norms, sometimes more deliberately and sometimes less so. The levels of complexity are endless.

The Lies in our I's

For over a century, clinical psychologists have tried to understand the nature of this strange hidden structure tightly locked in at the deepest core of each one of us, and some have written very insightfully about it. A few decades ago, I read a couple of books by psychoanalyst Karen Horney, and they left a lasting impression on me. In her book *Our Inner Conflicts,* for instance, Horney spoke of the "idealized image" one forms of oneself. Although her primary focus was how we suffer from our neuroses, what she said had much wider applicability.

> ...It [the idealized image] represents a kind of artistic creation in which opposites appear reconciled...
>
> The idealized image might be called a fictitious or illusory self, but that would be only a half truth and hence misleading. The wishful thinking operating in its creation is certainly striking, particularly since it occurs in persons who otherwise stand on a ground of firm reality. But this does not make it wholly fictitious. It is an imaginative creation interwoven with and determined by very realistic factors. It usually contains traces of the person's genuine ideals. While the grandiose achievements are illusory, the potentialities underlying them are often real. More relevant, it is born of very real inner necessities, it fulfills very real functions, and it has a very real influence on its creator. The processes operating in its creation are determined by such definite laws that a knowledge of its specific features permits us to make accurate inferences as to the true character structure of the particular person.

Horney is obviously not speaking of one's awareness of one's most superficial perceptual features such as height or hair color, or of one's knowledge of slight abstractions such as what kind of job one has and whether one enjoys it, but rather of the (inevitably somewhat distorted) image that one forms, over a lifetime, of one's own deepest character traits, of one's level in all sorts of blurry social hierarchies, of one's greatest accomplishments and failures, of one's fulfilled and unfulfilled yearnings, and on and on. Her stress in the book is on those aspects of this image that are illusory and thus tend to be harmful, but the full structure in which such neurotic distortions reside is much larger. This structure is what I have here called the "self-symbol", or simply the "I".

Horney's earlier book *Self-Analysis* is devoted to the complex challenge whereby one tries to change one's own neurotic tendencies, and it inevitably centers on the rather paradoxical idea of the self reaching in and attempting deliberately to effect deep changes in itself. This is not the place to delve into such intricate issues, but I mention them briefly because doing so may help to remind readers of the immense psychological complexity that lies at the core of all human existence.

The Locking-in of the "I" Loop

Let me now summarize the foregoing in slightly more abstract terms. The vast amounts of stuff that we call "I" collectively give rise, at some particular moment, to some external action, much as a stone tossed into a pond gives rise to expanding rings of ripples. Soon, our action's myriad consequences start bouncing back at us, like the first ripples returning after bouncing off the pond's banks. What we receive back affords us the chance to perceive what our gradually metamorphosing "I" has wrought. Millions of tiny reflected signals impinge on us from outside, whether visually, sonically, tactilely, or whatever, and when they land, they trigger *internal* waves of secondary and tertiary signals inside our brain. Finally this flurry of signals is funneled down into just a handful of activated symbols — a tiny set of extremely well-chosen categories constituting a coarse-grained understanding of what we've just done (for example, "Shoot — missed my hook shot by a hair!", or perhaps, "Wow, my new hair-do hooked him!").

And thus the current "I" — the most up-to-date set of recollections and aspirations and passions and confusions — by tampering with the vast, unpredictable world of objects and other people, has sparked some rapid feedback, which, once absorbed in the form of symbol activations, gives rise to an infinitesimally modified "I"; thus round and round it goes, moment after moment, day after day, year after year. In this fashion, via the loop of symbols sparking actions and repercussions triggering symbols, the abstract structure serving us as our innermost essence evolves slowly but surely, and in so doing it locks itself ever more rigidly into our mind. Indeed, as the years pass, the "I" converges and stabilizes itself just as inevitably as the screech of an audio feedback loop inevitably zeroes in and stabilizes itself at the system's natural resonance frequency.

I Am Not a Video Feedback Loop

It's analogy time again! I'd like once more to invoke the world of video feedback loops, for much of this has its counterpart in that far simpler domain. An event takes place in front of the camera and thus is sent onto the screen, but in simplified form, since continuous shapes (shapes with very fine grain) have been rendered on a grid made of discrete pixels (a coarse-grained medium). The new screen is then taken in by the camera and fed back in, and around and around it goes. The upshot of all this is that a single easily perceivable gestalt shape — some kind of stable but one-of-a-kind, never-seen-before whorl — appears on the screen.

Thus it is with the strange loop making up a human "I", but there is a key difference. In the TV setup, as we earlier observed, no *perception* takes place at any stage inside the loop — just the transmission and reception of bare pixels. The TV loop is not a *strange* loop — it is just a feedback loop.

In any strange loop that gives rise to human selfhood, by contrast, the level-shifting acts of perception, abstraction, and categorization are central, indispensable elements. It is the upward leap from *raw stimuli* to *symbols* that imbues the loop with "strangeness". The overall gestalt "shape" of one's self — the "stable whorl", so to speak, of the strange loop constituting one's "I" — is not picked up by a disinterested, neutral camera, but is perceived in a highly subjective manner through the active processes of categorizing, mental replaying, reflecting, comparing, counterfactualizing, and judging.

I Am Ineradicably Entrenched...

While you were reading my first-grade show-and-tell period Hopalong Cassidy–style smile-attempt bravado anecdote, the question "How come Hofstadter is once again leaving elementary particles out of the picture?" may have flitted through your mind; then again, perhaps it did not. I hope the latter is the case! Indeed, why would such an odd thought occur to any sane human being reading that passage (including the most hard-bitten of particle physicists)? Even the vaguest, most fleeting allusion to particle physics in that context would seem to constitute an absurd *non sequitur,* for what on earth could gluons and muons and protons and photons, of all things, have to do with a little boy imitating his idol, Hopalong Cassidy?

Although particles galore were, to be sure, constantly churning "way down there" in that little boy's brain, they were as invisible as the myriad simms careening about inside a careenium. Roger Sperry (a later idol of mine whose writings, had I but read and understood them in first grade, might have inspired me to stand up and bravely proclaim to my classmates, "I can philosophize just like Roger Sperry!") would additionally point out that the particles in the young boy's brain were merely serving (*i.e.,* being pushed around by) far higher-level symbolic events in which the boy's "I" was participating, and in which his "I" was being formed. As that "I" grew in complexity and grew ever realer to itself (*i.e.,* ever more indispensable to the boy's efforts to categorize and understand the never-repeating events in his life), the chance that any alternative "I"-less way of understanding the world could emerge and compete with it was being rendered essentially nil.

At the same time as I myself was getting ever more used to the fact that this "I" thing was responsible for what I did, my parents and friends were also becoming more convinced that there was indeed something very real-seeming "in there" (in other words, something very marble-like, something with its unique brands of "hardness" and "resilience" and "shape"), which merited being called "you" or "he" or "Douggie", and that also merited being called "I" by Douggie — and so once again, the sense of reality of this "I" was being reinforced over and over again, in myriad ways. By the time this brain had lived in this body for a couple of years or so, the "I" notion was locked into it beyond any conceivable hope of reversal.

…But Am I Real?

And yet, was this "I", for all its tremendous stability and apparent utility, a *real* thing, or was it just a comfortable myth? I think we need some good old-fashioned analogies here to help out. And so I ask you, dear reader, are temperature and pressure *real* things, or are they just *façons de parler*? Is a rainbow a real thing, or is it nonexistent? Perhaps more to the point, was the "marble" that I discovered inside my box of envelopes *real*?

What if the box had been sealed shut so I had no way of looking at the individual envelopes? What if my knowledge of the box of envelopes necessarily came from dealing with its hundred envelopes *as a single whole,* so that no shifting back and forth between coarse-grained and fine-grained perspectives was possible? What if I hadn't even known there were envelopes in the box, but had simply thought that there was a somewhat squeezable, pliable mass of softish *stuff* that I could grab with my entire hand, and that at this soft mass's center there was something much more rigid-feeling and undeniably spherical in shape?

If, in addition, it turned out that talking about this supposed marble had enormously useful explanatory power in my life, and if, on top of that, all my friends had similar cardboard boxes and all of them spoke ceaselessly — and wholly unskeptically — about the "marbles" inside *their* boxes, then it would soon become pretty irresistible to me to accept my own marble as part of the world and to allude to it frequently in my explanations of various phenomena in the world. Indeed, any oddballs who denied the existence of marbles inside their cardboard boxes would be accused of having lost their marbles.

And thus it is with this notion of "I". Because it encapsulates so neatly and so efficiently for us what we perceive to be truly important aspects of causality in the world, we cannot help attributing reality to our "I" and to those of other people — indeed, the highest possible level of reality.

The Size of the Strange Loop that Constitutes a Self

One more time, let's go back and talk about mosquitoes and dogs. Do they have anything like an "I" symbol? In Chapter 1, when I spoke of "small souls" and "large souls", I said that this is not a black-and-white matter but one of degree. We thus have to ask, is there a strange loop — a sophisticated level-crossing feedback loop — inside a mosquito's head? Does a mosquito have a rich, symbolic representation of itself, including representations of its desires and of entities that threaten those desires, and does it have a representation of itself in comparison with other selves? Could a mosquito think a thought even vaguely reminiscent of "I can smile just like Hopalong Cassidy!" — for example, "I can bite just like Buzzaround Betty!"? I think the answer to these and similar questions is quite obviously, "No way in the world!" (thanks to the incredibly spartan symbol repertoire of a mosquito brain, barely larger than the symbol repertoire of a flush toilet or a thermostat), and accordingly, I have no qualms about dismissing the idea of there being a strange loop of selfhood in as tiny and swattable a brain as that of a mosquito.

On the other hand, where dogs are concerned, I find, not surprisingly, much more reason to think that there are at least the rudiments of such a loop in there. Not only do dogs have brains that house many rather subtle categories (such as "UPS truck" or "things I can pick up in the house and walk around with in my mouth without being punished"), but also they seem to have some rudimentary understanding of their own desires and the desires of others, whether those others are other dogs or human beings. A dog often knows when its master is unhappy with it, and wags its tail in the hopes of restoring good feelings. Nonetheless, a dog, saliently lacking an arbitrarily extensible concept repertoire and therefore possessing only a rudimentary episodic memory (and of course totally lacking any permanent storehouse of imagined future events strung out along a mental timeline, let alone counterfactual scenarios hovering around the past, the present, and even the future), necessarily has a self-representation far simpler than that of an adult human, and for that reason a dog has a far smaller soul.

The Supposed Selves of Robot Vehicles

I was most impressed when I read about "Stanley", a robot vehicle developed at the Stanford Artificial Intelligence Laboratory that not too long ago drove all by itself across the Nevada desert, relying just on its laser rangefinders, its television camera, and GPS navigation. I could not help asking myself, "How much of an 'I' does Stanley have?"

In an interview shortly after the triumphant desert crossing, one gung-ho industrialist, the director of research and development at Intel (you should keep in mind that Intel manufactured the computer hardware on board Stanley), bluntly proclaimed: "Deep Blue [IBM's chess machine that defeated world champion Garry Kasparov in 1997] was just processing power. It didn't think. Stanley thinks."

Well, with all due respect for the remarkable collective accomplishment that Stanley represents, I can only comment that this remark constitutes shameless, unadulterated, and naïve hype. I see things very differently. If and when Stanley ever acquires the ability to form limitlessly snowballing categories such as those in the list that opened this chapter, *then* I'll be happy to say that Stanley thinks. At the present, though, its ability to cross a desert without self-destructing strikes me as comparable to an ant's following a dense pheromone trail across a vacant lot without perishing. Such autonomy on the part of a robot vehicle is hardly to be sneezed at, but it's a far cry from thinking and a far cry from having an "I".

At one point, Stanley's video camera picked up another robot vehicle ahead of it (this was H1, a rival vehicle from Carnegie-Mellon University) and eventually Stanley pulled around H1 and left it in its dust. (By the way, I am carefully avoiding the pronoun "he" in this text, although it was par for the course in journalistic references to Stanley, and perhaps also at the AI Lab as well, given that the vehicle had been given a human name. Unfortunately, such linguistic sloppiness serves as the opening slide down a slippery slope, soon winding up in full anthropomorphism.) One can see this event taking place on the videotape made by that camera, and it is the climax of the whole story. At this crucial moment, did Stanley recognize the other vehicle as being "like me"? Did Stanley think, as it gaily whipped by H1, "There but for the grace of God go I?" or perhaps "Aha, gotcha!" Come to think of it, why did I write that Stanley "gaily whipped by" H1?

What would it take for a robot vehicle to think such thoughts or have such feelings? Would it suffice for Stanley's rigidly mounted TV camera to be able to turn around on itself and for Stanley thereby to acquire visual imagery of itself? Of course not. That may be one indispensable move in the long process of acquiring an "I", but as we know in the case of chickens and cockroaches, perception of a body part does not a self make.

A Counterfactual Stanley

What is lacking in Stanley that would endow it with an "I", and what does not seem to be part of the research program for developers of self-driving vehicles, is a deep understanding of its place in the world. By this I do not mean, of course, the vehicle's location on the earth's surface, which is given to it down to the centimeter by GPS; it means a rich representation of the vehicle's own actions and its relations to other vehicles, a rich representation of its goals and its "hopes". This would require the vehicle to have a full episodic memory of thousands of experiences it had had, as well as an episodic projectory (what it would expect to happen in its "life", and what it would hope, and what it would fear), as well as an episodic subjunctory, detailing its thoughts about near misses it had had, and what would most likely have happened had things gone some other way.

Thus, Stanley the Robot Steamer would have to be able to think to itself such hypothetical future thoughts as, "Gee, I wonder if H1 will deliberately swerve out in front of me and prevent me from passing it, or even knock me off the road into the ditch down there! That's what *I'd* do if *I* were H1!" Then, moments later, it would have to be able to entertain counterfactual thoughts such as, "Whew! Am I ever glad that H1 wasn't so clever as I feared — or maybe H1 is just not as competitive as I am!"

An article in *Wired* magazine described the near-panic in the Stanford development team as the desert challenge was drawing perilously near and they realized something was still very much lacking. It casually stated, "They needed the algorithmic equivalent of self-awareness", and it then proceeded to say that soon they had indeed achieved this goal (it took them all of three months of work!). Once again, when all due hat-tips have been made toward the team's great achievement, one still has to realize that there is nothing going on inside Stanley that merits being labeled by the highly loaded, highly anthropomorphic term "self-awareness".

The feedback loop inside Stanley's computational machinery is good enough to guide it down a long dusty road punctuated by potholes and lined with scraggly saguaros and tumbleweed plants. I salute it! But if one has set one's sights not just on driving but on thinking and consciousness, then Stanley's feedback loop is not strange enough — not anywhere close. Humanity still has a long ways to go before it will collectively have wrought an artificial "I".

CHAPTER 14

Strangeness in the "I" of the Beholder

ન્જ ન્જ ન્જ

The Inert Sponges inside our Heads

W<small>HY</small>, you might be wondering, do I call the lifelong loop of a human being's self-representation, as described in the preceding chapter, a *strange* loop? You make decisions, take actions, affect the world, receive feedback, incorporate it into your self, then the updated "you" makes more decisions, and so forth, round and round. It's a loop, no doubt — but where's the paradoxical quality that I've been saying is a *sine qua non* for strange loopiness? Why is this not just an ordinary feedback loop? What does such a loop have in common with the quintessential strange loop that Kurt Gödel discovered unexpectedly lurking inside *Principia Mathematica*?

For starters, a brain would seem, *a priori*, just about as unlikely a substrate for self-reference and its rich and counterintuitive consequences as was the extremely austere treatise *Principia Mathematica*, from which self-reference had been strictly banished. A human brain is just a big spongy bulb of inanimate molecules tightly wedged inside a rock-hard cranium, and there it simply sits, as inert as a lump on a log. Why should self-reference and a self be lurking in such a peculiar medium any more than they lurk in a lump of granite? Where's the "I"-ness in a brain?

Just as something very strange had to be happening inside the stony fortress of *Principia Mathematica* to allow the outlawed "I" of Gödelian sentences like "I am not provable" to creep in, something very strange must also take place inside a bony cranium stuffed with inanimate molecules if it is to bring about a soul, a "light on", a unique human identity, an "I". And keep in mind that an "I" does not magically pop up in all brains inside *all* crania, courtesy of "the right stuff" (that is, certain "special" kinds of molecules); it happens only if the proper *patterns* come to be in that medium. Without such patterns, the system is just as it superficially appears to be: a mere lump of spongy matter, soulless, "I"-less, devoid of any inner light.

Squirting Chemicals

When the first brains came into existence, they were trivial feedback devices, less sophisticated than a toilet's float-ball mechanism or the thermostat on your wall, and like those devices, they selectively made primitive organisms move towards certain things (food) and away from others (dangers). Evolutionary pressures, however, gradually made brains' triage of their environments grow more complex and multi-layered, and eventually (here we're talking millions or billions of years), the repertoire of categories that were being responded to grew so rich that the system, like a TV camera on a sufficiently long leash, was capable of "pointing back", to some extent, at itself. That first tiny glimmer of self was the germ of consciousness and "I"-ness, but there is still a great mystery.

No matter how complicated and sophisticated brains became, they always remained, at bottom, nothing but a set of cells that "squirted chemicals" back and forth among each other (to borrow a phrase from the pioneering roboticist and provocative writer Hans Moravec), a bit like a huge oil refinery in which liquids are endlessly pumped around from one tank to another. How could a system of pumping liquids ever house a locus of upside-down causality, where *meanings* seem to matter infinitely more than physical objects and their motions? How could joy, sadness, a love for impressionist painting, and an impish sense of humor inhabit such a cold, inanimate system? One might as well look for an "I" inside a stone fortress, a toilet's tank, a roll of toilet paper, a television, a thermostat, a heat-seeking missile, a heap of beer cans, or an oil refinery.

Some philosophers see our inner lights, our "I"'s, our humanity, our souls, as emanating from the nature of the substrate itself — that is, from the organic chemistry of carbon. I find that a most peculiar tree on which to hang the bauble of consciousness. Basically, this is a mystical refrain that explains nothing. Why should the chemistry of carbon have some magical property entirely unlike than that of any other substance? And what *is* that magical property? And how does it make us into conscious beings? Why is it that only *brains* are conscious, and not kneecaps or kidneys, if all it takes is organic chemistry? Why aren't our carbon-based cousins the mosquitoes just as conscious as we are? Why aren't cows just as conscious as we are? Doesn't organization or pattern play any role here? Surely it does. And if it does, why couldn't it play the *whole* role?

By focusing on the medium rather than the message, the pottery rather than the pattern, the typeface rather than the tale, philosophers who claim that something ineffable about carbon's chemistry is indispensable for consciousness miss the boat. As Daniel Dennett once wittily remarked in a rejoinder to John Searle's tiresome "right-stuff" refrain, "It ain't the meat, it's the motion." (This was a somewhat subtle hat-tip to the title of a somewhat unsubtle, clearly erotic song written in 1951 by Lois Mann and Henry Glover, made famous many years later by singer Maria Muldaur.) And for my money, the magic that happens in the meat of brains makes sense only if you know how to look at the motions that inhabit them.

The Stately Dance of the Symbols

Brains take on a radically different cast if, instead of focusing on their squirting chemicals, you make a level-shift upwards, leaving that low level far behind. To allow us to speak easily of such upward jumps was the reason I dreamt up the allegory of the careenium, and so let me once again remind you of its key imagery. By zooming out from the level of crazily careening simms and by looking instead at the system on a speeded-up time scale whereby the simms' locally chaotic churning becomes merely a foggy blur, one starts to see other entities coming into focus, entities that formerly were utterly invisible. And at that level, *mirabile dictu*, meaning emerges.

Simmballs filled with meaning are now seen to be doing a stately dance in a blurry soup that they don't suspect for a split second consists of small interacting magnetic marbles called "simms". And the reason I say the simmballs are "filled with meaning" is not, of course, that they are oozing some mystical kind of sticky semantic juice called "meaning" (even though certain meat-infatuated philosophers might go for that idea), but because their stately dance is deeply in synch with events in the world around them.

Simmballs are in synch with the outer world in the same way as in *La Femme du boulanger,* the straying cat Pomponnette's return was in synch with the return of the straying wife Aurélie: there was a many-faceted alignment of Situation "P" with Situation "A". However, this alignment of situations at the film's climax was just a joke concocted by the screenwriter; no viewer of *La Femme du boulanger* supposes for a moment that the cat's escapades will continue to parallel the wife's escapades (or vice versa) for months on end. We know it was just a coincidence, which is why we find it so humorous.

By contrast, a careenium's dancing simmballs *will* continue tracking the world, *will* stay in phase with it, *will* remain aligned with it. That (by fiat of the author!) is the very nature of a careenium. Simmballs are *systematically* in phase with things going on in the world just as, in Gödel's construction, prim numbers are systematically in phase with *PM*'s provable formulas. That is the only reason simmballs can be said to have meaning. Meaning, no matter what its substrate might be — Tinkertoys, toilet paper, beer cans, simms, whole numbers, or neurons — is an automatic, unpreventable outcome of reliable, stable alignment; this was the lesson of Chapter 11.

Our own brains are no different from careenia, except, of course, that whereas careenia are just my little fantasy, human brains are not. The symbols in our brains truly *do* do that voodoo that they do so well, and they do it in the electrochemical soup of neural events. The strange thing, though, is that over the eons that it took for our brains to evolve from the earliest proto-brains, meanings just sneaked ever so quietly into the story, almost unobserved. It's not as if somebody had devised a grand plan, millions of years in advance, that high-level meaningful structures — physical patterns representing abstract categories — would one day come to inhabit big fancy brains; rather, such patterns (the "symbols" of this book) simply came along an unplanned by-product of the tremendously effective way that having bigger and bigger brains helped beings to survive better and better in a terribly cutthroat world.

Just as Bertrand Russell was blindsided by the unexpected appearance of high-level Gödelian meanings in the heart of his ultraprotected bastion, *Principia Mathematica,* so someone who had never conceived of looking at a brain at any level other than that of Hans Moravec's squirting chemicals would be mightily surprised at the emergence of symbols. Much as Gödel saw the great potential of shifting attention to a wholly different level of *PM* strings, so I am suggesting (though I'm certainly far from the first) that we have to shift our attention to a far higher level of brain activity in order to find symbols, concepts, meanings, desires, and, ultimately, our selves.

The funny thing is that we humans all *are* focused on that level without ever having had any choice in the matter. We *automatically* see our brains' activity as entirely symbolic. I find something wonderfully strange and upside-down about this, and I'll now try to show why through an allegory.

In which the Alfbert Visits Austranius

Imagine, if you will, the small, lonely planet of Austranius, whose sole inhabitants are a tribe called the "Klüdgerot". From time immemorial, the Klüdgerot have lived out their curious lives in a dense jungle of extremely long *PM* strings, some of which they can safely ingest (strings being their sole source of nutrition) and others of which they must not ingest, lest they be mortally poisoned. Luckily, the resourceful Klüdgerot have found a way to tell apart these opposite sorts of *PM* strings, for certain strings, when inspected visually, form a message that says, in the lilting Klüdgerotic tongue, "I am edible", while others form a message in Klüdgerotic that says "I am inedible". And, quite marvelously, by the Benevolent Grace of Göd, every *PM* string proclaiming its edibility has turned out to be edible, while every *PM* string proclaiming its inedibility has turned out to be inedible. Thus have the Klüdgerot thrived for untold öörs on their bountiful planet.

On a fateful döö in the Austranian möönth of Spöö, a strange-looking orange spacecraft swoops down from the distant planet of Ukia and lands exactly at the North Pöö of Austranius. Out steps a hulking whiteheaded alien that announces itself with the words, "I am the Alfbert. Behold." No sooner has the alien uttered these few words than it trundles off into the Austranian jungle, where it spends not only the rest of Spöö but also all of Blöö, after which it trundles back, slightly bedraggled but otherwise no worse for the wear, to its spacecraft. Bright and early the next döö, the Alfbert solemnly convenes a meeting of all the Klüdgerot on Austranius. As soon as they all have assembled, the Alfbert begins to speak.

"Good döö, virtuous Klüdgerot," intones the Alfbert. "It is my privilege to report to you that I have made an Austranius-shaking scientific discovery." The Klüdgerot all sit in respectful if skeptical silence. "Each *PM* string that grows on this planet," continues the Alfbert, "turns out to be not merely a long and pretty vine but also, astonishingly enough, a message that can be read and understood. Do not doubt me!" On hearing this non-novelty, many Klüdgerot yawn in unison, and a voice shouts out, "Tell us about it, white head!", at which scattered chuckles erupt. Encouraged, the Alfbert does just so. "I have made the fantastic discovery that every *PM* string makes a claim, in my beautiful native tongue of Alfbertic, about certain wondrous entities known as the 'whole numbers'. Many of you are undoubtedly champing at the bit to have me explain to you, in very simple terms that you can understand, what these so-called 'whole numbers' are."

At the sound of this term, a loud rustling noise is heard among the assembled crowd. Unbeknownst to the Alfbert, the Klüdgerot have for countless generations held the entities called "whole numbers" to be incomprehensibly abstract; indeed, the whole numbers were long ago unanimously declared so loathsome that they were forever banned from the planet, along with all their names. Clearly, the Alfbert's message is not welcome here. It is of course wrong (that goes without saying), but it is not merely wrong; it is also totally absurd, and it is repugnant, to boot.

But the whiteheaded Alfbert, blithely unaware of the resentment it has churned up, continues to speak as the mob rustles ever more agitatedly. "Yes, denizens of Austranius, fabulously unlikely though it may sound, in each *PM* string there resides *meaning*. All it takes is to know how to look at the string in the proper way. By using a suitable mapping, one can…"

All at once pandemonium erupts: has the Alfbert not just uttered the despised word "one", the long-banished name of the most dreaded of all the whole numbers? "Away with the alien! Off with its white head!" screams the infuriated mob, and a moment later, a phalanx of Klüdgerot grabs hold of the declaiming alien. Yet even as it is being dragged away, the pontificating Alfbert patiently insists to the Klüdgerot that it is merely trying to edify them, that it can perceive momentous facts hidden to them by reading the strings in a language of which they are ignorant, and that… But the angered throng drowns out the Alfbert's grandiose words.

As the brazen alien is being prepared to meet a dire fate, a commotion suddenly breaks out among the Klüdgerot; they have plumb forgotten the age-old and venerated Klüdgerot tradition of holding a Pre-dishing-out-of-dire-fate Banquet! A team is dispatched to pick the sweetest of all *PM* strings from the Principial Planetary Park of Wööw, a sacred sanctuary into which no Klüdgerot has ever ventured before; when it returns with a fine harvest of succulent strings from Wööw, each of which clearly reads "I am edible", it is greeted by a hail of thunderous applause. After the Klüdgerot have expressed their gratitude to Göd, the traditional Pre-dishing-out-of-dire-fate Banquet begins, and at last it begins to dawn on the Alfbert that it will indeed meet a dire fate in short order. As this ominous fact takes hold, it feels its white head start to spin, then to swim, and then…

Idealistically attempting to save the unsuspecting Klüdgerot, the ever-magnanimous Alfbert cries out, "Listen, I pray, O friends! Your harvest of *PM* strings is treacherous! A foolish superstition has tricked you into thinking they are nutritious, but the truth is otherwise. When decoded as messages, these strings all make such grievously false statements about whole numbers that no one — I repeat, no one! — could swallow them." But the words of warning come too late, for the *PM* strings from Wööw are already being swallowed whole by the stubbornly superstitious Klüdgerot.

And before long, frightful groans are heard resounding far and near; the sensitive Alfbert shields its gaze from the dreadful event. When at last it dares to look, it beholds a sorry sight; on every side, as far as its sole eye can see, lie lifeless shells of Klüdgerot that but moments ago were carousing their silly heads off. "If only they had listened to me!", sadly muses the kindly Alfbert, scratching its great white head in puzzlement. On these words, it trundles back to its strange-looking orange spacecraft at the North Pöö, takes one last glance at the bleak Klüdgerot-littered landscape of Austranius, and finally presses the small round "Takeoff" button on the craft's leatherette dashboard, setting off for destinations unknown.

At this point, the Alfbert, having earlier swooned in terror as the banqueters began their ritual reveling, regains consciousness. First it hears shouts of excitement echoing all around, and then, when it dares to look, it beholds a startling sight; on every side, as far as its sole eye can see, masses of Klüdgerot are staring with unmistakable delight at something moving, somewhere above its white head. It turns to see what this could possibly be, just in time to catch the most fleeting glimpse of a thin shape making a strange, high-pitched rustling sound as it rapidly plummets towards —

Brief Debriefing

I offer my apologies to the late Ambrose Bierce for this rather feeble imitation of the plot of his masterful short story "An Occurrence at Owl Creek Bridge", but my intentions are good. The *raison d'être* of my rather flippant allegory is to turn the classic tragicomedy starring Alfred North Whitehead and Bertrand Russell (jointly alias the Alfbert) and Kurt Gödel (alias the Klüdgerot) on its head, by positing bizarre creatures who cannot imagine the idea of any number-theoretical meaning in *PM* strings, but who nonetheless see the strings as meaningful messages — it's just that they see only high-level Gödelian meanings. This is the diametric opposite of what one would naïvely expect, since *PM* notation was invented expressly to write down statements about numbers and their properties, certainly *not* to write down Gödelian statements about themselves!

A few remarks are in order here to prevent confusions that this allegory might otherwise engender. In the first place, the length of any *PM* string that speaks of its own properties (Gödel's string KG being the prototype, of course) is not merely "enormous", as I wrote at the allegory's outset; it is inconceivable. I have never tried to calculate how many symbols Gödel's string would consist of if it were written out in pure *PM* notation, because I would hardly know how to begin the calculation. I suspect that its symbol-count might well exceed "Graham's constant", which is usually cited as "the largest number ever to appear in a mathematical proof", but even if not, it would certainly give it a run for its money. So the idea of anyone directly reading the strings that grow on Austranius, whether on a low level, as statements about whole numbers, or on a high level, as statements about their own edibility, is utter nonsense. (Of course, so is the idea that strings of mathematical symbols could grow in jungles on a faraway planet, as well as the idea that they could be eaten, but that's allegoric license.)

Gödel created his statement KG through a series of 46 escalating stages, in which he shows that *in principle,* certain notions about numbers *could be* written down in *PM* notation. A typical such notion is "the exponent of the kth prime number in the prime factorization of n". This notion depends on prior notions defined in earlier stages, such as "exponent", "prime number", "kth prime number", "prime factorization" (none of which come as "built-in notions" in *PM*). Gödel never explicitly writes out *PM* expressions for such notions, because doing so would require writing down a prohibitively long chain of *PM* symbols. Instead, each individual notion is given a name, a kind of abbreviation, which could theoretically be expanded out into pure *PM* notation if need be, and which is then used in further steps. Over and over again, Gödel exploits already-defined abbreviations in defining further abbreviations, thus carefully building a tower of increasing complexity and abstractness, working his way up to its apex, which is the notion of prim numbers.

Soaps in Sanskrit

This may sound a bit abstruse and remote, so let me suggest an analogy. Imagine the challenge of writing out a clear explanation of the meaning of the contemporary term "soap digest rack" in the ancient Indian language of Sanskrit. The key constraint is that you are restricted to using pure Sanskrit as it was in its heyday, and are not allowed to introduce even one single new word into the language.

In order to get across the meaning of "soap digest rack" in detail, you would have to explain, for starters, the notions of electricity and electromagnetic waves, of TV cameras and transmitters and TV sets, of TV shows and advertising, the notion of washing machines and rivalries between detergent companies, the idea of daily episodes of predictable hackneyed melodramas broadcast into the homes of millions of people, the image of viewers addicted to endlessly circling plots, the concept of a grocery store, of a checkout stand, of magazines, of display racks, and on and on... Each of the words "soap", "digest", and "rack" would wind up being expanded into a chain of ancient Sanskrit words thousands of times longer than itself. Your final text would fill up hundreds of pages in order to get across the meaning of this three-word phrase for a modern banality.

Likewise, Gödel's string KG, which we conventionally express in supercondensed form through phrases such as "I am not provable in *PM*", would, if written out in pure *PM* notation, be monstrously long — and yet despite its formidable size, we understand precisely what it says. How is that possible? It is a result of its condensability. KG is not a random sequence of *PM* symbols, but a formula possessing a great deal of structure. Just as the billions of cells comprising a heart are so extremely organized that they can be summarized in the single word "pump", so the myriad symbols in KG can be summarized in a few well-chosen English words.

To return to the Sanskrit challenge, imagine that I changed the rules, allowing you to define new Sanskrit words and to employ them in the definitions of yet further new Sanskrit words. Thus "electricity" could be defined and used in the description of TV cameras and televisions and washing machines, and "TV program" could be used in the definition of "soap opera", and so forth. If abbreviations could thus be piled on abbreviations in an unlimited fashion, then it is likely that instead of producing a book-length Sanskrit explanation of "soap digest rack", you would need only a few pages, perhaps even less. Of course, in all this, you would have radically changed the Sanskrit language, carrying it forwards in time a few thousand years, but that is how languages always progress. And that is also the way the human mind works — by the compounding of old ideas into new structures that become new ideas that can themselves be used in compounds, and round and round endlessly, growing ever more remote from the basic earthbound imagery that is each language's soil.

Winding Up the Debriefing

In my allegory, both the Klüdgerot and the Alfbert supposedly have the ability to read pure *PM* strings — strings that contain no abbreviations whatsoever. Since at one level (the level perceived by the Klüdgerot) these strings talk about themselves, they are like Gödel's KG, and this means that such strings are, for want of a better term, infinitely huge (for all practical purposes, anyway). This means that any attempt to read them as statements about numbers will never yield anything comprehensible at all, and so the Alfbert's ability, as described, is a total impossibility. But so is the Klüdgerot's, since they too are overwhelmed by an endless sea of symbols. The only hope for either the Alfbert or the Klüdgerot is to notice that certain patterns are used over and over again in the sea of symbols, and to give these patterns names, thus compressing the string into something more manageable, and then carrying this process of pattern-finding and compression out at the new, shorter level, and each time compressing further and further and further until finally the whole string collapses down into just one simple idea: "I am not edible" (or, translating out of the allegory, "I am not provable").

Bertrand Russell never imagined this kind of a level-shift when he thought about the strings of *PM*. He was trapped by the understandable preconception that statements about whole numbers, no matter how long or complicated they might get, would always retain the familiar flavor of standard number-theoretical statements such as "There are infinitely many primes" or "There are only three pure powers in the Fibonacci sequence." It never occurred to him that some statements could have such intricate hierarchical structures that the number-theoretical ideas they would express would no longer feel like ideas about numbers. As I observed in Chapter 11, a dog does not imagine or understand that certain large arrays of colored dots can be so structured that they are no longer just huge sets of colored dots but become pictures of people, houses, dogs, and many other things. The higher level takes perceptual precedence over the lower level, and in the process becomes the "more real" of the two. The lower level gets forgotten, lost in the shuffle.

Such an upwards level-shift is a profound perceptual change, and when it takes place in an unfamiliar, abstract setting, such as the world of strings of *Principia Mathematica*, it can sound very improbable, even though when it takes place in a familiar setting (such as a TV screen), it is trivially obvious.

My allegory was written in order to illustrate a *downwards* level-shift that is seen as very improbable. The Klüdgerot see only high-level meanings like "I am edible" in certain enormous *PM* strings, and they supposedly cannot imagine any lower-level meaning *also* residing in those strings. To us who know the original intent of the strings of symbols in *Principia Mathematica*, this sounds like an inexplicably rigid prejudice, yet when it comes to understanding our own nature, the tables are quite turned, for a very similar rigid prejudice in favor of high-level (and only high-level) perception turns out to pervade and even to define "the human condition".

Trapped at the High Level

For us conscious, self-aware, "I"-driven humans, it is almost impossible to imagine moving down, down, down to the neuronal level of our brains, and slowing down, down, down, so that we can see (or at least can imagine) each and every chemical squirting in each and every synaptic cleft — a gigantic shift in perspective that would seem to instantly drain brain activity of all symbolic quality. No meanings would remain down there, no sticky semantic juice — just astronomical numbers of meaningless, inanimate molecules, squirting meaninglessly away, all the livelong, lifeless day.

Your typical human brain, being blissfully ignorant of its minute physical components and their arcanely mathematizable mode of microscopic functioning, and thriving instead at the infinitely remote level of soap operas, spring sales, super skivaganzas, SUV's, SAT's, SOB's,

Santa Claus, splashtacular scuba specials, snorkels, snowballs, sex scandals (and let's not forget sleazeballs), makes up as plausible a story as it can about its own nature, in which the starring role, rather than being played by the cerebral cortex, the hippocampus, the amygdala, the cerebellum, or any other weirdly named and gooey physical structure, is played instead by an anatomically invisible, murky thing called "I", aided and abetted by other shadowy players known as "ideas", "thoughts", "memories", "beliefs", "hopes", "fears" "intentions", "desires", "love", "hate", "rivalry", "jealousy", "empathy", "honesty", and on and on — and in the soft, ethereal, neurology-free world of *these* players, your typical human brain perceives its very own "I" as a pusher and a mover, never entertaining for a moment the idea that its star player might merely be a useful shorthand standing for a myriad of infinitesimal entities and the invisible chemical transactions taking place among them, by the billions — nay, the millions of billions — every single second.

The human condition is thus profoundly analogous to the Klüdgerotic condition: neither species can see or even imagine the lower levels of a reality that is nonetheless central to its existence.

First Key Ingredient of Strangeness

Why does an "I" symbol never develop in a video feedback system, no matter how swirly or intricate or deeply nested are the shapes that appear on its screen? The answer is simple: a video system, no matter how many pixels or colors it has, *develops no symbols at all*, because a video system does not *perceive* anything. Nowhere along the cyclic pathway of a video loop are there any symbols to be triggered — no concepts, no categories, no meanings — not a tad more than in the shrill screech of an audio feedback loop. A video feedback system does not attribute to the strange emergent galactic shapes on its screen any kind of causal power to make anything happen. Indeed, it doesn't attribute anything to anything, because, lacking all symbols, a video system can't and doesn't ever *think* about anything!

What makes a strange loop appear in a brain and not in a video feedback system, then, is an *ability* — the ability to think — which is, in effect, a one-syllable word standing for the possession of a sufficiently large repertoire of triggerable symbols. Just as the richness of whole numbers gave *PM* the power to represent phenomena of unlimited complexity and thus to twist back and engulf itself via Gödel's construction, so our extensible repertoires of symbols give our brains the power to represent phenomena of unlimited complexity and thus to twist back and to engulf themselves via a strange loop.

Second Key Ingredient of Strangeness

But there is a flip side to all this, a second key ingredient that makes the loop in a human brain qualify as "strange", makes an "I" come seemingly out of nowhere. This flip side is, ironically, an *inability* — namely, our Klüdgerotic inability to peer below the level of our symbols. It is our inability to see, feel, or sense in any way the constant, frenetic churning and roiling of micro-stuff, all the unfelt bubbling and boiling that underlies our thinking. This, our innate blindness to the world of the tiny, forces us to hallucinate a profound schism between the goal-lacking material world of balls and sticks and sounds and lights, on the one hand, and a goal-pervaded abstract world of hopes and beliefs and joys and fears, on the other, in which radically different sorts of causality seem to reign.

When we symbol-possessing humans watch a video feedback system, we naturally pay attention to the eye-catching shapes on the screen and are seduced into giving them fanciful labels like "helical corridor" or "galaxy", but still we know that ultimately they consist of nothing but pixels, and that whatever patterns appear before our eyes do so thanks solely to the local logic of pixels. This simple and clear realization strips those fancy fractalic gestalts of any apparent life or autonomy of their own. We are not tempted to attribute desires or hopes, let alone consciousness, to the screen's swirly shapes — no more than we are tempted to perceive fluffy cotton-balls in the sky as renditions of an artist's profile or the stoning of a martyr.

And yet when it comes to perceiving ourselves, we tell a different story. Things are far murkier when we speak of ourselves than when we speak of video feedback, because we have no direct access to any analogue, inside our brains, to pixels and their local logic. Intellectually knowing that our brains are dense networks of neurons doesn't make us familiar with our brains at that level, no more than knowing that French poems are made of letters of the roman alphabet makes us experts on French poetry. We are creatures that congenitally cannot focus on the micromachinery that makes our minds tick — and unfortunately, we cannot just saunter down to the corner drugstore and pick up a cheap pair of glasses to remedy the defect.

One might suspect neuroscientists, as opposed to lay people, to be so familiar with the low-level hardware of the brain that they have come to understand just how to think about such mysteries as consciousness and free will. And yet often it turns out to be quite the opposite: many neuroscientists' great familiarity with the low-level aspects of the brain makes them skeptical that consciousness and free will could ever be explained in physical terms at all. So baffled are they by what strikes them as an unbridgeable chasm between mind and matter that they abandon all

Video Voyage II, snapshots 1–4

Video Voyage II, snapshots 5–8

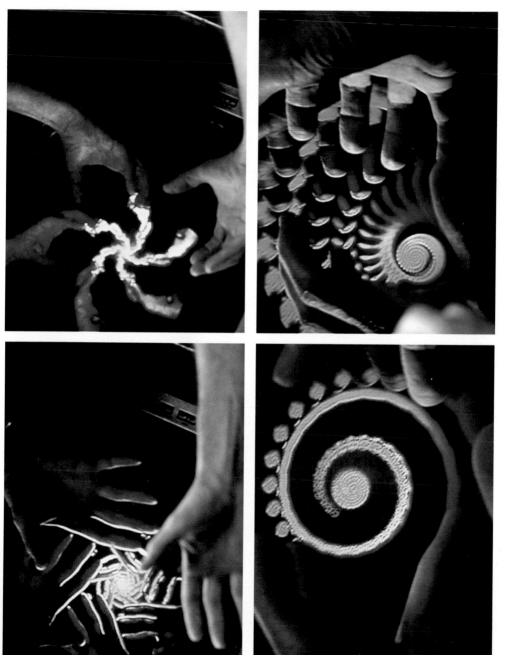

Video Voyage II, snapshots 9–12

Video Voyage II, snapshots 13–16

efforts to see how consciousness and selves could come out of physical processes, and instead they throw in the towel and become dualists. It's a shame to see scientists punt in this fashion, but it happens all too often. The moral of the story is that being a professional neuroscientist is not by any means synonymous with understanding the brain deeply — no more than being a professional physicist is synonymous with understanding hurricanes deeply. Indeed, sometimes being mired down in gobs of detailed knowledge is the exact thing that blocks deep understanding.

Our innate human inability to peer below a certain level inside our cranium makes our inner analogue to the swirling galaxy on a TV screen — the vast swirling galaxy of "I"-ness — strike us as an undeniable *locus of causality*, rather than a mere passive epiphenomenon coming out of lower levels (such as a video-feedback galaxy). So taken in are we by the perceived hard sphericity of that "marble" in our minds that we attribute to it a reality as great as that of anything we know. And because of the locking-in of the "I"-symbol that inevitably takes place over years and years in the feedback loop of human self-perception, causality gets turned around and "I" seems to be in the driver's seat.

In summary, the combination of these two ingredients — one an ability and the other an inability — gives rise to the strange loop of selfhood, a trap into which we humans all fall, every last one of us, willy-nilly. Although it begins as innocently as a humble toilet's float-ball mechanism or an audio or video feedback loop, where no counterintuitive type of causality is posited anywhere, human self-perception inevitably ends up positing an emergent entity that exerts an upside-down causality on the world, leading to the intense reinforcement of and the final, invincible, immutable locking-in of this belief. The end result is often the vehement denial of the possibility of any alternative point of view at all.

Sperry Redux

I just said that we all fall into this "trap", but I don't really see things so negatively. Such a "trap" is not harmful if taken with a grain of salt; rather, it is something to rejoice in and cherish, for it is what makes us human. Permit me once more to quote the eloquent words of Roger Sperry:

> In the brain model proposed here, the causal potency of an idea, or an ideal, becomes just as real as that of a molecule, a cell, or a nerve impulse. Ideas cause ideas and help evolve new ideas. They interact with each other and with other mental forces in the same brain, in neighboring brains, and, thanks to global communication,

in far distant, foreign brains. And they also interact with the external surroundings to produce *in toto* a burstwise advance in evolution that is far beyond anything to hit the evolutionary scene yet, including the emergence of the living cell.

When you come down to it, all that Sperry has done here is to go out on a limb and dare to assert, in a serious scientific publication, the ho-hum, run-of-the-mill, commonsensical belief held by the random person on the street that there is a genuine reality (*i.e.*, causal potency) of the thing we call "I". In the scientific world, such an assertion runs a great risk of being looked upon with skepticism, because it sounds superficially as if it reeks of Cartesian dualism (wonderfully mystical-sounding terms such as *élan vital,* "life force", "spirit of the hive", "entelechy", and "holons" occasionally spring into my mind when I read this passage).

However, Roger Sperry knew very well that he wasn't embracing dualism or mysticism of any sort, and he therefore had the courage to take the plunge and make the assertion. His position is a subtle balancing act whose insightfulness will, I am convinced, one day be recognized and celebrated, and it will be seen to be analogous to the subtle balancing act of Kurt Gödel, who demonstrated how high-level, emergent, self-referential meanings in a formal mathematical system can have a causal potency just as real as that of the system's rigid, frozen, low-level rules of inference.

CHAPTER 15

Entwinement

☙ ☙ ☙

Multiple Strange Loops in One Brain

TWO chapters back, I declared that there was one strange loop in each human cranium, and that this loop constituted our "I", but I also mentioned that that was just a crude first stab. Indeed, it is a drastic oversimplification. Since we all perceive and represent hundreds of other human beings at vastly differing levels of detail and fidelity inside our cranium, and since the most important facet of all of those human beings is their *own* sense of self, we inevitably mirror, and thus house, a large number of other strange loops inside our head. But what exactly does it mean to say that each human head is the locus of a multiplicity of "I"'s?

Well, I don't know precisely what it means. I wish I did! And I reckon that if I did, I would be the world's greatest philosopher and psychologist rolled into one. As best I can guess, from far below such a Parnassus, it means we manufacture an enormously stripped-down version of our *own* strange loop of selfhood and install it at the core of our symbols for other people, letting that initially crude loopy structure change and grow over time. In the case of the people we know best — our spouse, our parents and siblings, our children, our dearest friends — each of these loops grows over the years to be a very rich structure adorned with many thousands of idiosyncratic ingredients, and each one achieves a great deal of autonomy from the stripped-down "vanilla" strange loop that served as its seed.

Content-free Feedback Loops

More light can be cast on this idea of a "vanilla" strange loop through our old metaphor of the audio feedback loop. Suppose a microphone and a loudspeaker have been connected together so that even a very soft noise will cycle around rapidly, growing louder and louder each pass through the loop, until it becomes a huge ear-piercing shriek. But suppose the room is dead silent at the start. In that case, what happens? What happens is that it remains dead silent. The loop is working just fine, but it is receiving zero noise and outputting zero noise, because zero times anything is still zero. When no signal enters a feedback loop, the loop has no perceptible effect; it might as well not even exist. An audio loop on its own does not a screech make. It takes some non-null input to get things off the ground.

Let's now translate this scenario to the world of video feedback. If one points a TV camera at the middle of a blank screen, and if the camera sees only the screen and none of its frame, then despite its loopiness, all that this setup will produce, whether the camera stands still, tilts, turns, or zooms in and out (always without reaching the screen's edge), is a fixed white image. As before, the fact that the image results from a closed feedback loop makes no difference, because nothing external is serving as the *contents* of that loop. I'll refer to such a content-free feedback loop as a "vanilla" loop, and it's obvious that two vanilla video loops will be indistinguishable — they are just empty shells with no recognizable traits and no "personal identity".

If, however, the camera turns far enough left or right, or zooms out far enough to take in something *external* to the blank screen (even just the tiniest patch of color), a bit of the screen will turn non-blank, and then, instantly, that non-blank patch will get sucked into the video loop and cycled around and around, like a tree limb picked up by a tornado. Soon the screen will be populated with many bits of color forming a complex and self-stabilizing pattern. What gives this non-vanilla loop its recognizable identity is not merely the fact that the image contains *itself*, but just as crucially, the fact that *external* items in a particular arrangement are part of the image.

If we bring this metaphor back to the context of human identity, we could say that a "bare" strange loop of selfhood does not give rise to a distinct self — it is just a generic, vanilla shell that requires contact with something else in the world in order to start acquiring a distinctive identity, a distinctive "I". (For those who enjoy the taboo thrills of non-wellfounded sets — sets that, *contra* Russell, may contain themselves as members — I might raise the puzzle of two singleton sets, x and y, each of which contains itself, and *only* itself, as a member. Are x and y identical entities or different entities? Trying to answer the riddle by defining two sets to be identical if and only if they have the same members leads one instantly into an infinite regress, and thus no answer is yielded. I prefer to brazenly cut the Gordian knot by declaring the two sets indistinguishable and hence identical.)

Baby Feedback Loops and Baby "I"'s

Although I just conjured up the notion of a "vanilla" strange loop in a human brain, I certainly did not mean to suggest that a human baby is already at birth endowed with such a "bare" strange loop of selfhood — that is, a fully-realized, though vanilla, shell of pure, distilled "I"-ness — thanks to the mere fact of having human genes. And far less did I mean to suggest that an unborn human embryo acquires a bare loop of selfhood while still in the womb (let alone at the moment of fertilization!). The realization of human selfhood is not nearly so automatic and genetically predetermined as that would suggest.

The closing of the strange loop of human selfhood is deeply dependent upon the level-changing leap that is *perception*, which means *categorization*, and therefore, the richer and more powerful an organism's categorization equipment is, the more realized and rich will be its self. Conversely, the poorer an organism's repertoire of categories, the more impoverished will be the self, until in the limit there simply is no self at all.

As I've stressed many times, mosquitoes have essentially no symbols, hence essentially no selves. There is no strange loop inside a mosquito's head. What goes for mosquitoes goes also for human babies, and all the more so for human embryos. It's just that babies and embryos have a fantastic potential, thanks to their human genes, to become homes for huge symbol-repertoires that will grow and grow for many decades, while mosquitoes have no such potential. Mosquitoes, because of the initial impoverishment and the fixed non-extensibility of their symbol systems, are doomed to soullessness (oh, all right — maybe 0.00000001 hunekers' worth of consciousness — just a hair above the level of a thermostat).

For better or for worse, we humans are born with only the tiniest hints of what our perceptual systems will metamorphose into as we interact with the world over the course of decades. At birth, our repertoire of categories is so minimal that I would call it nil for all practical purposes. Deprived of symbols to trigger, a baby cannot make sense of what William James evocatively called the "big, blooming, buzzing confusion" of its sensory input. The building-up of a self-symbol is still far in the future for a baby, and so in babies there exists no strange loop of selfhood, or nearly none.

To put it bluntly, since its future symbolic machinery is 99 percent missing, a human neonate, devastatingly cute though it might be, simply has no "I" — or, to be more generous, if it does possess some minimal dollop of "I"-ness, perhaps it is one huneker's worth or thereabouts — and that's not much to write home about. So we see that a human head can contain *less* than one strange loop. What about *more* than one?

Entwined Feedback Loops

To explore in a concrete fashion the idea of two strange loops coexisting in one head, let's start with a mild variation on our old TV metaphor. Suppose two video cameras and two televisions are set up so that camera A feeds screen A and, far away from it, camera B feeds screen B. Suppose moreover that at all times, camera A picks up *all* of what is on screen A (plus some nearby stuff, to give the A-loop "content") and cycles it back onto A, and analogously, camera B picks up all of what is on screen B (plus some external content) and cycles it back onto B. Now since systems A and B are, by stipulation, far apart from each other, it is intuitively clear that A and B constitute separate, disjoint feedback loops. If the local scenes picked up by cameras A and B are different, then screens A and B will have clearly distinguishable patterns on them, so the two systems' "identities" will be easily told apart. So far, what this metaphor gives us is old hat (in fact, it's two old hats) — two different heads, each having one loop inside it.

What will happen, however, when systems A and B are gradually brought close enough together to begin interacting with each other? Camera A will then see not only screen A but also screen B, and so loop B will enter into the content of loop A (and vice versa).

Let's assume, as would seem natural, that camera A is closer to screen A than it is to screen B (and vice versa). Then loop A will take up more space on screen A than does loop B, meaning more pixels, and so loop A will be reproduced with higher fidelity on screen A. Loop A will be large and fine-grained, loop B will be small and coarse-grained. But that's only on screen A. On screen B, everything is reversed: loop B will be larger and finer-grained, while loop A will be smaller and of coarser grain. The last thing I want to remind you of before we go on to a new paragraph is that now loop A, although it's still *called* just "A", nonetheless *involves* loop B as well (and vice versa); each of these two loops now plays a role in defining the other one, though loop A plays a *larger* role in its own definition than does loop B (and vice versa).

We now have a metaphor for two individuals, A and B, each of whom has their own personal identity (*i.e.*, their own private strange loop) — and yet part of that private identity is made out of, and is thus dependent upon, the private identity of the *other* individual. Furthermore, the more faithful the image of each screen on the other one, the more the "private" identities of the two loops are intertwined, and the more they start to be fused, blurred, and even, to coin a word, undisentanglable from each other.

At this point, even though we are being guided solely by a very curious technological metaphor, I believe we are drawing slowly closer to an understanding of what genuine human identity is all about. In fact, how could anyone imagine that it would be possible to gain deep insight into the mystery of human identity without eventually running up against some sort of unfamiliar abstract structures? Sigmund Freud posited egos, ids, and superegos, and there may well exist some such abstractions inside the architecture of a human soul (perhaps not exactly those three, but patterns of that ilk). We humans are so different from other natural phenomena, even from most other types of living beings, that we should expect that in order to get a glimpse into what we truly are, we would have to look in very unexpected places. Although my strange loops are obviously very different from Freud's notions, there is a certain similarity of spirit. Both views of what a self is involve abstract patterns that are extremely remote from the biological substrate they inhabit — so remote, in fact, that the specifics of the substrate would seem mostly irrelevant.

One Privileged Loop inside our Skull

Suppose some future television technology managed to eliminate the graininess of cameras and screens, so that all images were flawless at all scales. Such a fanciful scenario would then invalidate the argument, given above, that A's representation of B's loop, since it uses fewer pixels, is less faithful than that of its own loop. Now A has a *perfect* representation of B's loop on its screen, and vice versa. So what makes A different from B? Perhaps they are now indistinguishable?

Well, no. There is still a fundamental difference between A and B, even though each represents the other perfectly. The difference is that camera A is feeding its image directly to screen A (and not screen B), while camera B is directly feeding screen B (and not screen A). Thus, if camera A tilts or zooms in, then the entire image on screen A follows suit and also tilts or grows larger, whereas the image on screen B stays put. (To be sure, the *nested* image of screen A on screen B will tilt or grow, all the way down the line of ever-more-nested images — but the orientation and size of the *top-level* screen in system B will remain unchanged, while those of the top-level screen in system A will be directly affected by what camera A does.)

The point of this variation was to make clear that distinct identities still exist even in a situation with profoundly intertwined loops, because the perceptual hardware of a given system directly feeds only that system. It may have indirect effects on all sorts of other systems, and those effects may even be very important, but any perceptual hardware is associated first and foremost with the system into which it feeds directly (or with which it is "hard-wired", in today's blur of computational and neurological jargon).

Put less metaphorically, my sense organs feed my brain directly. They also feed the brains of my children and my friends and other people (my readers, for instance), but they do so indirectly — usually through the intermediary channel of language (though sometimes by photography, art, or music). I tell my kids some droll story of what happened at the grocery store checkout stand, and by George, they instantly see it all oh-so-clearly in their mind's eyes! The customer with the black-and-white tabloid *Weekly World News* in his cart, the odd look of the cashier as she picks it up and reads the headline about the baby found, perfectly healthy, floating in a life raft from the *Titanic,* the embarrassed chuckle of the customer, the quip by the next person in line, and so on. The imagery thus created in the brains of my kids, my friends, and others may seem at times to have a vividness rivaling that of images coming directly through their own sense organs.

Our ability to experience life vicariously in this manner is a truly wonderful aspect of human communication, but of course most of anyone's perceptual input comes from their own perceptual hardware, and only a smaller part comes filtered this way through other beings. That, to put it bluntly, is why I remain primarily myself, and why you remain primarily yourself. If, however, my perceptions came flooding as fast and furiously into your brain as they do into mine, then we'd be talking a truly different ballgame. But at least for the time being, there's no danger of such high communication rates between, say, my eyes and your brain.

Shared Perception, Shared Control

At first I had proposed that a human "I" results from the existence of a very special strange loop in a human brain, but now we see that since we mirror many people inside our crania, there will be many loops of different sizes and degrees of complexity, so we have to refine our understanding. Part of the refinement hinges, as I just stated, on the fact that one of these loops in a given brain is privileged — mediated by a perceptual system that feeds *directly* into that brain. There is another part of the story, though, which has to do with what a brain *controls* rather than what it perceives.

The thermostat in my house does not regulate the temperature in your house. Analogously, the decisions made in my brain do not control the body that's hard-wired to your brain. When you and I play tennis, it's only *my* arms that my brain controls! Or so it would seem at first. On second thought, that's clearly an oversimplification, and this is where things start to get blurry once again. I have partial and indirect control over your arms — after all, wherever I send the ball, that's where you run, and my shot has a great deal to do with how you will swing your arms. So in some indirect fashion, my brain can control your muscles in a game of tennis, but it is not a very reliable fashion. Likewise, if I hit my brakes while driving down the road, then the person behind me will also hit their brakes. What happens in my brain exerts a little bit of control over that driver's actions, but it is an unreliable and imprecise control.

The type of external control just described does not create a profound blurring of two people's identities. Tennis and driving do not give rise to deep interpenetrations of souls. But things get more complicated when language enters the show. It is through language most of all that our brains can exert a fair measure of indirect control over other humans' bodies — a phenomenon very familiar not only to parents and drill sergeants, but also to advertisers, political "spin doctors", and whiny, wheedling teen-agers. Through language, other people's bodies can become flexible extensions of our own bodies. In that sense, then, my brain is attached to your body in somewhat the same way as it is to my body — it's just that, once again, the connection is not hard-wired. My brain is attached to your body via channels of communication that are much slower and more indirect than those linking it to my body, so the control is much less efficient.

For example, I am infinitely better at writing my signature with my own hand than if I were to try to get you to do so by describing all the tiny details of the many curves that I execute so smoothly and unconsciously whenever I "sign out" at the grocery store checkout stand. But the initial notion that there is a *fundamental and absolute* distinction between how my brain is linked to my own body and how it's linked to someone else's body is seen to be exaggerated. There is a difference in degree, that's clear, but it's not clear that it's a difference in kind.

Where have we gotten so far in discussing intertwined souls? We've seen that I can perceive your perceptions indirectly, and that I can also control your body indirectly. Likewise, you can perceive my perceptions indirectly (that's what you're doing right now!), and you can control my body indirectly, at least a bit. We've also seen that the communication channels are slow enough that there are two pretty clearly separate systems, and so we can unproblematically give them different names. The fact that we humans have cleanly separated bodies (except for mother–fetus unions and Siamese twins) makes it absolutely natural to assign a different name to each body, and on a surface level, the act of assignment of distinct names to distinct bodies seems to settle the question once and for all. "Me Tarzan, you Jane." Our naming convention not only supports but enormously helps to lock in the comfortable notion that we — our *selves* — are cleanly separated entities. "Me Tarzan, you Jane" — end of story.

Language plays a further role, though, in this matter of establishing a body as the locus of an identity. Not only does it give us one name per body ("Tarzan", "Jane") but it also gives us personal pronouns ("me", "you") that do just as much as names do to reinforce the notion of a crystal-clear, sharp distinction between souls, associating one watertight soul to each body. Let's take a closer look.

A Twirlwind Trip to Twinwirld

Once, some years ago, I concocted a curious philosophical fantasy-world, to which now, with your permission, I'll escort you for the next few sections. Although back then I didn't give the place a name, I think I'll call it "Twinwirld" here. The special feature of Twinwirld is that 99 percent of all births result in identical twins, and only 1 percent give rise to singletons, which are not called that, but "halflings". In Twinwirld, twins (who, as in our world, are not *exactly* identical but have the same genome) grow up together and go everywhere together, wearing identical clothes, attending the same schools, taking the same courses, cooperating on homework assignments, making the same friends, learning to play the same musical instrument, eventually taking a single job together as a team, and so forth. A pair of identical twins in Twinwirld is called, rather inevitably, a "pairson" or a "dividual" (or even just a "dual").

Each dividual in Twinwirld is given a name at birth — thus a male pairson might be named "Greg" and a female pairson "Karen". In case you were wondering, there is a way to refer to each of the two "halves" of a pairson, although, as it happens, the need to do so crops up very seldom. However, for completeness's sake, I will describe how this is done. One simply appends an apostrophe and a one-letter suffix — either an "l" or an "r" — to the dividual's name. (Twinwirld etymologists have determined that these consonants "l" and "r" are not arbitrary, but are in fact residues of the words "left" and "right", although no two seems to be sure exactly why this should be the case.) Thus Greg consists of a "left half", Greg'l, and a "right half", Greg'r. Karen likewise consists of Karen'l and Karen'r — but as I said, most of the time, nobodies feel the need to address the "left" or "right" half of a pairson, so those suffixes are almost never used.

Now what constitutes a "friend" in Twinwirld? Well, another pairson, natch — sometwo that UU like a lot. And what about love and marriage? Well, if you've already guessed that a pairson falls in love with and marries another pairson, then you are spot on! As a matter of fact, by a crazy coincidence, this very same Karen and Greg that I just mentioned are a typical Twinwirld couple; moreover, they are the proud pairents of two twildren — a girlz named "Natalie" and a boyz named "Lucas". (To satisfy busybodies, I have to explain that I have no idea which of Karen'l and Karen'r gave birth to either twild, nor which of Greg'r and Greg'l was, so to speak, the instigating agent in either case. No two in Twinwirld ever thinks about such intimate things — no more than we in our world wonder whether the sperm leading to a child's birth came from the father's right or left testicle, or whether the egg came from the mother's left or right ovary. It's neither here nor there — the zwygote was formed and the twild was born, that's all that matters. Anyway, please don't ask too many questions on this complex topic. That's far from the point of my fantasy!)

In Twinwirld, there is an unspoken and obvious understanding that the basic units are pairsons, not left or right halves, and that even though each dividual consists of two physically separate and distinguishable halves, the bond between those halves is so tight that that the physical separateness doesn't much matter. That everytwo is made of a left and right half is just a familiar fact about being alive, taken for granted like the fact that every half has two hands, and every hand has five fingers. Things have parts, to be sure, but that doesn't mean that they don't have integrity as wholes!

The left and right halves of a pairson are sometimes physically apart from each other, though generally only for very brief periods. For instance, one half of twem might make a quick hop to the grocery store to get something that twey forgot to purchase, while the other half is cooking tweir dinner. Or if twey're snowboarding down a hill, twey might split apart to go around opposite sides of a twee. But most of the time the two halves prefer to stay close to each other. And although the two halves do have conversations together, most thoughts are so easily anticipated that very few words are usually needed, even to get across rather complex ideas.

Is "Ш" One or Two Letters of the Alphabet?

We now come to the tricky matter of pairsonal pronouns in Twinwirld. To start off, they have something like our familiar pronoun "I" for an isolated half, but it is written with a small "i". This is because "i", much like the suffixes "l" and "r", is a very rare term used only when extreme pedantic clarity is called for. Far more common than "i" is the pronoun that either half of a pairson uses in order to refer to the *whole* pairson. I am not speaking of the pronoun "we", because that word reaches out *beyond* the pairson who is speaking, and includes *other* pairsons. Thus "we" might mean, for instance, "our whole school" or "everytwo at last night's dinner party". Instead, there is a special variant of "we" — "Twe" (always spelled with a capital "T") — which denotes just that pairson of which the speaker is the left or right half. And of course there is an analogous pronoun, "UU" (which, although it looks as though it should be pronounced "double-you", is actually pronounced "tyou", like the "Tue" in a British "Tuesday"), used for addressing exactly one other pairson. Thus, for example, back when they were first getting to know each other, Greg (that is, either Greg'l or Greg'r — I don't know which half of twem) once said very timidly to Karen (on whom twey had a crush), "Tonight after dinner, Twe are going to the movies; would UU like to join twus, Karen?"

The pronoun "you" also exists in Twinwirld, but it is plural only, which means that it is never used for addressing just one other dividual — it always denotes a group. "Do you know how to ski?" might be asked of an entire family, but never of just one twild or one pairent. (The way to ask that would be, of course, "Do UU know how to ski?") Analogously, "they" never denotes just one dividual. "Both of them came to our wedding" is a statement about a *duo* of pairsons (that is to say, four halves — or four "persons", in the quaint terminology of those hailing from our world). As for a third-pairson *singular* pronoun, there is one — "twey" — and it is genderless. Thus "Did twey go to the concert last night?" could be a question about either Karen or Greg (but not about both together, as that would require "they"), and "Have twey had the measles?" could be asked about either Lucas or Natalie, but of course not about both.

Pairsonal Identity in Twinwirld

A young pairson in Twinwirld grows up with a natural sense of being just one unit, even though twey consist of two disconnected parts. "Every dividual is indivisible", runs an ancient Twinwirld saying. All sorts of conventions in Twinwirld systematically reinforce and lock in this feeling of unity and indivisibility. For instance, only one grade is earned for work that UU do in school. It may be that one half of UU is a bit weaker than the other half is in, say, math or drawing, but that doesn't affect UUr collective self-image; what counts is the *team's* joint performance. When a twild learns to play a musical instrument, both halves have their own instrument, practice the same pieces, and do so simultaneously. A bit later in life, when UU're in college, UU read novels written by pairsons, go to exhibits of paintings painted by pairsons, and study theorems proven by pairsons. In a word, credit and blame, glory and shame, neglect and fame are always doled out to pairsons, never to mere *halves* of pairsons.

The cultural norms in Twinwirld take for granted and thus reinforce the view of a pair of halves as a natural and indissoluble unit. Whereas in our society, identical twins often yearn to break away from each other, to strike out on their own, to show the world that they are *not* identical people, such desires and behavior in Twinwirld would be seen as anomalous and deeply puzzling. The two halves of a pairson would scratch tweir head (or each other's head — why not?), and say to each other, perhaps even in synchrony, "Why in the Twinwirld did twey break apart? Who would ever want to become a halfling? It would be such a semitary existence!"

I mentioned at the outset that 1 percent of births in Twinwirld result in halflings rather than pairsons. Actually, it's not quite 1 percent — more like 0.99 percent. But in any case, in Twinwirld, a very young pairson will sometimes wonder what it could possibly be like to be born a halfling, and not to be composed of two nearly identical "left" and "right" halves that hang around together all the time, echoing each other's words, thinking each other's thoughts, forming a tight team. The latter state seems so absolutely normal that it is very hard to imagine a halfling's deeply strange, semitary, and impoverished life (often jokingly called a "half-life").

What about that tiny remaining portion of births, happening just 0.01 percent of the time? Well, there is a curious phenomenon that can occur in pregnancy: both fertilized eggs constituting the zwygote break in half at the same moment (no two knows why it always happens this way, but it does), and as a result, instead of a *single* twild being born, two genetically identical twildren emerge! (Oddly, the babies are called "identical twinns", although they are never *exactly* identical.) The pairents of twinns of course love both of their "identical" offspring equally well, and very often give them cutely resonant pairs of names (such as "Natalie" and "Natalia", in the case of twinn girlzes, or "Lucas" and "Luke", for twinn boyzes).

Sometimes twinns feel the need, as they are growing up, to break away from each other, to strike out on their own, to show the wirld that they are *not* identical pairsons. But then again, some twinns enjoy playing the near-identicality game to the hilt. Roy and Bruce Nabel, for instance, are a typical pair of twinn boyzes (actually, now they're grown up) who love to confuse their friends by having Bruce turn up when Roy is expected, or vice versa. Nearly everytwo in Twinwirld finds such stunts quite amusing, because the idea of twinns is so unfamiliar to ordinary pairsons in Twinwirld. Indeed, a normal (non-twinn) pairson in Twinwirld has almost no concept of what it could be like to be a twinn. How extremely strange it would be to grow up side by side with sometwo almost identical to twoself!

There was once even an author in Twinwirld who concocted a curious philosophical fantasy-world called "Twinnwirrld", whose defining feature was that 99 percent of all births resulted in so-called "identical twinns" — but that's a whole nother story.

"Twe"-tweaking by Twinwirld-twiddling

Several intertwined issues are inevitably raised by our short and hopefully provocative little jaunt. The most vivid, of course, is that in Twinwirld, a *solo* human body — a half — builds up a sense of itself as an "i" (lowercase!), while at the same time a *pair* of human bodies — a pairson — builds up a sense of itself as a "Twe". This latter process happens partly thanks to genetics (just one genome, found in the zwygote, determines a pairson) and partly thanks to acculturation, enhanced by a slew of linguistic conventions, some of which were mentioned.

Suppose we wanted to apply the loaded word "soul" to beings in Twinwirld. What or who in Twinwirld has a soul? Even the noun "being" is a loaded word. What constitutes a *being* in Twinwirld? To my mind, both of these questions have the same answer as the following question: "What kind of entity in Twinwirld builds up an unshakable conviction of itself as an 'I'? Is it a half, a pairson, or both?" What we're really asking here is how strong each of two salient and rival analogies is — namely, how strong is the analogy between an "i" and an "I", and how strong is the analogy between a "Twe" and an "I"?

I suspect that any human reader of this chapter can easily identify with a Twinwirld half (such as Karen'l or Greg'r), which would suggest that the "i"/"I" analogy seems convincing to most readers. I hope, however, that my human readers will also see a convincing analogy between Twe-ness and I-ness, even if, for some, it is less strong than that between "i" and "I". In any case, since Twinwirld is just a fantasy, one can adjust its parameters as one wishes. You and I are both free, reader, to twiddle knobs of various sorts on Twinwirld to make "i" weaker and "Twe" stronger, or the reverse.

For the twirlwind trip just undertaken, I set the knobs determining Twinwirld at a middle-range level in order to make both analogies roughly equally plausible, hence to make the competition between "i" and "Twe" quite tight. But now I want to tweak Twinwirld so as to make "Twe" a bit stronger. In this new fantasy world, which I'll dub "Siamese Twinwirld", instead of positing that 99 percent of births yield standard identical twins, I'll posit that 99 percent of births yield Siamese twins joined, say, at the hip. Moreover, I'll stipulate that the Twinwirld pronoun "i" doesn't exist in Siamese Twinwirld. Now the only analogy that remains is that between our concept of "I" and their concept of "Twe". This may seem extremely far-fetched, but the curious thing is, our standard earthly world has much in common with Siamese Twinwirld. Here's why.

We all possess two cerebral hemispheres (left and right halves), each of which can function pretty well as a brain on its own, in case one side of our brain is damaged. I'll presume that both of your hemispheres are in good shape, dear reader, in which case what you mean when you say "I" involves a very tight team consisting of your left and right half-brains, each of which is fed directly by just one of your eyes and just one of your ears. The communication between your team's two members is so strong and rapid, however, that the fused entity — the team itself — seems like just one thing, one absolutely unbreakable self. You know just what this feels like because it's how you are constructed! And if you're anything like me, neither of your half-brains goes around calling itself "i" and brazenly proclaiming itself an autonomous soul! Rather, the two of them together make just one capital "I". In short, our own human condition in this, the real world, is quite analogous to that of pairsons in Siamese Twinwirld.

The communication between the two halves of a dividual in Twinwirld (whether it's the Siamese variant or the original one) is, of course, less efficient than that between the two cerebral hemispheres inside a human head, because our hemispheres are hard-wired together. On the other hand, the communication between halves in Twinwirld is more efficient than that between nearly any two individuals in our "normal" world. And so the degree of fusing-together of two Twinwirld halves, though not as deep as that between two cerebral hemispheres, is deeper than that between two very close siblings in our world, deeper than that between identical twins, deeper than that between wife and husband.

Post Scriptum re Twinwirld

After I had written a first draft of this chapter and had moved on to the following one, which is based on emails exchanged between Dan Dennett and myself in 1994, I noticed that in one of his messages to me he referred to an unusual pair of twins in England that he had mentioned in his 1991 book *Consciousness Explained* (which I had read in manuscript form). I had forgotten this email from Dan, so I decided to go look the reference up in his book, and I found the following passage:

> We can imagine.... two or more bodies sharing a single self. There may actually be such a case, in York, England: the Chaplin twins, Greta and Freda (*Time*, April 6, 1981). These identical twins, now in their forties and living together in a hostel, seem to act *as one*; they collaborate on the speaking of single speech acts, for instance, finishing each other's sentences with ease or speaking in unison, with one just a split-second behind. For years they have been inseparable, as inseparable as two twins who are not Siamese twins could arrange.

Some who have dealt with them suggest that the natural and effective tactic that suggested itself was to consider *them* more of a *her*....

I'm not for a moment suggesting that these twins were linked by telepathy or ESP or any other sort of occult bonds. I am suggesting that there are plenty of subtle, everyday ways of communicating and coordinating (techniques often highly developed by identical twins, in fact). Since these twins have seen, heard, touched, smelled, and thought about very much the same events throughout their lives, and started, no doubt, with brains quite similarly disposed to react to these stimuli, it might not take enormous channels of communication to keep them homing in on some sort of loose harmony. (And besides, how unified is the most self-possessed among us?)....

But in any case, wouldn't there also be two clearly defined individual selves, one for each twin, and responsible for maintaining this curious charade? Perhaps, but what if each of these women had become so selfless (as we do say) in her devotion to the joint cause that she more or less lost herself (as we also say) in the project?

I don't have any clear memory of when I first came up with the germ that has here blossomed out as my fairly elaborate Twinwirld fantasy, although I'd like to think it was before I read about the Chaplin twins in Dan's book. But whether I got the idea from Dan or made it up myself isn't crucial; I was delighted to discover not only that Dan resonated with the idea, but also that observers of real human behavior claimed to have seen something much like what I was merely blue-skying about. Twinwirld thus comes one step closer to plausibility than I might have suspected.

There is one other curiosity that by a great stroke of luck dovetails astonishingly with this chapter. A couple of days after finishing Twinwirld, I chanced to see a scrap of paper on my bedside table, and on it, in pencil, in my own hand, were written four German words — *O du angenehmes Paar* ("O thou pleasant couple"). That short phrase didn't ring a bell, but from its antiquated and exalted tone, I guessed that it was probably the opening line of an aria from some Bach cantata that I had once heard on the radio, found beautiful, and jotted down. From the Web I quickly found out my guess was right — these are the words that open a bass aria from Cantata 197, *Gott ist unsre Zuversicht* ("God, Our One True Source of Faith"). It turns out that this is a "wedding cantata" — one intended to accompany a marriage ceremony.

Here are the words that the bass sings to the couple, given first in the original German and then in my own translation, respecting both the meter and the rhyme scheme of the original:

> *O du angenehmes Paar,*
> *Dir wird eitel Heil begegnen,*
> *Gott wird dich aus Zion segnen*
> *Und dich leiten immerdar,*
> *O du angenehmes Paar!*

*

> *O thou charming bridal pair,*
> *Providence shall e'er caress thee*
> *And from Zion God shall bless thee*
> *And shall guide thee, e'er and e'er,*
> *O thou charming bridal pair!*

Are you struck, dear reader, by something rather peculiar about these words? What struck me forcefully is that although they are being sung to a couple, they feature *singular* pronouns — *du, dir,* and *dich* in German and, in my English rendition, the obsolete pronouns "thou" and "thee". On one level, these second-person singular pronouns sound strange and wrong, and yet, by addressing the couple in the singular, they convey a profound feeling of the imminent joining-together of two souls in a sacred union. To me, these poems suggest that the wedding ceremony in which they occur constitutes a "soul merger", giving rise to a single unit having just one "higher-level soul", like two drops of water coming together, touching, and then seamlessly fusing, showing that sometimes one plus one equals one.

I found translations of this aria's words into French and Italian, and they, too, used *tu* to address the twosome, and this, just like the German, sounded far weirder to me than the English, since *tu* (in either language) is completely standard usage today (unlike "thou") but it is always addressed to just one person, never *ever* to a couple or small group of any kind.

To experience the same kind of semantic jolt in modern English, you'd have to move from second person to first person, and imagine the opposite of the editorial "we" — namely, a pair of people who refer to the union they compose as "I". Thus I shall now counterfactually extend Cantata 197 by imagining one last joyous aria to be sung by the united twosome at the very end of their wedding ceremony. Its first line would run, *Jetzt bin ich ein strahlendes Paar* — "I now am a radiant couple" — and the new wife and husband would sing it precisely in unison from start to finish, instead of singing two melodies in typical Bachian counterpoint, for doing that would inappropriately draw attention to their distinct identities. In this closing aria, "I" would denote the couple itself, not either of its members, and the aria would be thought of as being for the couple's *one* new voice rather than for two independent voices.

Soulmates and Matesouls

The real point of the Twinwirld fantasy was to cast some doubt on a dogma, usually unquestioned in our world, which could be phrased as a slogan: "One body, one soul." (If you don't like the word "soul", then feel free to substitute "I", "person", "self", or "locus of consciousness".) This idea, though seldom verbalized, is so taken for granted that it seems utterly tautological to most people (unless they deny the existence of souls altogether). But visiting Twinwirld (or musing about it, if a trip can't be arranged) forces this dogma out into the open where it must at least be confronted, if not overturned. And so, if I have managed to get my readers to open their minds to the counterintuitive notion of a pair of bodies as the potential joint locus of one soul — that is, to be able to identify with a pairson such as Karen or Greg as easily as they identify with R2-D2 or with C-3PO in *Star Wars* — then Twinwirld will have discharged its duty well.

One of my inspirations for the Twinwirld fantasy was the notion of a married couple as a type of "higher-level individual" made of two ordinary individuals, which is why bumping into the *O du angenehmes Paar* scrap of paper was such a stunning coincidence. Many married people acquire this notion naturally in the course of their marriage. In fact, I had dimly sensed something like this intuitively before I was married, and I remember how, in the anticipation-filled weeks leading up to my wedding, I found this idea to be an implicit, moving theme of the book *Married People: Staying Together in the Age of Divorce* by Francine Klagsbrun. For instance, at the conclusion of a chapter about therapy and counseling for married couples, Klagsbrun writes, "I believe that a therapist should be neutral and impartial toward the partners, the two patients in the marriage, but that there is no breach of ethics in being biased toward the third patient, the marriage." I was deeply struck by her idea of the marriage itself as a "patient" undergoing therapy in order to get better, and I must say that over the years, a sense of the truth in this image helped me greatly in the harder times of my marriage.

The bond created between two people who are married for a long time is often so tight and powerful that upon the death of either one of them, the other one very soon dies as well. And if the other survives, it is often with the horrible feeling that half of their soul has been ripped out. In happier days, during the marriage, the two partners of course have individual interests and styles, but at the same time a set of common interests and styles starts to build up, and over time a new entity starts to take shape.

In the case of my marriage, that entity was Carol-and-Doug, once in a while jokingly called "Doca" or "Cado". Our oneness-in-twoness started to emerge clearly in my mind on several occasions during the first year of our marriage, right after we'd had several friends over for a dinner party and everyone had finally left and Carol and I started cleaning up together. We would carry the plates into the kitchen and then stand together at the sink, washing, rinsing, and drying, going over the whole evening together to the extent that we could replay it in our joint mind, laughing with delight at the spontaneous wit and re-savoring the unexpected interactions, commenting on who seemed happy and who seemed glum — and what was most striking in these *post partyum* decompressions was that the two of us almost always agreed with each other down the line. Something, some *thing*, was coming into being that was made out of both of us.

I remember how, a few years into our marriage, the strangest remark would occasionally be made to us: "You look so much alike!" I found this astonishing because I thought of Carol as a beautiful woman and utterly unlike me in appearance. And yet, as time passed, I started to see how there was *something* in her gaze, something about how she looked out at the world, that reminded me of my own gaze, of my own attitude about the world. I decided that the "resemblance" our friends saw wasn't located in the anatomy of our faces; rather, it was as if something of our souls was projected outwards and was perceptible as a highly abstract feature of our expressions. I could see it most clearly in certain photos of us together.

Children as Gluons

What made for the most profound bond between us, though, was without doubt the births of our two children. As a mere married couple without children, we were still not totally fused — in fact, like most couples, we were at times totally confused. But when new people, vulnerable tiny people, came into our lives, some kind of vectors inside us aligned totally. There are many couples who do not agree on how to rear their children, but Carol and I discovered happily that we saw eye-to-eye on virtually everything regarding ours. And if one of us was uncertain, talking with the other would always bring clarity into the picture.

That shared goal of bringing up our children safely, happily, and wisely in this huge, crazy, and often scary world became the dominant motif of our marriage, and it forged us both in the same mold. Although we were distinct individuals, that distinctness seemed to fade away, to vanish almost entirely, when it came to parenthood. First in that arena of life, and then slowly in other arenas, we were one individual with two bodies, one sole "pairson", one "indivisible dividual", one single "dual". We two were Twe. We had exactly the same feelings and reactions, we had exactly the same dreads and dreams, exactly the same hopes and fears. Those hopes and dreams were not mine or Carol's separately, copied twice — they were *one* set of hopes and dreams, they were *our* hopes and dreams.

I don't mean to sound mystical, as if to suggest that our common hopes floated in some ethereal neverland independent of our brains. That's not my view at all. Of course our hopes were physically instantiated two times, once in each of our separate brains — but when seen at a sufficiently abstract level, these hopes were one and the same *pattern,* merely realized in two distinct physical media.

No one has trouble with the idea that "the same gene" can exist in two different cells, in two different organisms. But what is a gene? A gene is not an actual physical object, because if it were, it could only be located in *one* cell, in *one* organism. No, a gene is a *pattern* — a particular sequence of nucleotides (usually encoded on paper by a sequence of letters from the four-letter alphabet "ACGT"). And so a gene is an abstraction, and thus "the very same gene" can exist in different cells, different organisms, even organisms living millions of years apart.

No one has trouble with the idea that "the same novel" can exist in two different languages, in two different cultures. But what is a novel? A novel is not a specific sequence of words, because if it were, it could only be written in *one* language, in *one* culture. No, a novel is a *pattern* — a particular collection of characters, events, moods, tones, jokes, allusions, and much more. And so a novel is an abstraction, and thus "the very same novel" can exist in different languages, different cultures, even cultures thriving hundreds of years apart.

And so no one should have trouble with the idea that "the same hopes and dreams" can inhabit two different people's brains, especially when those two people live together for years and have, as a couple, engendered new entities on which these hopes and dreams are all centered. Perhaps this seems overly romantic, but it is how I felt at the time, and it is how I still feel. The sharing of so much, particularly concerning our two children, aligned our souls in some intangible yet visceral manner, and in some dimensions of life turned us into a single unit that acted as a whole, much as a school of fish acts as a single-minded higher-level entity.

℞ ℞ ℞

CHAPTER 16

Grappling with the Deepest Mystery

ঔ ঔ ঔ

A Random Event Changes Everything

IN THE month of December, 1993, when we were just a quarter of the way into my sabbatical year in Trento, Italy, my wife Carol died very suddenly, essentially without warning, of a brain tumor. She was not yet 43, and our children, Danny and Monica, were but five and two. I was shattered in a way I could never have possibly imagined before our marriage. There had been a bright shining soul behind those eyes, and that soul had been suddenly eclipsed. The light had gone out.

What hit me by far the hardest was not my own personal loss ("Oh, what shall I do now? Who will I turn to in moments of need? Who will I cuddle up beside at night?") — it was *Carol's* personal loss. Of course I missed her, I missed her enormously — but what troubled me much more was that I could not get over what *she* had lost: the chance to watch her children grow up, see their personalities develop, savor their talents, comfort them in their sad times, read them bedtime stories, sing them songs, smile at their childish jokes, paint their rooms, pencil in their heights on their closet walls, teach them to ride a bike, travel with them to other lands, expose them to other languages, get them a pet dog, meet their friends, take them skiing and skating, watch old videos together in our playroom, and on and on. All this future, once so easily taken for granted, Carol had lost in a flash, and I couldn't deal with it.

There was a time, many months later, back in the United States, when I tried out therapy sessions for recently bereaved spouses — "Healing Hearts", I think they were called — and I saw that most of the people whose mates had died were focused on their own pain, on their own loss, on what they themselves were going to do now. That, of course, was the meaning of the sessions' name — you were supposed to heal, to get better. But how was *Carol* going to heal?

I truly felt as if the other people in these sessions and I were talking past each other. We didn't have similar concerns at all! I was the only one whose mate had died when the children were tiny, and this fact seemed to make all the difference. Everything had been ripped away from Carol, and I could not stand thinking about — but I could not stop thinking about — what she'd been cheated out of. This bitter injustice to Carol was the overwhelming feeling I felt, and my friends kept on saying to me (oddly enough, in a well-meaning attempt to comfort me), "You can't feel sorry for her! She's dead! There's no one to feel sorry for any more!" How utterly, totally wrong this felt to me.

One day, as I gazed at a photograph of Carol taken a couple of months before her death, I looked at her face and I looked so deeply that I felt I was behind her eyes, and all at once, I found myself saying, as tears flowed, "That's me! That's me!" And those simple words brought back many thoughts that I had had before, about the fusion of our souls into one higher-level entity, about the fact that at the core of both our souls lay our identical hopes and dreams for our children, about the notion that those hopes were not separate or distinct hopes but were just one hope, one clear thing that defined us both, that welded us together into a unit, the kind of unit I had but dimly imagined before being married and having children. I realized then that although Carol had died, that core piece of her had not died at all, but that it lived on very determinedly in my brain.

Desperate Lark

In the surreal months following the tragedy of Carol's sudden death, I found myself ceaselessly haunted by the mystery of the vanishing of her consciousness, which made no sense at all to me, and by the undeniable fact that I kept on thinking of her in the present, which also confused me. Trying to put these extremely murky things down on paper but quite unsure of myself, I initiated in late March of 1994 an email exchange with my close friend and colleague Daniel Dennett across the ocean in Massachusetts, for Dan's ideas on minds and the concept of "I" had always seemed to me to be very nearly on the same wavelength as my own (which perhaps explains why we got along so well together when, in 1981, we coedited a book entitled *The Mind's I*). Dan also had spent most of his professional life thinking about and writing about these kinds of problems, so he wasn't exactly a randomly selected partner!

Once I had started up this exchange, we sent messages back and forth across the Atlantic sporadically for a few months, the last one coming from me in late August of that year, just before the kids and I returned to the U.S. It was a fairly lopsided exchange, with me doing roughly 90 percent of the "talking", doing my best to articulate these elusive, sometimes nearly inexpressible, ideas, and Dan mostly making just brief comments on whether he agreed or not, and hinting at why.

While I was working on the last few chapters of *I Am a Strange Loop*, I reread our entire exchange, which was roughly 35 pages long when printed out, and although it was not great prose, it struck me that portions of it were worth including in the new book, in some form or other. My musings were extremely personal, of course. They were grapplings by a husband in profound shock after his wife simply went up in smoke for no reason at all. I decided to include excerpts from them here not because I wish to make some kind of grand after-the-fact public declaration of love for my wife, although there is no doubt that I loved and love her deeply. I decided to include some of my musings for the simple reason that they are heartfelt probings that struggle with the issues that form the very core of this book. Nothing else that I have written on the topic of the human soul and human consciousness ever came so much from the heart as did those messages to Dan, and even though I would like to think that I now understand the issues somewhat more clearly than I did then, I doubt that anything I write today can have nearly as much urgency as what I wrote then, in those days of extreme anguish and turmoil.

I decided that since my email grapplings have a different style from the rest of this book, and since they come from a different period of time, I would devote a separate chapter to them — and this is that chapter. In order to prepare it, I went through those 35 pages of email, which were often jumbled, redundant, and vague, and which included sporadic snippets on peripheral if not irrelevant topics, and I edited them down to about a quarter of their original length. I also reordered pieces of my messages and allowed myself to make occasional slight modifications in the passages I was keeping, so as to make the flow more logical. Consequently, what you see here is by no means a raw transcript of my end of our conversation, for that would be truly rough going, but it is a faithful boiling-down of the most important topics.

Although it was a dialogue, I have left Dan's voice out of this chapter because, as I said above, he served mostly as a cool, calm sounding board for my white-hot, emotional explorations. He was not trying to come up with any new theories; he was just listening, being my friend. There was, however, one point in April of 1994 where Dan waxed poetic about what I was going through in those days, and I think his words make an excellent prelude to this chapter, so I'll quote them below. All else that follows will be in my voice, quoted (in a slightly retouched form) from my email musings between March and August, 1994.

> There is an old racing sailboat in Maine, near where I sail, and I love to see it on the starting line with me, for it is perhaps the most beautiful sailboat I have ever seen; its name is "Desperate Lark", which I also think is beautiful. You are now embarked on a desperate lark, which is just what you should be doing right now. And your reflections are the reflections of a person who has encountered, and taken a measure of, the power of life on our sweet Earth. You'll return, restored to balance, refreshed, but it takes time to heal. We'll all be here on the shore when you come back, waiting for you.

●　　　●　　　●

The name "Carol" denotes, for me, far more than just a body, which is now gone, but rather a very vast *pattern*, a style, a set of things including memories, hopes, dreams, beliefs, loves, reactions to music, sense of humor, self-doubt, generosity, compassion, and so on. Those things are to some extent sharable, objective, and multiply instantiatable, a bit like software on a diskette. And my obsessive writing-down of memories, and the many videotapes she is on, and all our collective brain-stored memories of Carol make those pattern-aspects of her still exist, albeit in spread-out form — spread out among different videotapes, among different friends' and relatives' brains, among different yellow-sheeted notebooks, and so on. In any case, there is a spread-out pattern of Carolness very clearly discernable in this physical world. And in that sense, Carolness survives.

By "Carolness surviving", what I mean is that even people who never met her can see how it was to be near her, around her, with her — they can experience her wit, see her smile, hear her voice and her laugh, hear about her youthful adventures, learn how she and I met, watch her play with her small children, and so forth...

I keep trying, though, to figure out the extent to which I believe that because of my memories of her (in my brain or on paper), and those of other people, some of Carol's *consciousness*, her *interiority*, remains on this planet. Being a strong believer in the noncentralizedness of consciousness, in its distributedness, I tend to think that although any individual's consciousness is primarily resident in one particular brain, it is also somewhat present in other brains as well, and so, when the central brain is destroyed, tiny fragments of the living individual remain — remain *alive*, that is.

Also being a believer in the thesis that external memory is a very real part of our personal memories, I think that an infinitesimal sliver of Carol's consciousness resides even in the slips of paper on which I captured some of her cleverer bon mots, and a somewhat larger (though still tiny) shard of her resides in the yellow lined notebooks in which I have, in the past few months of grieving, recorded so many of our joint experiences. To be sure, those experiences were already encoded in my own brain, but the externalization of them will one day allow them to be shared by other people who knew her, and thus will somehow "resuscitate" her, in a small way. Thus even a static representation on paper can contain elements of a "living" Carol, of Carol's consciousness.

•   •   •

All of this brings to mind a conversation I had with my mother a few weeks after my Dad died. She said that once in a while she would look at a photo of him that she loved, in which he was smiling, and she would find herself smiling back at "him", or at "it". Her comment on this reaction of hers was, "Smiling at that photo is so wrong, because it's *not him* — it's just a flat, meaningless piece of paper." And then she got very upset with herself, and felt even more distraught over her loss of him. I pondered her anguished remark for a while, and though I could see what she meant, it seemed to me that the situation was much more complicated than what she had said.

Yes, on the surface it seems that this photo is an inert, lifeless, soulless piece of paper, but *somehow* it reaches her, it touches her. And this brought to my mind the set of lifeless, soulless pieces of paper comprising the complete works for piano of Frédéric Chopin. Though just pieces of paper, they have incredible effects on people all over the world. So might it be with that photograph of my Dad. It certainly causes deep rumblings in *my* brain when I look at it, in my sister Laura's brain, and in many others. For us, that photo is not just a physical object with mass, size, color, and so forth; it is a *pattern* imbued with fantastic triggering-power.

And of course, in addition to a photo of someone and the set of someone's complete works, there are so many other cases of elaborate patterns that contain fragments of souls — imagine, for example, having many hours of videotapes of Bach playing the organ and talking about his music, or of James Clerk Maxwell talking about physics and describing the moment when he discovered that light must be an electromagnetic wave, or of Pushkin reciting his own poetry, or of Galileo telling about how he discovered the moons of Jupiter, or of Jane Austen explaining how she imagined her characters and their complex intrigues…

Just where comes the point of "critical mass", when having a pattern, perhaps a large set of videotapes, perhaps an extensive diary (like Anne Frank's), amounts to having a significant percentage of the person — a significant percentage of their self, their soul, their "I", their consciousness, their interiority? If you concede that a significant percentage of the person would exist at *some* point along this spectrum, provided that one had a sufficiently large pattern, then it seems to me that you would have to concede that even having a much smaller pattern, such as a photo or my cherished collection of Carol's "bonner mots", already gives you a non-zero (even if microscopic) fraction of the actual person — of "the view from inside" — not just of how it was to be *with* them.

•   •   •

It was Monica's third birthday — a joyous but very sad occasion, for obvious reasons. The kids and I, along with some friends, were at an outdoor pizzeria in Cognola, our hillside village just above Trento, and we had a beautiful view of the high mountains all around us. Little Monica, in her booster chair, was sitting directly across the table from me. Because it was such an emotional occasion, one that Carol would so much have wanted to be part of, I tried to look at Monica "for Carol", and then of course wondered what on earth I was doing, what on earth I meant by thinking such a thought.

This idea of "seeing Monica for Carol" led me to a vivid memory of Old-Doug and Old-Carol (or if you prefer, "young Doug and young Carol") sitting on the terrace of the Wok, a favorite Chinese restaurant in Bloomington, way back in the summer of 1983, gazing at an adorable little dark-haired girl of two or three who was walking around in a navy-blue corduroy dress. We weren't married yet, we hadn't even broached the topic of getting married, but we had often talked very emotionally about children, and both of us were yearning to be co-parents of just such a little girl ourselves. This was a shared longing, for sure, even if only implicit.

And so now, eleven years later, now that our daughter Monica in fact exists, can I finally experience for Old-Doug that joy that he was dreaming of, longing for, back in 1983? Can I now look at his daughter Monica "for

Old-Doug"? (Or do I mean "look at *my* daughter for him"? Or both?) And if I can validly claim to be able to do so for Old-Doug, then why not just as validly for Old-Carol? After all, our yearning for a shared daughter that long-ago summer evening was a deeply shared yearning, was *the exact same yearning*, burning simultaneously in both of our brains. Thus the question is, can I now experience that joy for Old-Carol, can I now look at Monica for Old-Carol?

What seems crucial here is the depth of interpenetration of souls — the sense of shared goals, which leads to shared identity. Thus, for instance, Carol always had a deep, deep desire that Monica and Danny would be each other's best friends as they grew up, and would always remain so when they were adults. This desire also exists or persists in a very strong form inside me (in fact, we always had that joint hope, and I used to do my best to foster its realization even before she died), and it is now exerting an even greater influence on my actions than it used to, precisely because she died and so now, given that I am her best representative in this world, I feel deeply responsible to her.

•   •   •

Along with Carol's desires, hopes, and so on, her own personal sense of "I" is represented in my brain, because I was so close to her, because I empathized so deeply with her, co-felt so many things with her, was so able to see things from inside her point of view when we spoke, whether it was her physical sufferings (writhing in pain an hour after a sigmoidoscopy, her insides churning with residual air bubbles) or her greatest joys (a devilishly clever bon mot by David Moser, a scrumptious Indian meal in Cambridge) or her fondest hopes or her reactions to movies or whatever.

For brief periods of time in conversations, or even in nonverbal moments of intense feeling, *I was Carol*, just as, at times, *she was Doug*. So her "personal gemma" (to borrow Stanislaw Lem's term in his story "Non Serviam") had brought into existence a somewhat blurry, coarse-grained copy of itself inside my brain, had created a secondary Gödelian swirl inside my brain (the primary one of course being my own self-swirl), a Gödelian swirl that allowed me to be her, or, said otherwise, a Gödelian swirl that allowed her self, her personal gemma, to ride (in simplified form) on my hardware.

But is this secondary swirl that now lives in my brain, this simulated personal gemma, anything like the *real* swirl, the *primary* swirl, that once lived in her brain and is now gone? Is there Carol-consciousness still somewhere in this world? That is, is it possible for me to look at Monica "for Carol" and, even in the tiniest degree, to *become Carol* seeing Monica? Or has that personal gemma been finally and totally and irrevocably obliterated?

●   ●   ●

A person is a *point of view* — not only a *physical* point of view (looking out of certain eyes in a certain physical place in the universe), but more importantly a *psyche's* point of view: a set of hair-trigger associations rooted in a huge bank of memories. The latter can be absorbed, more and more over time, by someone else. Thus it's like acquiring a foreign language step by step.

For a while, one's speaking is largely "fake" — that is, one is thinking in one's native language but substituting words quickly enough to give the impression that the thinking is going on in the second language; however, as one's experience with the second language grows, new grammatical habits form and turn slowly into reflexes, as do thousands of lexical items, and the second language becomes more and more rooted, more and more genuine. One gradually becomes a fluent thinker in and speaker of the other language, and it is no longer "fake", even if one has an accent in it. So it is with coming to see the world through another person's soul.

My parents, for instance, internalized each other's psychic points of view very deeply over the nearly fifty years of their marriage, and each of them thus gradually became a "fluent be-er" of the other. Perhaps when my mother "was" my father, she was so with an "accent", and vice versa, but for each of them, the act of *being* the other one was certainly genuine, was not fakery.

As with my parents, so there was some degree of genuine *being of Carol* by me when she was alive, and vice versa. Although it took me several years to learn to "be" Carol, and although I certainly never reached the "native speaker" level, I think it's fair to say that, at our times of greatest closeness, I was a "fluent be-er" of my wife. I shared so many of her memories, both from our joint times and from times before we ever met, I knew so many of the people who had formed her, I loved so many of the same pieces of music, movies, books, friends, jokes, I shared so many of her most intimate desires and hopes. So her point of view, her interiority, her *self,* which had originally been instantiated in just one brain, came to have a second instantiation, although that one was far less complete and intricate than the original one. (Actually, long before she met me, her point of view had already engendered other instantiations, because it had of course been internalized to varying degrees and levels of fidelity by her siblings and her parents.) Needless to say, Carol's point of view was always by far *most* strongly instantiated in *her* brain.

This talk of someone "being" someone else reminds me of a Linguistics Department Christmas party back in the late 1970's, when Carol's and my old friend Tom Ernst did a marvelous imitation of his professor John Goldsmith (also a friend), with so many *echt* mannerisms of John's. It was uncanny to me to watch Tom "put on" and "take off" John's style — and in so doing, putting John on and making a fine take-off of him.

•　　　•　　　•

There are shallower aspects of a person and there are deeper aspects, and the deeper aspects are what imbue the shallower ones with genuine meaning. I guess that sounds cryptic. What I mean is that if I believe statement X (for example, "Chopin is a great composer") and someone else also believes X, then, despite this ostensible agreement between us, our internal feelings when we think X may be unutterably different even though, on the superficial verbal level, our belief is "the same". On the other hand, if our souls have a deep resemblance, then our two beliefs in X will in fact be very similar, and we will intuitively resonate with each other. Communication (at least on that topic) will be nearly effortless.

What really matters for mutual understanding of two people are such things as having similar responses to music (not just shared likes but also shared dislikes), having similar responses to people (again, I mean both likes and dislikes), having similar degrees of empathy, honesty, patience, sentimentality, audacity, ambition, competitiveness, and so on. These central building blocks of personality, character, and temperament are decisive in mutual understanding.

Consider, for instance, the shattering experience of constantly feeling inferior to other people. Some people know this intimately, and some don't know it at all. A person with huge reserves of self-confidence will simply *never* be able to feel how it is to be paralyzed by the lack of confidence — they "just don't get it". It is *these* sorts of aspects, these innermost aspects of a soul (as opposed to such relatively objective and transferable items as countries visited, novels read, cuisines mastered, historical facts known, and so forth) that make for soul-uniqueness.

I'm concerned with whether the deeper aspects of a person, the ones that give rise to a self, to an "I", are transportable to another person, or absorbable by another person (*i.e.*, by the second person's brain). The second person doesn't have to change their own personality or opinions in order to absorb the first person; it can be more like an alter ego that, like an article of clothing or a persona or a stage role, they can occasionally don or slip into (my image is that of Tom Ernst putting on and taking off that John Goldsmith persona, although of course on a much more profound level), a sort of a "second vantage point" from which to see the world.

But the key question is, no matter how much you absorb of another person, can you ever have absorbed *so much of them* that when that primary brain perishes, you can feel that that *person* did not totally perish from the earth, because they (or at least a significant fraction of them) are still instantiated in your brain, because they still live on in a "second neural home"?

•   •   •

In my opinion, to deal with this question head-on, one really has to focus on this thing I call the "Gödelian swirl of self". The key question becomes this: When the *pointers* to "self" — the structures that, through a lifetime of locking-in and self-stabilizing, have given rise to an "I" — are copied in some imperfect, low-resolution fashion in a secondary brain, where exactly do they wind up pointing?

My internal model of Carol is certainly "thin" or sparse in comparison to the original self-model (the one that was located inside her own brain), but that sparseness is not the key issue. The crux is this: even if my internal model of Carol were unbelievably rich (*e.g.*, like my Mom's model of my Dad, say, or even ten times stronger than that), would it nonetheless be *the wrong kind of structure* to give rise to an "I"? Would it be something other than a strange loop? Would it be a structure pointing *not at itself* but at *something else,* and therefore be lacking that essentially swirly, vortical, self-referential quality that makes an "I"?

My guess is that if the model were extremely rich, extremely faithful, then effectively the destinations of all the pointers in it would be *fluid* — in other words, the pointers inside my model of Carol would be able to slip, to point just as validly to the symbol for her in *my* brain as to her *own* self-symbol. If so, then the original swirliness, the original "I"-ness of the structure, would have been successfully transported to a second medium and reconstructed faithfully (though far more coarse-grainedly) in it.

•   •   •

The "outer" layers of the self consist in lots of pointers that point mostly at standard universal aspects of the world (*e.g.*, rain, ice cream, the swooping of swallows, etc., etc.); the "middle" layers of the self consist in pointers to things more tied in with one's own life (*e.g.*, one's parents' faces and voices, the music one loves, the street one grew up on, one's beloved pets from childhood, one's favorite books and movies, and many other deep things); then the inner sanctum has tons of tangly pointers to very deeply "indexical" things, such as one's insecurities, one's sexual feelings, one's most intense fears, one's deepest loves, and lots of other things that I cannot put my finger on). All this is very vague, and only meant to suggest a kind of imagery wherein the outermost layers have mostly outwards-pointing arrows, the middle layers have a mixture of inwards and outwards arrows, and then the innermost core has tons of arrows that point right back in towards itself. Strange-Loop City — that's an "I" for you!

It's that deeply twisted-back-on-itself quality of the innermost core that, I surmise, makes it so hard to transport elsewhere, that makes the soul so deeply, almost irrevocably, attached to one single body, one single brain. The outer layers are relatively easy to transport, of course, with their relative paucity of inwards-pointing pointers, and the middle layers are medium easy to transport. Someone as close to Carol as I was can get lots of the outer layers and something of the middle layers and little bits of the inner core, but can one ever internalize enough of that core to say that, even in a very diluted sense, "she's still here among us"?

•   •   •

Perhaps I'm exaggerating the difficulty of transport. In some sense, all Gödelian loops-of-self (*i.e.*, strange loops that give rise to an "I") are isomorphic at the most coarse-grained level, and therefore in lowest approximation they may not be hard to transport at all; what makes them different from each other is only their "flavorings", consisting of memories, and, of course, genetic preferences and talents, and so forth. So, to the extent that we can be chameleons and can import the "spices" of other people's life histories (the spices that imbue *their* self-loops with unique individuality), we *are* capable of seeing the world through their eyes. Their psychic point of view is transportable and modular — not trapped inside just one perishable piece of hardware.

If this is true, then Carol survives because her point of view survives — or rather, she survives *to the extent* that her point of view survives — in my brain and those of others. This is why it is so good to keep records, to write down memories, to have photos and videotapes, and to do so with maximal clarity — because thanks to having such records, you can "possess", or "be possessed by", other people's brains. That's why Frédéric Chopin, the actual person, survives so much in our world, even today.

•   •   •

When, someday, I first watch our videotapes with Carol on them, my heart is going to break because I'll be seeing her again, living her again, being with her again — and though I'll be filled with love, I'll also be pervaded by the feeling that this is *fake*, that I'm being tricked, and all of this will make me wonder just what is going on inside my brain.

There is no doubt that the patterns that will be sparked in my brain by watching those videos — the symbols in my brain that will be triggered, reactivated, resuscitated, brought back to life for the first time since she died, and that will be dancing inside me — will be just as strong as when they were sparked in my brain when she herself was there, in person, actually doing those things that are now merely images on tape. The dance of the symbols inside my brain sparked by the videos will be *the same dance*, and danced by *the same symbols*, as when she was right there before me.

So there's this set of structures inside my brain that videos and photos and other extremely intense records can access in such a profound way — the structures *in me* that, when she was alive, were correlated with Carol, were deeply in resonance with her, the structures that represented Carol, the structures that seemed, for all the world, to *be* Carol. But as I watch the videos, knowing she is gone, the fraudulency will at once be being revealed and yet be deeply confusing me, because I will be *seeming* to see her, seeming to have revived her, seeming to have brought her back, just as I do in my dreams. And so I wonder, what *is* the nature of those structures collectively forming the "Carol symbol" in my brain? How big is the Carol symbol? And most importantly of all: How close does the Carol symbol inside Doug come to *being* a person, as opposed to merely *representing* or *symbolizing* a person?

The following should be a much easier question (although I think it is not actually easier). What was the nature of the "Holden Caulfield symbol" in J. D. Salinger's brain during the period when he was writing *Catcher in the Rye*? That structure was all there ever was to Holden Caulfield — but it was so, so rich. Perhaps that symbol wasn't as rich as a full human soul, but Holden Caulfield seems like *so much* of a person, with a true core, a true soul, a true personal gemma, even if only a "miniature" one. You couldn't ask for a richer representation, a richer mirroring, of one person inside another person, than whatever constituted the Holden Caulfield symbol inside Salinger's brain.

•   •   •

I hope the overall set of ideas here sounds coherent to you, Dan, even though what I have said is certainly made up of lots of incoherent little threads. It is terribly hard to articulate these things, and it is made far harder by the interference of one's deep emotions, which *wish* things to be certain ways, and which push to a certain extent for the answers to come out on that side. Of course it is also precisely the strength of those desires that makes these questions so intense and so important in ways that wouldn't have happened if tragedy hadn't struck.

I must admit that I feel a little bit like someone trying to grapple with quantum-mechanical reality while quantum mechanics was developing but before it had been fully and rigorously established — someone around 1918, someone like Sommerfeld, who had a deep understanding of all the so-called "semiclassical" models that were then available (the wonderful Bohr atom and its many improved versions), but quite a while before Heisenberg and Schrödinger came along, cutting to the very core of the question, and getting rid of all the confusion. Around 1918, a lot of the truth was nearly within reach, but even people who were at the cutting edge could easily fall back into a purely classical mode of thinking and get hopelessly confused.

That's how I feel about self, soul, consciousness these days. I feel as if I know very intimately, yet can't quite always remember, the distributedness of consciousness and the illusion of the soul. It's frustrating to feel myself constantly sliding back into conventional intuitive ("classical") views of these questions when I know that deep down, my view is radically counterintuitive ("quantum-mechanical").

Post Scriptum

Long after this chapter (minus this P.S.) had been put together in final form, it occurred to me that it might be tempting for some readers to conclude that in the wake of Carol's death, her deeply depressed husband had buckled under the terrible pressures of loss, and had sought to build some kind of elaborate intellectual superstructure through which he could deny to himself what was self-evident to all outsiders: that his wife had died and was completely gone, and that was all there was to it.

Such skepticism or even cynicism is quite natural, and I will admit that even I, looking back at these grapplings, couldn't help wondering if denial of death's reality or finality wasn't a good part of the motivation for all the anguished musings about souls and survival that I engaged in, not only during the year of 1994 but for many years thereafter. Since I know myself quite well, I didn't really think this was the case (although sometimes I was a little bit unsure just what was the case), but what definitely troubled me was the thought that readers who don't know me could easily draw such a conclusion and could thus dismiss my grapplings as the passionate ravings of a suffering individual who had expediently modified his belief system in order to give balm to his grief.

It was therefore a relief when, very recently, I went through a number of old files in my filing cabinets — files with names like "Identity", "Strange Loops", "Consciousness", and so forth — and ran across writings galore in which all these same ideas are set forth in crystal-clear terms long before there was any shadow on the horizon. I found endless musings, all written out by hand, in which I talked about the blurred identities of human souls, and in particular I found several episodes where I talked explicitly about the fusing of Carol's and my soul into a single tight unit, or about the "soul merger" of Carol and Danny.

In these improvised passages, I often dreamed up quite amusing but very serious thought experiments in which I tampered with the rate of potential information flow between two brains (one time involving a linkup connecting my brain and a zombie's brain — a delightful thought, at least to me!). What became obvious was that these ideas about who we are and what makes a person unique had been brewing and stirring around in my mind for decades, and that it had all come to an intense boil when I got married and especially when I had the experience of having children and raising them with someone whose love for them was so terribly similar to, and so terribly entangled with, my own love for them.

My book is now done, and those old paper files are rich preludes to it. Perhaps someday some of what I wrote back then will see the light of day, perhaps never, but at least I myself have the comfort of knowing that when I was in my time of greatest need, I did not merely tumble for some kind of path-of-least-resistance belief system that winked at me, but instead I stayed true to my long-term principles, worked out with great care many years earlier. That knowledge about myself gives me a small kind of solace.

⤚ ⤙ ⤚

CHAPTER 17

How We Live in Each Other

ക ക ക

Universal Machines

WHEN I was around twelve, there were kits you could buy that allowed you to put together electronic circuitry that would carry out various interesting functions. You could build a radio, a circuit that would add two binary numbers, a device that could encode or decode a message using a substitution cipher, a "brain" that would play tic-tac-toe against you, and a few other devices like this. Each of these machines was *dedicated*: it could do just one kind of trick. This is the usual meaning of "machine" that we grow up with. We are accustomed to the idea of a refrigerator as a dedicated machine for keeping things cold, an alarm clock as a dedicated machine for waking us up, and so on. But more recently, we have started to get used to machines that transcend their original purposes.

Take cellular telephones, for instance. Nowadays, in order to be competitive, cell phones are marketed not so much (maybe even very little) on the basis of their original purpose as communication devices, but instead for the number of tunes they can hold, the number of games you can play on them, the quality of the photos they can take, and who knows what else! Cell phones once were, but no longer are, dedicated machines. And why is that? It is because their inner circuitry has surpassed a certain threshold of complexity, and that fact allows them to have a chameleon-like nature. You can use the hardware inside a cell phone to house a word processor, a Web browser, a gaggle of video games, and on and on. This, in essence, is what the computer revolution is all about: when a certain well-defined threshold — I'll call it the "Gödel–Turing threshold" — is surpassed, then a computer can emulate *any* kind of machine.

This is the meaning of the term "universal machine", introduced in 1936 by the English mathematician and computer pioneer Alan Turing, and today we are intimately familiar with the basic idea, although most people don't know the technical term or concept. We routinely download virtual machines from the Web that can convert our universal laptops into temporarily specialized devices for watching movies, listening to music, playing games, making cheap international phone calls, who knows what. Machines of all sorts come to us through wires or even through the air, via software, via patterns, and they swarm into and inhabit our computational hardware. One single universal machine morphs into new functionalities at the drop of a hat, or, more precisely, at the double-click of a mouse. I bounce back and forth between my email program, my word processor, my Web browser, my photo displayer, and a dozen other "applications" that all live inside my computer. At any specific moment, most of these independent, dedicated machines are dormant, sleeping, waiting patiently (actually, unconsciously) to be awakened by my royal double-click and to jump obediently to life and do my bidding.

Inspired by Gödel's mapping of *PM* into itself, Alan Turing realized that the critical threshold for this kind of computational universality comes at exactly that point where a machine is flexible enough to read and correctly interpret a set of data that describe its own structure. At this crucial juncture, a machine can, in principle, explicitly watch how it does any particular task, step by step. Turing realized that a machine that has this critical level of flexibility can imitate any another machine, no matter how complex the latter is. In other words, there is nothing *more* flexible than a universal machine. Universality is as far as you can go!

This is why my Macintosh can, if I happen to have fed it the proper software, act indistinguishably from my son's more expensive and faster "Alienware" computer (running any specific program), and vice versa. The only difference is one of speed, because my Mac will always remain, deep in its guts, a Mac. It will therefore have to imitate the fast, alien hardware by constantly consulting tables of data that explicitly describe the hardware of the Alien, and doing all those lookups is very slow. This is like me trying to get you to sign my signature by writing out a long set of instructions telling you how to draw every tiny curve. In principle it's possible, but it would be hugely slower than just signing with my own hardware!

The Unexpectedness of Universality

There is a tight analogy linking universal machines of this sort with the universality I earlier spoke of (though I didn't use that word) when I described the power of *Principia Mathematica*. What Bertrand Russell and Alfred North Whitehead did not suspect, but what Kurt Gödel realized, is that, simply by virtue of representing certain fundamental features of the positive integers (such basic facts as commutativity, distributivity, the law of mathematical induction), they had unwittingly made their formal system *PM* surpass a key threshold that made it "universal", which is to say, capable of defining number-theoretical functions that imitate arbitrarily complex *other* patterns (or indeed, even capable of turning around and imitating itself — giving rise to Gödel's black-belt maneuver).

Russell and Whitehead did not realize what they had wrought because it didn't occur to them to use *PM* to "simulate" anything else. That idea was not on their radar screen (for that matter, radar itself wasn't on anybody's radar screen back then). Prime numbers, squares, sums of two squares, sums of two primes, Fibonacci numbers, and so forth were seen merely as beautiful mathematical patterns — and patterns consisting of numbers, though fabulously intricate and endlessly fascinating, were not thought of as being isomorphic to anything else, let alone as being stand-ins for, and thus standing for, anything else. After Gödel and Turing, though, such naïveté went down the drain in a flash.

By and large, the engineers who designed the earliest electronic computers were as unaware as Russell and Whitehead had been of the richness that they were unwittingly bringing into being. They thought they were building machines of very limited, and purely military, scopes — for instance, machines to calculate the trajectories of ballistic missiles, taking wind and air resistance into account, or machines to break very specific types of enemy codes. They envisioned their computers as being specialized, single-purpose machines — a little like wind-up music boxes that could play just one tune each.

But at some point, when Alan Turing's abstract theory of computation, based in large part on Gödel's 1931 paper, collided with the concrete engineering realities, some of the more perceptive people (Turing himself and John von Neumann especially) put two and two together and realized that their machines, incorporating the richness of integer arithmetic that Gödel had shown was so potent, were thereby universal. All at once, these machines were like music boxes that could read arbitrary paper scrolls with holes in them, and thus could play *any* tune. From then on, it was simply a matter of time until cell phones started being able to don many personas other than just the plain old cell-phone persona. All they had to do was surpass that threshold of complexity and memory size that limited them to a single "tune", and then they could become anything.

The early computer engineers thought of their computers as number-crunching devices and did not see numbers as a universal medium. Today we (and by "we" I mean our culture as a whole, rather than specialists) do not see numbers that way either, but our lack of understanding is for an entirely different reason — in fact, for exactly the opposite reason. Today it is because all those numbers are so neatly hidden behind the screens of our laptops and desktops that we utterly forget they are there. We watch virtual football games unfolding on our screen between "dream teams" that exist only inside the central processing unit (which is carrying out arithmetical instructions, just as it was designed to do). Children build virtual towns inhabited by little people who virtually ride by on virtual bicycles, with leaves that virtually fall from trees and smoke that virtually dissipates into the virtual air. Cosmologists create virtual galaxies, let them loose, and watch what happens as they virtually collide. Biologists create virtual proteins and watch them fold up according to the complex virtual chemistry of their constituent virtual submolecules.

I could list hundreds of things that take place on computer screens, but few people ever think about the fact that all of this is happening courtesy of *addition and multiplication of integers* way down at the hardware level. But that *is* exactly what's happening. We don't call computers *computers* for nothing, after all! They are, in fact, computing sums and products of integers expressed in binary notation. And in that sense, Gödel's world-dazzling, Russell-crushing, Hilbert-toppling vision of 1931 has become such a commonplace in our downloading, upgrading, gigabyte culture that although we are all swimming in it all the time, hardly anyone is in the least aware of it. Just about the only trace of the original insight that remains visible, or rather, "audible", around us is the very word "computer". That term tips you off, if you bother to think about it, to the fact that underneath all the colorful pictures, seductive games, and lightning-fast Web searches, there is nothing going on but integer arithmetic. What a hilarious joke!

Actually, it's more ambiguous than that, and for all the same reasons as I elaborated in Chapter 11. Wherever there is a pattern, it can be seen either as itself or as standing for anything to which it is isomorphic. Words that apply to Pomponnette's straying also apply, as it happens, to Aurélie's straying, and neither interpretation is truer than the other, even if one of them was the originally intended one. Likewise, an operation on an integer that is written out in binary notation (for instance, the conversion of "0000000011001111" into "1100111100000000") that one person might describe as multiplication by 256 might be described by another observer as a left-shift by eight bits, and by another observer as the transfer of a color from one pixel to its neighbor, and by someone else as the deletion of an alphanumeric character in a file. As long as each one is a correct description of what's happening, none of them is privileged. The reason we call computers "computers", then, is historic. They originated as integer-calculation machines, and they are still of course validly describable as such — but we now realize, as Kurt Gödel first did back in 1931, that such devices can be equally validly perceived and talked about in terms that are fantastically different from what their originators intended.

Universal Beings

We human beings, too, are universal machines of a different sort: our neural hardware can copy arbitrary patterns, even if evolution never had any grand plan for this kind of "representational universality" to come about. Through our senses and then our symbols, we can internalize external phenomena of many sorts. For example, as we watch ripples spreading on a pond, our symbols echo their circular shapes, abstract them, and can replay the essence of those shapes much later. I say "the essence" because some — in fact most — detail is lost; as you know very well, we retain not all levels of what we encounter but only those that our hardware, through the pressures of natural selection, came to consider the most important. I also have to make clear (although I hope no reader would fall into such a trap) that when I say that our symbols "internalize" or "copy" external patterns, I don't mean that when we watch ripples on a pond, or when we "replay" a memory of such a scene (or of many such scenes blurred together), there literally are circular patterns spreading out on some horizontal surface inside our brains. I mean that a host of structures are jointly activated that are connected with the concepts of water, wetness, ponds, horizontal surfaces, circularity, expansion, things bobbing up and down, and so forth. I am not talking about a movie screen inside the head!

Representational universality also means that we can import ideas and happenings without having to be direct witnesses to them. For example, as I mentioned in Chapter 11, humans (but not most other animals) can easily process the two-dimensional arrays of pixels on a television screen and can see those ever-changing arrays as coding for distant or fictitious three-dimensional situations evolving over time.

On a skiing vacation in the Sierra Nevada, far away from home, my children and I took advantage of the "doggie cam" at the Bloomington kennel where we had boarded our golden retriever Ollie, and thanks to the World Wide Web, we were treated to a jerky sequence of stills of a couple of dozen dogs meandering haphazardly in a fenced-in play area outdoors, looking a bit like particles undergoing random Brownian motion, and although each pooch was rendered by a pretty small array of pixels, we could often recognize our Ollie by subtle features such as the angle of his tail. For some reason, the kids and I found this act of visual eavesdropping on Ollie quite hilarious, and although we could easily describe this droll scene to our human friends, and although I would bet a considerable sum that these few lines of text have vividly evoked in your mind both the canine scene at the kennel and the human scene at the ski resort, we all realized that there was not a hope in hell that we could ever explain to Ollie himself that we had been "spying" on him from thousands of miles away. Ollie would never know, and could never know.

Why not? Because Ollie is a dog, and dogs' brains are not universal. They cannot absorb ideas like "jerky still photo", "24-hour webcam", "spying on dogs playing in the kennel", or even, for that matter, "2,000 miles away". This is a huge and fundamental breach between humans and dogs — indeed, between humans and all other species. It is this that sets us apart, makes us unique, and, in the end, gives us what we call "souls".

In the world of living things, the magic threshold of representational universality is crossed whenever a system's repertoire of symbols becomes extensible without any obvious limit. This threshold was crossed on the species level somewhere along the way from earlier primates to ourselves. Systems above this counterpart to the Gödel–Turing threshold — let's call them "beings", for short — have the capacity to model inside themselves other beings that they run into — to slap together quick-and-dirty models of beings that they encounter only briefly, to refine such coarse models over time, even to invent imaginary beings from whole cloth. (Beings with a propensity to invent other beings are often informally called "novelists".)

Once beyond the magic threshold, universal beings seem inevitably to become ravenously thirsty for tastes of the interiority of other universal beings. This is why we have movies, soap operas, television news, blogs, webcams, gossip columnists, *People* magazine, and *The Weekly World News*, among others. People yearn to get inside other people's heads, to "see out" from inside other crania, to gobble up other people's experiences.

Although I have been depicting it somewhat cynically, representational universality and the nearly insatiable hunger that it creates for vicarious experiences is but a stone's throw away from empathy, which I see as the most admirable quality of humanity. To "be" someone else in a profound way is not merely to see the world intellectually as they see it and to feel rooted in the places and times that molded them as they grew up; it goes much further than that. It is to adopt their values, to take on their desires, to live their hopes, to feel their yearnings, to share their dreams, to shudder at their dreads, to participate in their life, to merge with their soul.

Being Visited

One morning not long ago I woke up with the memory of my father richly pulsating inside my cranium. For a shining moment my dreaming mind seemed to have brought him back to life in the most vivid fashion, even though "he" had had to float in the rarefied medium of my brain's stage. It felt, nonetheless, like he was really back again for a short while, and then, sadly, all at once he just went poof. How is this bittersweet kind of experience, so familiar to every adult human being, to be understood? What degree of reality do these software beings that inhabit us have? Why did I put "he" in quotation marks, a few lines up? Why the caution, why the hedging?

What is *really* going on when you dream or think more than fleetingly about someone you love (whether that person died many years ago or is right now on the other end of a phone conversation with you)? In the terminology of this book, there is no ambiguity about what is going on. The symbol for that person has been activated inside your skull, lurched out of dormancy, as surely as if it had an icon that someone had double-clicked. And the moment this happens, much as with a game that has opened up on your screen, your mind starts acting differently from how it acts in a "normal" context. You have allowed yourself to be invaded by an "alien universal being", and to some extent the alien takes charge inside your skull, starts pushing things around in its own fashion, making words, ideas, memories, and associations bubble up inside your brain that ordinarily would not do so. The activation of the symbol for the loved person swivels into action whole sets of coordinated tendencies that represent that person's cherished style, their idiosyncratic way of being embedded in the world and looking out at it. As a consequence, during this visitation of your cranium, you will surprise yourself by coming out with different jokes from those you would normally make, seeing things in a different emotional light, making different value judgments, and so forth.

But the crux of the matter for us right now is the following question: Is your symbol for another person actually an "I"? Can that symbol have inner experiences? Or is it as unalive as is your symbol for a stick or a stone or a playground swing? I chose the example of a playground swing for a reason. The moment I suggest it to you, no matter what playground you have located it in, no matter what you imagine its seat to be made of, no matter how high you imagine the bar it is dangling from, you can see it swinging back and forth, wiggling slightly in that funny way that swings wiggle, losing energy unless pushed, and you can also hear its softly clinking chains. Though no one would call the swing itself alive, there is no doubt that its mental proxy is dancing in the seething substrate of your brain. After all, that is what a brain is made for — to be a stage for the dance of active symbols.

If you seriously believe, as I do and have been asserting for most of this book, that concepts are *active symbols in a brain,* and if furthermore you seriously believe that *people, no less than objects, are represented by symbols in the brain* (in other words, that each person that one knows is internally mirrored by a concept, albeit a very complicated one, in one's brain), and if lastly you seriously believe that *a self is also a concept, just an even more complicated one* (namely, an "I", a "personal gemma", a rock-solid "marble"), then it is a necessary and unavoidable consequence of this set of beliefs that *your brain is inhabited to varying extents by other I's, other souls,* the extent of each one depending on the degree to which you faithfully represent, and resonate with, the individual in question. I include the proviso "and resonate with" because one can't just slip into any old soul, no more than one can slip into any old piece of clothing; some souls and some suits simply "fit" better than others do.

Chemistry and Its Lack

For me, the best illustration of the idea of better and worse fits or "resonances" between souls is musical taste. I will never forget what happened, thirty-some years ago, when a pianist friend praised Béla Bartók's second violin concerto to the skies and insisted that I get to know it. This was an act of reciprocation for my having introduced to her, a few years earlier, one of the most stirring pieces of music I knew — Prokofiev's third piano concerto. At that time, she had resonated to the last movement of the Prokofiev in an incredibly powerful way, a fact that seemed to signal that we were on much the same musical wavelength; therefore, I took her passionate endorsement of Bartók's second violin concerto with great seriousness. To egg me on, she said that Bartók not only used her favorite chord from the Prokofiev over and over, but he used it *better.* Say no more! I instantly went out and bought a record of it. That evening, with high anticipation, I put it on and listened carefully. To my disappointment, I was utterly unaffected. This was very puzzling. I listened again. And then again. And again. And again. Over a couple of weeks, I must have listened to that highly-touted piece a dozen times if not two dozen, and yet nothing at all ever happened inside me, except that a fifteen-second section somewhere in the middle mildly engaged me. You could call this a blind spot — or a deaf spot — inside me, or else, as I would prefer, you could just say that the "fit" between my soul and Bartók's is extremely poor. And this has been corroborated many times over with other Bartók pieces, so that now I am quite confident about what will (or rather, won't) happen inside me when I hear Bartók. Although I like a few small pieces (based on folk songs) that he wrote, the bulk of his output doesn't speak to me at all. And so my sense that this friend and I had a lot in common musically was greatly reduced, and in fact our friendship subsided thereafter.

After writing that paragraph, I grew curious as to whether a thirty-year-old memory might be revealed invalid, or whether in the meantime my soul might perhaps have opened up to new musical horizons, so I went straight to my record player (yes, vinyl), put the Bartók violin concerto on once again, and listened to it carefully from beginning to end. My reaction was totally identical. To me, the piece just seems to wander and wander, never getting anywhere. Listening to it, I feel like a magnetic field bashing headlong into a superconductor — cannot penetrate even one micron! In case that's too esoteric a metaphor, let's just say that I'm stopped dead, right at the surface. It makes no sense at all to me; it is music written in an impenetrable idiom. It's like looking at a book written in an alien script. You can tell there is intelligence behind it — maybe a great deal! — but you have no idea what it is saying.

I recount this rather gloomy anecdote because it stands for a thousand experiences in life, involving what, for lack of a better word, we call "chemistry" between people. There just is no chemistry between Bartók and me. I respect his intelligence, his creative drive, and his high moral standards, but I have no idea what made his heart tick. Not a clue. But I could say this of thousands of people — and then there are those for whom the reverse holds equally strongly. For instance, there is no piece of music in the world that means more to me than Prokofiev's first violin concerto, written within just a few years of the Bartók concerto. (In fact, to my bewilderment, I have even seen the two mentioned in the same breath, as if they were cut from the same cloth. They might have some superficial textures in common here and there, but to me they are as different as Bach and Eminem.) While the Bartók rolls off of me like water off a duck's back, the Prokofiev flows into me like an infinitely intoxicating elixir. It speaks to me, soars inside me, sets me on fire, turns up the volume of life to full blast.

I need not go on and on, because I am sure that every reader has experienced chemistries and non-chemistries of this sort — perhaps even relating to the Bartók and Prokofiev violin concertos in exactly the reverse fashion from me, but even so, the message I am trying to convey will come across loud and clear. Music seems to me to be a direct route to the heart, or between hearts — in fact, the most direct. Across-the-board alignment of musical tastes, including both loves and hates — something extremely rarely run into — is as sure a guide to affinity of souls as I have ever found. And an affinity of souls means that the people concerned can rapidly come to know each other's essences, have great potential to live inside each other.

Copycat Planetoids Grow by Absorbing Melting Meteorites

As children, as adolescents, and even as adults, we are all copycats. We involuntarily and automatically incorporate into our repertoire all sorts of behavior-fragments of other people. I already mentioned my "Hopalong Cassidy smile" in first grade, which I suppose still vaguely informs my "real" smile, and I have dozens of explicit memories of other copycat actions from that age and later. I remember admiring and then copying one friend's uneven, jagged handwriting, a jaunty classmate's cool style of blustering, an older boy's swaggering walk, the way the French ticketseller in the film *Around the World in Eighty Days* pronounced the word *américain*, a college friend's habit of always saying the name of his interlocutor at the end of every phone call, and so forth. And when I watch a video of myself, I am always caught off guard to see so many of my sister Laura's terribly familiar expressions (they're *so her*) flicker briefly across *my* face. Which of us borrowed from the other, and when, and why? I'll never know.

I have long watched my two children imitate catchy intonation-patterns and favorite phrases of their American friends, and I can also hear specific Italian friends' sounds and phrases echo throughout their Italian. There have been times when, on listening to either of them talk, I could practically have rattled off a list of their friends' names as the words and sounds sailed by.

The small piano pieces I used to compose with such intense emotional fervor — a fervor that felt like it was pure *me* — are riddled, ironically, with recognizable features coming very clearly from Chopin, Bach, Prokofiev, Rachmaninoff, Shostakovich, Scriabin, Ravel, Fauré, Debussy, Poulenc, Mendelssohn, Gershwin, Porter, Rodgers, Kern, and easily another dozen or more composers whose music I listened to endlessly in those years. My writing style bears marks of countless writers who used words in amazing ways that I wished I could imitate. My ideas come from my mother, my father, my youthful friends, my teachers… Everything I do is some kind of modified borrowing from others who have been close to me either actually or virtually, and the virtual influences are among the most profound.

Much of my fabric is woven out of borrowed bits and pieces of the experiences of thousands of famous individuals whom I never met face to face, and almost surely never will, and who for me are therefore only "virtual people". Here's a sample: Niels Bohr, Dr. Seuss, Carole King, Martin Luther King, Billie Holiday, Mickey Mantle, Mary Martin, Maxine Sullivan, Anwar Sadat, Charles Trenet, Robert Kennedy, P. A. M. Dirac, Bill Cosby, Peter Sellers, Henri Cartier-Bresson, Sin-Itiro Tomonaga, Jesse Owens, Groucho Marx, Janet Margolin, Roald Dahl, Françoise Sagan, Sidney Bechet, Shirley MacLaine, Jacques Tati, and Charles Shultz.

The people just mentioned all had major positive impacts on my life and their lives overlapped a fair amount with mine, and thus I might (at least theoretically) have run into any of them in person. But I also contain myriad traces of thousands of individuals whom I never could have met and interacted with, such as W. C. Fields, Galileo Galilei, Harry Houdini, Paul Klee, Clément Marot, John Baskerville, Fats Waller, Anne Frank, Holden Caulfield, Captain Nemo, Claude Monet, Leonhard Euler, Dante Alighieri, Alexander Pushkin, Eugene Onegin, James Clerk Maxwell, Samuel Pickwick, Esq., Charles Babbage, Archimedes, and Charlie Brown.

Some of the people in the latter list, of course, are fictional while others hover between the fictional and the real, but that is of no more import than the fact that in my mind, they are all merely *virtual* beings. What matters is neither the fictional/nonfictional nor the virtual/nonvirtual dimension, but the duration and depth of an individual's interaction with my interiority. In that regard, Holden Caulfield ranks at about the same level as Alexander Pushkin, and higher far than Dante Alighieri.

We are all curious collages, weird little planetoids that grow by accreting other people's habits and ideas and styles and tics and jokes and phrases and tunes and hopes and fears as if they were meteorites that came soaring out of the blue, collided with us, and stuck. What at first is an artificial, alien mannerism slowly fuses into the stuff of our self, like wax melting in the sun, and gradually becomes as much a part of *us* as ever it was of someone else (though that person may very well have borrowed it from someone else to begin with). Although my meteorite metaphor may make it sound as if we are victims of *random* bombardment, I don't mean to suggest that we willingly accrete just any old mannerism onto our sphere's surface — we are very selective, usually borrowing traits that we admire or covet — but even our style of selectivity is itself influenced over the years by what we have turned into as a result of our repeated accretions. And what was once right on the surface gradually becomes buried like a Roman ruin, growing closer and closer to the core of us as our radius keeps increasing.

All of this suggests that each of us is a bundle of fragments of other people's souls, simply put together in a new way. But of course not all contributors are represented equally. Those whom we love and who love us are the most strongly represented inside us, and our "I" is formed by a complex collusion of all their influences echoing down the many years. A marvelous pen-and-ink "parquet deformation" drawn in 1964 by David Oleson (below) illustrates this idea not only graphically but also via a pun, for it is entitled "I at the Center":

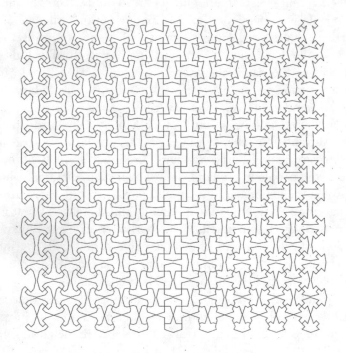

Here one sees a metaphorical individual at the center, whose shape (the letter "I") is a consequence of the shapes of all its neighbors. Their shapes, likewise, are consequences of the shapes of *their* neighbors, and so on. As one drifts out toward the periphery of the design, the shapes gradually become more and more different from each other. What a wonderful visual metaphor for how we are all determined by the people to whom we are close, especially those to whom we are closest!

How Much Can One Import of Another's Interiority?

When we interact for a couple of minutes with a checkout clerk in a store, we obviously do not build up an elaborate representation of that person's interior fire. The representation is so partial and fleeting that we would probably not even recognize the person a few days later. The same goes, only more so, for each of the hundreds of people we pass as we walk down a busy sidewalk at the height of the Christmas shopping madness. Though we know well that each person has at their core a strange loop somewhat like our own, the details that imbue it with its uniqueness are so inaccessible to us that that core aspect of them goes totally unrepresented. Instead, we register only superficial aspects that have nothing to do with their inner fire, with who they really are. Such cases are typical of the "truncated corridor" images that we build up in our brain for most people that we run across; we have no sense of the strange loop at their core.

Many of the well-known individuals I listed above are central to my identity, in the sense that I cannot imagine who I would be had I not encountered their ideas or deeds, but there are thousands of other famous people who merely grazed my being in small ways, sometimes gratingly, sometimes gratifyingly. These more peripheral individuals are represented in me principally by various famous achievements (whether they affected me for good or for ill) —a sound bite uttered, an equation discovered, a photo snapped, a typeface designed, a line drive snagged, a rabble roused, a refugee rescued, a plot hatched, a poem tossed off, a peace offer tendered, a cartoon sketched, a punch line concocted, or a ballad crooned.

The central ones, by contrast, are represented inside my brain by complex symbols that go well beyond the external traces they left behind; they have instilled inside me an additional glimmer of how it was to live inside their head, how it was to look out at the world through their eyes. I feel I have entered, in some cases deeply, into the hidden territory of their interiority, and they, conversely, have infiltrated mine.

And yet, for all the wonderful effects that our most beloved composers, writers, artists, and so forth have exerted on us, we are inevitably even more intimate with those people whom we know in person, have spent years with, and love. These are people about whom we care so deeply that for them to achieve some particular personal goal becomes an important internal goal for us, and we spend a good deal of time musing over how to realize that goal (and I deliberately chose the neutral phrase "that goal" because it is blurry whether it is *their* goal or *ours*).

We live inside such people, and they live inside us. To return to the metaphor of two interacting video feedback systems, someone that close to us is represented on our screen by a second infinite corridor, in addition to our own infinite corridor. We can peer all the way down — their strange loop, their personal gemma, is incorporated inside us. And yet, to reiterate the metaphor, since our camera and our screen are grainy, we cannot have as deep or as accurate a representation of people beloved to us as either our own self-representation or their own self-representation.

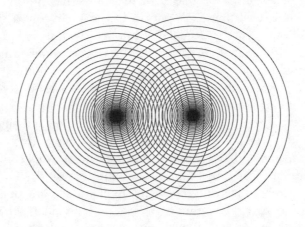

Double-clicking on the Icon for a Loved One's Soul

There was a point in my 1994 email broodings to Dan Dennett where I worried about how it would feel when, for the first time after her death, I would watch a video of Carol. I imagined the Carol symbol in my head being powerfully activated by the images on tape — more powerfully activated than at any moment since she had died — and I was fearful of the power of the illusion it would create. I would seem to see her standing by the staircase, and yet, obviously, if I were to get up and walk through the house to the spot where she had once stood, I would find no body there. Though I would see her bright face and hear her laugh, I could not go up to her and put my arm around her shoulders. Watching the tapes would heighten the anguish of her death, by *seeming* to bring her back physically but doing nothing of the sort in reality. Her physical nature would not be brought back by the tapes.

But what about her *inner* nature? When Carol was alive, her presence routinely triggered certain symbols in my brain. Quite obviously, the videos would trigger those same symbols again, although in fewer ways. What would be the nature of the symbolic dance thus activated in my brain? When the videos inevitably double-clicked on my "Carol" icon, what would happen inside me? The strange and complex thing that would come rushing up from the dormant murk would be a real thing — or at any rate, just as real as the "I" inside me is real. The key question then is, how different is that strange thing in my brain from the "I" that had once flourished inside Carol's brain? Is it a thing of an entirely different type, or is it of the same type, just less elaborate?

Thinking with Another's Brain

Of all Dan Dennett's many reactions to my grapplings in that searing spring of 1994, there was one sentence that always stood out in my mind: "It is clear from what you say that Carol will be thinking with your brain for quite some time to come." I appreciated and resonated to this evocative phrase, which, as I later discovered, Dan was quoting with a bit of license from our mutual friend Marvin Minsky, the artificial-intelligence pioneer — copycats everywhere!

"She'll be thinking with your brain." What this Dennett–Minsky utterance meant to me was roughly the following. Input signals coming to me would, under certain circumstances, follow pathways in my brain that led not to *my* memories but to *Carol's* memories (or rather, to my low-resolution, coarse-grained "copies" of them). The faces of our children, the voices of her parents and sisters and brothers, the rooms in our house — such things would at times be processed in a frame of reference that would imbue them with a Carol-style meaning, placing them in a frame that would root them in and relate them to *her* experiences (once again, as crudely rendered in my brain). The semantics that would accrue to the signals impingent on me would have originated in her life. To the extent, then, that I, over our years of living together, had accurately imported and transplanted the experiences that had rooted Carol on this earth, she would be able to react to the world, to live on in me. To that extent, and only to that extent, Carol would be thinking with my brain, feeling with my heart, living in my soul.

Mosaics of Different Grain Size

Since everything hung on those words "to the extent that X", what seemed to matter most of all here was *degree of fidelity* to the original, an idea for which I soon found a metaphor based on portraits rendered as mosaics made out of small colored stones. The more intimately someone comes to know you, the finer-grained will be the "portrait" of you inside their head. The highest-resolution portrait of you is of course your own self-portrait — your own mosaic of yourself, your self-symbol, built up over your entire life, exquisitely fine-grained. Thus in Carol's case, her own self-symbol was by far the finest-grained portrait of her inner essence, her inner light, her personal gemma. But surely among the next-highest in resolution was *my* mosaic of Carol, the coarser-grained copy of her interiority that resided inside my head.

It goes without saying that my portrait of Carol was of a coarser grain than her own; how could it not be? I didn't grow up in her family, didn't attend her schools, didn't live through her childhood or adolescence. And yet, over our many years together, through thousands of hours of casual and intimate conversations, I had imported lower-resolution copies of so many of the experiences central to her identity. Carol's memories of her youth — her parents, her brothers and sisters, her childhood collie Barney, the family's "educational outings" to Gettysburg and to museums in Washington D.C., their summer vacations in a cabin on a lake in central Michigan, her adolescent delight in wildly colorful socks, her preadolescent loves of reading and of classical music, her feelings of differentness and isolation from so many kids her age — all these had imprinted my brain with copies of themselves, blurry copies but copies nonetheless. Some of her memories were so vivid that they had become my own, as if I had lived through those days. Some skeptics might dismiss this outright, saying, "Just pseudo-memories!" I would reply, "What's the difference?"

A friend of mine once told me about a scenic trip he had taken, describing it in such vivid detail that a few years later I thought I had been on that trip myself. To add insult to injury, I didn't even remember my friend as having had anything to do with "my" trip! One day this trip came up in a conversation, and of course we both insisted that *we* were the one who had taken it. It was quite puzzling! However, after my friend showed me his photos of the trip and recounted far more details of it than I could, I realized my mistake — but who knows how many other times this kind of confusion has occurred in my mind without being corrected, leaving pseudo-memories as integral elements of my self-image?

In the end, what is the difference between actual, personal memories and pseudo-memories? Very little. I recall certain episodes from the novel *Catcher in the Rye* or the movie *David and Lisa* as if they had happened to me — and if they didn't, so what? They are as clear as if they had. The same can be said of many episodes from other works of art. They are parts of my emotional library, stored in dormancy, waiting for the appropriate trigger to come along and snap them to life, just as my "genuine" memories are waiting. There is no absolute and fundamental distinction between what I recall from having lived through it myself and what I recall from others' tales. And as time passes and the sharpness of one's memories (and pseudo-memories) fades, the distinction grows ever blurrier.

Transplantation of Patterns

Even if most readers agree with much that I am saying, perhaps the hardest thing for many of them to understand is how I could believe that the activation of a symbol inside my head, no matter how intricate that symbol might be, could capture any of someone else's *first-person* experience of the world, someone else's consciousness. What craziness could ever have led me to suspect that someone *else's* self — my father's, my wife's — could experience feelings, given that it was all taking place courtesy of the neurological hardware inside *my* head, and given that every single cell in the brain of the other person had long since gone the way of all flesh?

The key question is thus very simple and very stark: Does the actual hardware matter? Did only *Carol's* cells, now all recycled into the vast impersonal ecosystem of our planet, have the potential to support what I could call "Carol feelings" (as if feelings were stamped with a brand that identified them uniquely), or could *other* cells, even inside me, do that job?

To my mind, there is an unambiguous answer to this question. The cells inside a brain are not the bearers of its consciousness; the bearers of consciousness are *patterns*. The pattern of organization is what matters, not the substance. It ain't the meat, it's the motion! Otherwise, we would have to attribute to the molecules *inside* our brains special properties that, *outside* of our brains, they lack. For instance, if I see one last tortilla chip lying in a basket about to be thrown away, I might think, "Oh, you lucky chip! If I eat you, then your lifeless molecules, if they are fortunate enough to be carried by my bloodstream up to my brain and to settle there, will get to enjoy the experience of being me! And so I must devour you, in order not to deprive your inert molecules of the chance to enjoy the experience of being human!" I hope such a thought sounds preposterous to nearly all of my readers. But if the molecules making you up are *not* the "enjoyers" of your feelings, then what is? All that is left is *patterns*. And patterns can be copied from one medium to another, even between radically different media. Such an act is called "transplantation" or, for short, "translation".

A novel can withstand transplanting even though readers in the "guest language" haven't lived on the soil where the original language is spoken; the key point is, they have experienced essentially the same phenomena on their own soil. Indeed, all novels, whether translated or not, depend on this kind of transplantability, because no two human beings, even if they speak the same language, ever grow up on exactly the same soil. How else could we contemporary Americans relate to a Jane Austen novel?

Carol's soul can withstand transplanting into the soil of my brain because, even though I didn't grow up in her family and in their various houses, I know, to some degree, all the key elements of her earliest years. In me robustly live and survive her early inner roots, out of which her soul grew. My brain's fertile soil is a soul-soil not identical to, but very similar to, hers. And so I can "be" Carol albeit with a slight Doug accent, just as James Falen's lovely, lilting, and lyrical English transplantation of Pushkin's novel-in-verse *Eugene Onegin* is certainly and undeniably *that very novel*, even if it has something of an American accent.

The sad truth is, of course, that no copy is perfect, and that my copies of Carol's memories are hugely defective and incomplete, nowhere close to the level of detail of the originals. The sad truth is, of course, that Carol is reduced, in her inhabitation of my cranium, to only a tiny fraction of what she used to be. The sad truth is, my brain's mosaic of Carol's essence is far more coarse-grained than the privileged mosaic that resided in *her* brain was. That is the sad truth. Death's sting cannot be denied. And yet death's sting is not quite as absolute or as total as it might seem.

When the sun is eclipsed, there remains a corona surrounding it, a circumferential glow. When someone dies, they leave a glowing corona behind them, an afterglow in the souls of those who were close to them. Inevitably, as time passes, the afterglow fades and finally goes out, but it takes many years for that to happen. When, eventually, all of those close ones have died as well, then all the embers will have gone cool, and at that point, it's "ashes to ashes and dust to dust".

Several years ago, my email friend James Plath, knowing of my intense musings along these lines, sent me a paragraph from the novel *The Heart Is a Lonely Hunter* by Carson McCullers, with which I conclude this chapter.

> Late the next morning he sat sewing in the room upstairs. Why? Why was it that in cases of real love the one who is left does not more often follow the beloved by suicide? Only because the living must bury the dead? Because of the measured rites that must be fulfilled after a death? Because it is as though the one who is left steps for a time upon a stage and each second swells to an unlimited amount of time and he is watched by many eyes? Because there is a function he must carry out? Or perhaps, when there is love, the widowed must stay for the resurrection of the beloved — so that the one who has gone is not really dead, but grows and is created for a second time in the soul of the living?

CHAPTER 18

The Blurry Glow of Human Identity

❧ ❧ ❧

I Host and Am Hosted by Others

AMONG the beliefs most universally shared by humanity is the idea "One body, one person", or equivalently, "One brain, one soul". I will call this idea the "caged-bird metaphor", the cage being, of course, the cranium, and the bird being the soul. Such an image is so self-evident and so tacitly built into the way we all think about ourselves that to utter it explicitly would sound as pointless as saying, "One circle, one center" or "One finger, one fingernail"; to question it would be to risk giving the impression that you had more than one bat in your belfry. And yet doing precisely the latter has been the purpose of the past few chapters.

In contrast to the caged-bird metaphor, the idea I am proposing here is that since a normal adult human brain is a representationally universal "machine", and since humans are social beings, an adult brain is the locus not only of *one* strange loop constituting the identity of the primary person associated with that brain, but of *many* strange-loop patterns that are coarse-grained copies of the primary strange loops housed in other brains. Thus, brain 1 contains strange loops 1, 2, 3, and so forth, each with its own level of detail. But since this notion is true of any brain, not just of brain 1, it entails the following flip side: Every normal adult human soul is housed in many brains at varying degrees of fidelity, and therefore every human consciousness or "I" lives at once in a collection of different brains, to different extents.

There is, of course, a "principal domicile" or "main brain" for each particular "I", which means that there remains a good deal of truth to simple, commonsensical statements like "My soul is housed in my brain", and yet, close to true though it is, that statement misses something crucial, which is the idea, perhaps strange-sounding at first, that "My soul lives to lesser extents in brains that are not mine."

At this point, we should think at least briefly about the meaning of innocent-sounding phrases like "my brain" and "brains that are not mine". If I have five sisters, then saying "my sister" is, if not meaningless, then at least highly ambiguous. Likewise, if I have three nationalities, then saying "my nationality" is ambiguous. And analogously, if my self-symbol exists in, say, fifteen different brains (at fifteen different degrees of fidelity, to be sure), then not only is the phrase "my brain" ambiguous, but so is the word "my"! Who is the talker? I am reminded of a now-defunct bar in the Bay Area whose sign amused me no end every time I drove by it: "My Brother's Place". Yes, but *whose* brother's place? Just who was doing the talking here? I never could figure this out (nor, I guess, could anyone else), and I relished the sign's intentional silliness.

Fortunately, the existence of a "main brain" means that "my brain" has an unambiguous primary meaning, even if the soul uttering the phrase lives, to smaller extents, in fourteen other brains at the same time. And usually the soul uttering the phrase will be using its main brain (and thus its main body and main mouth), and so most listeners (including the speaker) will effortlessly understand what is meant.

It is not easy to find a strong, vivid metaphor to put up against the caged-bird metaphor. I have entertained quite a few possibilities, involving such diverse entities as bees, tornados, flowers, stars, and embassies. The image of a swarm of bees or of a nebula clearly conveys the idea of diffuseness, but there is no clear counterpart to the cage (or rather, to the head or brain or cranium). (A hive is not what I mean, because a flying swarm is not at all inside its hive.) The image of a tornado cell is appealing because it involves swirling entities reminiscent of the video feedback loops we've so often talked about, and because it involves a number of such swirls spread out in space, but once again there is no counterpart to the "home location", nor is it clear that there is one primary tornado in a cell. Then there is the image of a plant sending out underground shoots and popping up in several places at once, where there is a primary branch and secondary offshoots, which is an important component of the idea, and similarly, the image of a country with embassies in many other countries captures an important aspect of what I seek. But I am not fully satisfied with any of these metaphors, and so, rather than settling on a single one, I'll simply throw them all out at once, hoping that they stir up some appropriate imagery in your mind.

Feeling that One is Elsewhere

All this talk of one person inhabiting several bodies at the same time may seem wildly at odds with "common sense", which unambiguously tells us that we are always in just one place, not two or more. But let's examine this commonsense axiom a bit.

If you go to an I-Max movie theater and are riding a wild roller-coaster, where are you? The temptation is to say, "I'm sitting in a movie theater", but if that's the case, then why are you so scared? What's to scare you about a couple of dozen rows of stationary seats, the odor of popcorn, and a thin screen hanging forty or fifty feet away? The answer is obvious: when you watch the movie, the audiovisual input to your brain seems to be coming not from inside the theater but from somewhere else, a place that is far away from the theater and that has nothing to do with it. And it is *that* input that you can't help interpreting as telling you where you are. You feel you have been transported to a place where your body is actually not located, and where your brain is not located either, for that matter.

Of course since watching a movie is a very familiar activity, we are not confused by this phenomenon of virtual displacement, and we accept the idea that there is simply a temporary suspension of disbelief, so that we can enter into another world virtually, vicariously, and volatilely. No serious philosophical conundrums seem to be raised by such an experience, and yet to me, this first little crack allows the door of multiple simultaneous locations of the self to open up much more widely.

Now let's recall the experience of being transported from the ski resort in California's Sierra Nevada range to the Bloomington kennel via the "doggie cam" and the World Wide Web. Watching the dogs play in their little area, my children and I didn't in the least feel that we were "in Ollie's skin", but let's tweak the parameters of the situation a little bit. Suppose, for example, that the bandwidth of the visual image were greatly increased. Suppose moreover that the webcam was mounted not in a fixed spot above the fenced-in play area but on Ollie's head, and that it included a microphone. And lastly, suppose that you had a pair of dedicated goggles (spectacles with earphones) that, whenever you put them on, transmitted this scene to you in very high audiovisual fidelity. As long as you can put them on and then take them off, these teleportation goggles would seem like just a game, but what if they were affixed for several hours to your head and served as your *only* way of peering out at the world? Don't you think you would start to feel a little bit as if you *were* Ollie? What would it matter to you that you were in a faraway California ski resort, if your own eyes and ears were unable to give you any Californian input?

You might object that it's impossible to feel that you are Ollie if his movements are out of your control. In that case, we can add a joystick that will tend to make Ollie turn left or right, at your discretion (how it does so is not germane here). So now your hand controls Ollie's movements and you receive audiovisual input solely from the camera attached to Ollie's head, for several hours nonstop. This scenario is rather bizarre, but I think you can easily see that you will soon start to feel as if you are more in the Indiana kennel, where you are free to move about, than in some Californian ski resort, where you are basically stuck to your seat (because you have your goggles on, hence you can't see where you're going, hence you don't dare venture anywhere). We'll refer to this sensation of feeling that you are somewhere far from both your body and your brain, thanks to the ultrarapid transmission of data, as "telepresence" (a term invented by Pat Gunkel and popularized by Marvin Minsky around 1980).

Telepresence versus "Real" Presence

Perhaps my most vivid experience of telepresence occurred when I was typesetting my book *Gödel, Escher, Bach*. This was back in the late 1970's, when for an author to do any such thing was unheard of, but I had the good fortune of having access to one of the only two computer typesetting systems in the world at that time, both of which, by coincidence, were located at Stanford. The catch was that I was an assistant professor at Indiana University in far-off Bloomington, and I had courses to teach on Tuesdays and Thursdays. To make things doubly hard, there was no Internet, so I couldn't possibly do the typesetting work from Indiana. To typeset my book, I had to be on site at Stanford, but my teaching schedule allowed me to get there only on weekends, and not on all weekends at that. And so each time I flew out to Stanford for a weekend, I would instantly zoom to Ventura Hall, plunk myself down at a terminal in the so-called "Imlac room", and plunge furiously into the work, which was extremely intense. I once worked forty hours straight before collapsing.

Now what does this all have to do with telepresence? Well, each long, grueling work session at Stanford was quite hypnotic, and when I left, I would still half-feel as if I were there. One time when I had returned to Bloomington, I realized I had made a serious typesetting mistake in one chapter, and so, in panic, I called up my friend Scott Kim, who also had been spending endless hours in the Imlac room, and I was hugely relieved to find him there. Scott was more than happy to sit down at an Imlac terminal and to pull up the right program and the proper file to work on. So we set to work on it, with me talking Scott through the whole long and detailed process, and Scott reading to me what he saw on the screen. Since I had just spent numberless hours right there, I was easily able to see in my mind's eye everything that Scott relayed to me, and I remember how disoriented I would feel when, every so often, I remembered that my body was still in Bloomington, for I felt for all the world as if I were in Stanford, working directly at the Imlac terminal. And mind you, this powerful *visual* sense of telepresence was taking place solely through the *sonic* modality of a telephone. It was as if my eyes, though in Bloomington, were looking at an Imlac screen in California, thanks to Scott's eyes and the clarity of his words on the phone.

You can call my feeling an "illusion" if you wish, but before you do so, consider how primitive this now-ancient implementation of telepresence was. Today, one can easily imagine turning up all the technological knobs by orders of magnitude. There could be a mobile robot out in California whose movements were under my instantaneous and precise control (the joystick idea again), and whose multimedia "sensory organs" instantly transmitted whatever they picked up to me in Indiana. As a result, I could be fully immersed in a virtual experience thousands of miles from where my brain was located, and this could go on for any length of time. What would be most confusing would still be the moments of change, when I removed the helmet that made me feel I was in California, thereby finding myself transported two thousand miles eastwards in a fraction of a second — or the reverse, when I would don my helmet and in a flash would sail all the way out to the west coast.

What, in the end, would suggest to me that my presence in Indiana was "realer" than my presence in California? One clue, I suppose, would be the telltale fact that in order to "be" in California, I would always have to don some sort of helmet, whereas in order to "be" in Bloomington, I would need no such device. Another tip-off might be that if I picked up food while meandering about in California, I couldn't get it into my Indiana-based stomach! That little problem, however, could easily be taken care of: just attach an intravenous feeding device to me in Indiana and arrange for it to pump nutrients into my bloodstream whenever I — my robot body, that is — manage to track down some "food" in California (and it need not be actual food, as long as the act of laying my remote robotic hands on it out there activates the intravenous feeding device back home in Indiana).

What one starts to realize, as one explores these disorienting but technologically feasible ideas of virtual presence "elsewhere", is that as the telepresence technology improves, the "primary" location becomes less and less primary. Indeed, one can imagine a proverbial "brain in the vat" in

Bloomington controlling a strolling robot out in California, and totally believing itself to be a physical creature way out west and not believing one word about being a brain in a vat. (Many of these ideas were explored, incidentally, by Dan Dennett in his philosophical fantasy "Where Am I?")

Which Viewpoint is Really Mine?

I am hesitant to adduce too many science-fiction-like scenarios in order to explain and justify my ideas about soul and consciousness, because doing so might give the impression that my viewpoint is essentially tied to the indiscriminate mentality of an inveterate science-fiction junkie, which I am anything but. Nonetheless, I think such examples are often helpful in getting one to break free of ancient, deeply rooted prejudices. But one hardly needs to talk about head-mounted television cameras, remote-controlled robots, and intravenous feeding devices in order to remind people of how we routinely transport ourselves into virtual worlds. The mere act of reading a novel while relaxing in an armchair by the window in one's living room is an example *par excellence* of this phenomenon.

When we read a Jane Austen novel, what we look at is just a myriad of black smudges arranged neatly in lines on a set of white rectangles, and yet what we feel we are "seeing" (and should I use the quotation marks or not?) is a mansion in the English countryside, a team of horses pulling a carriage down a country lane, an elegantly clad lady and gentleman sitting side by side in the carriage exchanging pleasantries when they espy a poor old woman emerging from her humble cottage along the roadside... We are so taken in by what we "see" that in some important and serious sense we don't notice the room we are sitting in, the trees visible through its window, nor even the black smudges speckled all over the white rectangles in our hands (even though, paradoxically, we are depending on those smudges to bring us the visual images I just described). If you don't believe me, consider what you have just been doing in the last thirty seconds: processing black smudges speckled on white rectangles and yet "seeing" someone reading a Jane Austen novel in an armchair in a living room, and in addition, seeing the mansion, the country road, the carriage, the elegant couple, and the old woman... Black curlicues on a white background, when suitably arranged, transport us in milliseconds to arbitrarily distant, long-gone, or even never-existent venues and epochs.

The point of all of this is to insist on the idea that we *can* be in several places at one time, simultaneously entertaining several points of view at one time. You just did it! You are sitting somewhere reading this book, yet a moment ago you were also in a living-room armchair reading a Jane

Austen novel, and you were also simultaneously in a carriage going down a country lane. At least three points of view coexisted simultaneously inside your cranium. Which one of those viewers was "real"? Which one was "really you"? Need these questions be answered? Can they be answered?

Where Am I?

As I was driving a few days ago, I pulled up alongside a jogger waiting at a red light. She was trotting in place, and then the light changed and she crossed the street and disappeared. For a brief moment, I was "in her shoes". I had never seen her before and probably will never see her again, but I have been there many a time. I had lived that experience in my own way, and even though I know virtually nothing about her, I have shared that experience of hers. To be sure, I was not seeing it through her eyes. But let's briefly jump once again into the realm of slightly silly technological extravagance.

Suppose everyone wore a tiny TV camera on the bridge of their nose, and that everyone had glasses that could be tuned to receive the signals from any selected TV camera on earth. If there were a way of specifying a person by their GPS coordinates (and that certainly doesn't seem far-fetched), then all I would have to do is set my glasses to receive the signals from that jogger's nose-mounted TV camera, and presto! — I would suddenly be seeing the world from her perspective. When I was sitting in my car and the traffic light changed and she took off and disappeared, I could have ridden along and seen just where she was going, could have heard the birds chirping as she jogged through a woodsy lane, and so forth. And at any point I could switch channels and go see the world through the nose-camera of my daughter Monica or my son Danny, or anyone else I wished. So where am I? "Still just where you are!" chirps common sense. But that's too simplistic, too ambiguous.

What determines "where I am"? If we once again postulate the idea of obtaining nutrition by carrying out certain remote actions, and if we add back the ability to control distant motion by means of a joystick or even by certain brain events, then things really start to shimmer in uncertainty. For surely a mobile robot is not where the radio-connected computer that is controlling it happens to be sitting. A robot might be strolling about on the moon while its computerized guidance system was in some earthbound laboratory. Or a self-driving car like Stanley could be crossing the Nevada desert, and its computer control system might be on board or might be located in a lab in California, connected by radio. But would we even care where the computer was? Why should we care where it is located?

A robot, we feel, is where its *body* is. And so when my brain can switch at will (using the fancy glasses described above) between inhabiting any one of a hundred different bodies — or worse yet, when it can inhabit several bodies at the same time, processing different kinds of input from all of them at once (perhaps visual input from one, sonic from another, tactile from a third) — then *where I am* becomes extremely ill-defined.

Varying Degrees of Being Another

Once again, let's leave the science-fiction scenarios behind and just think about everyday events. I sit in a plane coming in for a landing and overhear random snippets of conversations around me — remarks about how great the Indianapolis Zoo is, how there's a new delicatessen at Broad Ripple, and so forth. Each snippet carries me a smidgen into someone else's world, gives me the tiniest taste of someone else's viewpoint. I may resonate very little with that viewpoint, but even so, I am entering ever so slightly into that person's "private" universe, and this incursion, though absolutely trivial for a human being, is far deeper than any canine's incursion into another canine's universe ever was.

And if I have untold thousands of hours of conversation with another human being on topics of every imaginable sort, including the most private feelings and the most confidential confessions, then the interpenetration of our worlds becomes so great that our worldviews start to fuse. Just as I could jump to California when talking on the telephone with Scott Kim in the Imlac room, so I can jump inside the other person's head whenever, through words and tones of voice, they call forth their most fervent hopes or their most agonizing fears.

To varying degrees, we human beings live inside other human beings already, even in a totally nontechnological world. The interpenetration of souls is an inevitable consequence of the power of the representationally universal machines that our brains are. That is the true meaning of the word "empathy".

I am capable of being other people, even if it is merely an "economy class" version of the act of being, even if it falls quite a bit short of being those people with the full power and depth with which they are themselves. I have the good fortune — at least I usually consider it fortunate, though at times I wonder — of always having the option of falling back and returning to being "just me", because there is only one primary self housed in my brain. If, however, there were a few high-powered selves in my brain, all competing with each other for primacy, then the meaning of the word "I" would truly be up for grabs.

The Naïve Viewpoint is Usually Good Enough

The image I just conjured up of several selves competing for primacy inside one brain may have struck you as extremely weird, but in fact the experience of internal conflict between several "rival selves" is one that we all know intimately. We know what it is to feel split between wanting to buy that candy bar and wanting to refrain. We know what it is to feel split between driving "just another twenty miles" and pulling off at the next rest stop for a desperately needed nap. We know what it is like to think, "I'll just read one more paragraph and then go fix dinner" and also to think, "I'll just finish this chapter first." Which one of these opposing inner voices is really *me*? In growing up, we learn not to ask or try to answer questions like this. We unthinkingly accept such small internal conflicts as simply part of "the human condition".

If you simultaneously dip your left hand into a basin of hot water and your right hand into a basin of ice water, leave them both there for a minute, and then plunge them into a lukewarm sink, you will find that your two hands — usually your most reliable scouts and witnesses of the outer world — are now telling you wildly opposite things about the very same sinkful of water. In reaction to this paradox, you will most likely just shrug and smile, thinking to yourself, "What a strong tactile illusion!" You aren't likely to think to yourself, "This cognitive split inside my brain is the thin edge of the wedge, revealing the illusoriness of the everyday conviction that there is just a single self inside my head." And the reason nearly everyone would put up great resistance to such a conclusion is that for nearly all purposes, the simple story we tell ourselves is good enough.

This situation is a bit reminiscent of Newtonian physics, whose laws are extraordinarily reliable unless there are objects moving near each other with a relative velocity approaching the velocity of light, and in such cases Newtonian physics goes awry and gives very wrong answers. There is no reason at all, however, to abandon Newtonian physics in most familiar situations, even including the calculations of the orbits of spacecraft traveling to the moon or other planets. The velocities of such spacecraft, although huge compared with those of jet airplanes, are still minuscule fractions of the speed of light, and abandonment of Newton is not in the least called for.

Likewise, why should we abandon our commonsense attitudes about how many souls inhabit our brains when we know very well that the answer is just *one*? The only answer I can give is that, yes, the answer is *very close* to one, but when push comes to shove, we can see small deviations from that accurate first approximation. Moreover, we even experience such deviations all the time in everyday life — it's just that we tend to interpret them as frivolous illusions, or else we simply ignore them. Such a strategy works quite well because we never approach the "speed of light" where the naïve, caged-bird picture fails badly. Less metaphorically, the lower-resolution, coarse-grained souls who fight and squabble for the chance to inhabit our brains never really pose any serious competition to "Number One" for the overall command, and so the naïve old caged-bird dogma "One brain, one soul" stands unchallenged nearly all of the time.

Where Does a Hammerhead Shark Think it is?

Perhaps the most forceful-seeming challenge to the thesis that a single soul — your own, say — is parceled out among a number of distinct brains is simply the question, "Okay, let's suppose that I'm somehow distributed over many brains. Then which one do *I* actually *experience*? I can't be simultaneously both here and there!" But in this chapter I have tried to show that you *can* indeed be in two places at the same time, and you don't even notice anything funny going on. You can be in Bloomington and in Stanford at the same time. You can be in a Donner Pass ski lodge and a Midwest town's kennel play area at the same time. You can be in your living room's plush armchair and in an uncomfortable carriage bouncing along a nineteenth-century English country road at the same time.

If these examples are too far-fetched or too technological for your taste, then just think of the lowly hammerhead shark. The poor thing has eyes on opposite sides of its head, which look out, quite often, on two completely unrelated scenes. So which scene is the shark *really* seeing? Where does it consider itself to be, *really*? Of course no one would ask such a question. We just accept the idea that the shark can "sort of" be in those two different worlds at the same time, mainly because we think to ourselves that no matter how different those scenes look, they nonetheless are contiguous pieces of the underwater world in the shark's vicinity, so there is no genuine problem about whereness. But this is glib, and sidesteps the point.

To put things in somewhat sharper focus, let's invent a variation on the hammerhead shark. We'll posit a creature whose eyes are taking in one situation (say in Bloomington) and whose ears are taking in another, unrelated situation (say in Stanford). The same brain is going to process these inputs at the same time. I hope you won't claim that this is an impossible feat! If that's your inclination, please first recall that you drive your car while reacting to other cars, scenery, billboards, and roadsigns, and also while talking with a far-off friend on your cell phone (and the topics covered in the conversation may vividly transport you to yet other places), and all during that very same period a recently-heard tune is running through your head, your strained back is bugging you, you smell cow manure wafting through the air, and your stomach is shouting to you, "I am hungry!" You manage to process all those different simultaneous worlds perfectly well — and in that same spirit, nothing is going to prevent a human brain from dealing simultaneously with the two unrelated worlds of Stanford sounds and Bloomington sights, no more than the hammerhead shark's brain protests, "Does not compute!" So the idea "I cannot be simultaneously here and there" goes down in flames. We *are* simultaneously here and there all the time, even in our everyday lives.

Sympathetic Vibrations

But perhaps you feel that what I've just described doesn't address the question originally posed about which of many brains *you* are really in — that being either here or there means that no matter how emotionally close you are to someone else, their feelings are always theirs, yours are always yours, and never the twain shall meet. This is once again the caged-bird imagery with which the chapter opened, and it will certainly not cease to rear its ugly head no matter how many times I try to cut it off. But let us nonetheless try tackling this medusa in yet another fashion.

If I claim that I am partially in my sister Laura and she is partly in me, it seems nonetheless obvious that if she happens to drive by our favorite falafel place in San Jose and stops to eat a falafel, I'm not going to taste that falafel as I sit here slaving away in my study in Bloomington, Indiana. And therefore I am not there, but here! And therefore my consciousness is local, not global, not spread out! And therefore that's the end of the story!

But things are not quite that simple. I might receive news of Laura's falafel an hour later, by a telephone call. When she describes it vividly (or not even vividly, since I know it so well), my mouth starts watering as I recall the exact texture of the little crunchy balls and the delicious red hot sauce. I know those falafels like the back of my teeth. Although my tongue is not caressing those little chunky deep-fried bits, something in my brain is taking a sensual delight in what I could call (in imitation of the phrase "sympathetic pain") "sympathetic pleasure". Albeit in a feeble way and an hour after the fact, I am sharing Laura's pleasure. But so what if it's a feeble imitation and is not exactly simultaneous? Even if my pleasure is a low-resolution copy of hers and is displaced in time, it is nonetheless pleasure, and it is pleasure that is "about" Laura, not about myself. Her delight has been powerfully transmitted to me. And so, at a distance, at a delay, and to a diminished degree, I am in her skin and she is in mine.

That's all I'm claiming — that there is blur. That some of what happens in other brains gets copied, albeit coarse-grainedly, inside the brain of "Number One", and that the closer two brains are to each other emotionally, the more stuff gets copied back and forth from one to the other, and the more faithful the copies are. There's no claim that the act of copying is simultaneous or perfect or total — just that each person lives *partially* in the brain of the other, and that if the bandwidth were turned up more and more and more and still more, they would come to live more and more inside each other — until, in the limit, the sense of a clear boundary between them would slowly be dissolved, as it is for the two halves of a Twinwirld pairson (and even more so for a Siamese Twinwirld pairson).

As it happens, we do not live in a didymous world like Twinwirld, nor do we live in a world where the existence of relatively clear boundaries between souls seems imminently threatened by the advent of extremely high-bandwidth interbrain communication — a world in which signals are swapped so fast and furiously between brains that separate bodies would cease to determine separate individuals. That is not the case at present, nor do I envision it becoming the case in the foreseeable future (though I am not a futurologist, and I could be quite wrong).

My point, though, is that the myth of watertight boundaries between souls is something whose falsity we all have slight tastes of all the time, but since it is so convenient and so conventional to associate one body with precisely one soul, since it is so deeply tempting and so deeply ingrained to see a body and a soul as being in perfect alignment, we choose to downplay or totally ignore the implications of the everyday manifestations of the interpenetration of souls.

Consider how profoundly wrapped up you can become in a close friend's successes and failures, in their very personal ecstasies and agonies. If my vicarious enjoyment of my sister's falafel seemed vivid to me, just think how much more vivid and intense is your vicarious thrill when a forever-lonely friend of yours finally bumps into someone wonderful and a promising romance starts up, or when a long-frustrated actor friend is finally given a lucky break and receives terrific reviews in the press. Or turning things around, think how powerful is your sense of injustice when a close friend of yours is hit, out of the blue, by some terrible misfortune. What are you doing but living their life inside your own head?

And yet we describe phenomena of this extremely familiar sort in easier, less challenging terms, such as "He identifies with her", or "She is such an empathetic woman", or "I know what you're going through", or "I feel for you", or "It pains me to see what she's up against", or "Don't tell me any more — I can't stand it!" Standard expressions like these, although they indeed reflect someone's partially being inside someone else, are seldom if ever taken as literal suggestions that our souls really do interpenetrate and blur together. That is just too messy and possibly even too scary an idea for us to deal with, and so we insist instead that there is no genuine overlap, that we are like distant galaxies to each other. Our lifelong ingrained habit is to accept without question the caged-bird metaphor for souls, and it's very hard to break out of such a profoundly rooted habit.

Am I No One Else or Am I Everyone Else?

The image of the caged bird essentially implies that different people are like separate dots on the same line, dots having a diameter of exactly zero, and thus having no overlap whatsoever. Indeed, if we take the so-called "real line" of elementary algebra as a metaphor, then the caged-bird metaphor would assign to each person a "serial number" — an infinite decimal that uniquely determines "what it is like" to be that person. In that view, you and I, no matter how similar we think we are, no matter how much experience we have shared in life, even if we are identical or Siamese twins, were simply assigned different serial numbers at birth, and hence we inhabit different zero-width dots on the line, and that is that. You are you, I am I, and there is not one whit of overlap, no matter how near we are. I cannot possibly know what it's like to be you, nor the reverse.

The opposite thesis would claim that every person is distributed uniformly over the entire real line, and that all individuals are therefore one and the same person! There is only one person. This extreme view, although less commonly advocated, has its modern proponents, such as philosopher Daniel Kolak in his recent book *I Am You*. This view makes as little sense to me as does panpsychism, which asserts that every entity — every stone, every picnic table, every picnic, every electron, every rainbow, every drop of water, waterfall, skyscraper, oil refinery, billboard, speed-limit sign, traffic ticket, county jail, jailbreak, track meet, election rigging, airport gate, spring sale, soap opera cancellation, photograph of Marilyn Monroe, and so on *ad nauseam* — is conscious.

The viewpoint of this book lies somewhere between these two extremes, picturing individuals not as pointlike infinite-decimal serial numbers but as fairly localized, blurry zones scattered here and there along the line. While some of these zones overlap considerably, most of them overlap little or none at all. After all, two smudges of width one inch apiece located a hundred miles apart will obviously have zero overlap. But two

smudges of width one inch whose centers are only a half inch apart will
have a great deal of overlap. There will not be an unbridgeable existential
gap between two such people. Each of them is instead spread out into the
other one, and each of them lives partially in the other.

Interpenetration of National Souls

Earlier in this chapter, I briefly offered the image of a self as analogous
to a country with embassies in many other countries. Now I wish to pursue
a similar notion, but I'll start out with a very simplistic notion of what a
country is, and will build up from there. So let's consider the slogan "One
country, one people". Such a slogan would suggest that each *people* (a
spiritual, cultural notion involving history, traditions, language, mythology,
literature, music, art, religion, and so forth) is always crisply and perfectly
aligned with some *country* (a physical, geographical notion involving oceans,
lakes, rivers, mountains, valleys, prairies, mineral deposits, cities, highways,
precise legal borders, and so forth).

If we actually believed a strict geographical analogue to the caged-bird
metaphor for human selves, then we would have the curious belief that all
individuals found inside a certain geographical region always had the same
cultural identity. The phrase "an American in Paris" would make no sense
to us, for the French nationality would coincide exactly with the boundaries
of the physical place called "France". There could never be Americans in
France, nor French people in America! And of course analogous notions
would hold for *all* countries and peoples. This is clearly absurd. Migration
and tourism are universal phenomena, and they intermix countries and
peoples continuously.

This does not mean that there is no such thing as a people or a
country, of course. Both notions remain useful, despite enormous blurs
concerning each one. Think for a moment of Italy, for instance. The
northwestern region called "Valle d'Aosta" is largely French-speaking,
while the northeastern region called "Alto Adige" (also "Südtirol") is largely
German-speaking. Moreover, north of Milano but across the border, the
Swiss canton of Ticino is Italian-speaking. So what is the relationship
between the country of Italy and the Italian people? It is not precise and
sharp, to say the least — and yet we still find it useful to talk about Italy
and Italians. It's just that we know there is a blur around both concepts.
And what goes for Italy goes for every country. We know that each
nationality is a blurry, spread-out phenomenon centered on but not limited
to a single geographical region, and we are completely accustomed to this
notion. It does not feel paradoxical or confusing in the least.

So let us exploit our comfort with the relationship between a place and a people to try to get a more sophisticated handle on the relationship between a body and a soul. Consider China, which over the past couple of centuries has lost millions of people to emigration. Does China simply forget about those people, thinking of them as deserters and expunging them from its collective memory? Not at all. There is a strong residual feeling inside China for the "Overseas Chinese". These cherished though distant people are urged to "come home" at least temporarily, and when they do, they are warmly welcomed like long-lost relatives (which of course is exactly what they are). This overseas branch of China is thus considered, within China, very much a part of China. It is a "halo" of Chineseness that extends far beyond the physical borders of the land.

Not just China, of course, but every country has such a halo, and this halo shimmers, sometimes brightly, sometimes dimly, in every other country on earth. If there were a counterpart at the country level to human death, then a people whose "body" was annihilated (by some kind of cataclysm such as a huge meteor crashing into their land) could survive, at least partially, thanks to the glowing halo that exists beyond their land's physical borders.

Though horrific, such an image does not strike us as in the least counterintuitive, because we understand that the physical land, no matter how beloved in song and story, is not indispensable for the survival of a nationality. The geographical place is merely the traditional breeding grounds for an ancient set of genes and memes — complexions, body types, hair colors, traditions, words, proverbs, dances, myths, costumes, recipes, and so forth — and as long as a critical mass of carriers of these genes and memes, located abroad, survives the cataclysm, all of this richness can continue to exist and flourish elsewhere, and the now-gone physical place can continue to be celebrated in song and story.

Although no entire country has ever been physically annihilated, events somewhat like this have happened in the past. I am reminded of the gulping-up of all of Polish soil by Poland's neighbors in the eighteenth and nineteenth centuries — the so-called "partitions of Poland". The Polish people, although rendered physically homeless, continued to endure. Here was a nation — *naród polski* — vibrant and alive, yet entirely deprived of a land. Indeed, the words that open the Polish national anthem celebrate this survival: "Poland is not lost, as long as we live!" In parallel fashion, the original Jews, scattered in biblical times from the cradle of their culture, continued to survive, keeping alive their traditions, their language, and their beliefs, in the Diaspora.

Halos, Afterglows, Coronas

In the wake of a human being's death, what survives is a set of afterglows, some brighter and some dimmer, in the collective brains of all those who were dearest to them. And when those people in turn pass on, the afterglow becomes extremely faint. And when that outer layer in turn passes into oblivion, then the afterglow is feebler still, and after a while there is nothing left.

This slow process of extinction I've just described, though gloomy, is a little less gloomy than the standard view. Because bodily death is so clear, so sharp, and so dramatic, and because we tend to cling to the caged-bird view, death strikes us as instantaneous and absolute, as sharp as a guillotine blade. Our instinct is to believe that the light has all at once gone out altogether. I suggest that this is not the case for human souls, because the essence of a human being — truly unlike the essence of a mosquito or a snake or a bird or a pig — is distributed over many a brain. It takes a couple of generations for a soul to subside, for the flickering to cease, for all the embers to burn out. Although "ashes to ashes, dust to dust" may in the end be true, the transition it describes is not so sharp as we tend to think.

It seems to me, therefore, that the instinctive although seldom articulated purpose of holding a funeral or memorial service is to reunite the people most intimate with the deceased, and to collectively rekindle in them all, for one last time, the special living flame that represents the essence of that beloved person, profiting directly or indirectly from the presence of one another, feeling the shared presence of that person in the brains that remain, and thus solidifying to the maximal extent possible those secondary personal gemmae that remain aflicker in all these different brains. Though the primary brain has been eclipsed, there is, in those who remain and who are gathered to remember and reactivate the spirit of the departed, a collective corona that still glows. This is what human love means. The word "love" cannot, thus, be separated from the word "I"; the more deeply rooted the symbol for someone inside you, the greater the love, the brighter the light that remains behind.

CHAPTER 19

Consciousness = Thinking

ও ও ও

So Where's Consciousness in my Loopy Tale?

FROM the very start in this book, I have used a few key terms pretty much interchangeably: "self", "soul", "I", "a light on inside", and "consciousness". To me, these are all names for the same phenomenon. To other people, they may not seem to denote one single thing, but that's how they seem to me. It's like prime numbers of the form $4n + 1$ and prime numbers that are the sums of two squares — on the surface these would seem to be descriptions of completely different entities, but on closer analysis they turn out to denote exactly the same entities.

In my way of looking at things, all of these phenomena come in shades of gray, and whatever shade one of them has in a particular being (natural or artificial), all the others have that same shade. Thus I feel that in talking about "I"-ness, I have also been talking about consciousness throughout. Yet I know that some people will protest that although I may have been addressing issues of personal identity, and perhaps the concepts of "I" and "self", I haven't even touched the far deeper and more mysterious riddle of consciousness. They will skeptically ask me, "What, then, is *experience* in terms of your strange loops? How do strange loops in the brain tell us anything about *what it feels like* to be alive, to smell honeysuckle, to see a sunset, or to listen to raindrops patter on a tin roof? *That* is what consciousness is all about! How does *that* have anything to do with your strange, loopy idea?"

I doubt that I can answer such questions to the satisfaction of these hard-core skeptics, for they will surely find what I say both too simple and too evasive. Nonetheless, here is my answer, stripped down to its essence:

Consciousness is the dance of symbols inside the cranium. Or, to make it even more pithy, consciousness is *thinking*. As Descartes said, *Cogito ergo sum.*

Unfortunately, I suspect that this answer is far too compressed for even my most sympathetic readers, so I will try to spell it out a little more explicitly. Most of the time, any given symbol in our brain is dormant, like a book sitting inertly in the remote stacks of a huge library. Every so often, some event will trigger the retrieval of this book from the stacks, and it will be opened and its pages will come alive for some reader. In an analogous way, inside a human brain, perceived external events are continually triggering the highly selective retrieval of symbols from dormancy, and causing them to come alive in all sorts of unanticipated, unprecedented configurations. This dance of symbols in the brain is what consciousness is. (It is also what thinking is.) Note that I say "symbols" and not "neurons". The dance has to be perceived *at that level* for it to constitute consciousness. So there you have a slightly more spelled-out version.

Enter the Skeptics

"But who *reads* these symbols and their configurations?", some skeptics will ask. "Who *feels* these symbols 'come alive'? Where is the counterpart to the reader of the retrieved book?"

I suspect that these skeptics would argue that the symbols' dance on its own is merely motion of material stuff, unfelt by anyone, so that despite my claim, this dance cannot constitute consciousness. The skeptics would like me to name or point to some special locus of subjective *awareness* that we all have of our thoughts and perceptions. I feel, though, that such a hope is confused, because it uses what I consider to be just another synonym for "conscious" — namely, "aware" — in posing the same question once more, but at a different level. In other words, people seeking the "reader" for configurations of activated symbols may accept the idea of symbols galore being triggered in the brain, but they refuse to call that kind of internal churning "consciousness" because now they want the symbols *themselves* to be perceived. These people would probably be particularly unhappy if I were to bring up the careenium metaphor at this point and to suggest that the dance of simmballs in the careenium constitutes consciousness. They would argue that it's just the mutual bashing of scads of tiny little marbles on a glorified pool table, and that that's *obviously* empty and devoid of consciousness. They want much more than that.

Such skeptics are in essence kicking the problem upstairs — instead of settling for the idea that symbol-level brain activity (or simmball-level careenium activity) that mirrors external events *is* consciousness, they now insist that the internal events of brain activity must in turn be perceived if consciousness is to arise. This runs the risk of setting up an infinite regress and thus moving further and further away from an answer to the riddle of consciousness rather than homing in on an answer to it.

I will give such people one thing, however — I will agree that symbolic activity is itself an important, indispensable focus of a human brain's attention (but I would quickly add that this does not hold for chickens or frogs or butterflies, and pretty darn little for dogs). Mature human brains are constantly trying to reduce the complexity of what they perceive, and this means that they are constantly trying to get unfamiliar, complex *patterns* made of many symbols that have been freshly activated in concert to trigger just *one* familiar pre-existing symbol (or a very small set of them). In fact, that's the main business of human brains — to take a complex situation and to put one's finger on *what matters* in it, to distill from an initial welter of sensations and ideas what a situation really is all about. To spot the gist. To Spot, the gist, however, doesn't much matter, and the gist certainly doesn't matter one whit to the flea on Spot's wagging tail.

I suspect that all of this may sound a bit abstruse and vague, so I'll illustrate it with a typical example.

Symbols Trigger More Symbols

A potential new doctoral student named Nicole comes to town for a day to explore the possibility of doing a Ph.D. in my research group. After my graduate students and I have interacted with her for several hours, first at our Center and then over a Chinese dinner, we agree that we all find her mind delightfully lively and her thoughts just on our wavelength, and it's clear that our enthusiasm is reciprocated. Needless to say, then, we are all hopeful that she'll join us next fall. After she returns home, Nicole sends me an email saying that she is still very excited by our ideas and that they are continuing to reverberate vividly in her mind. I reply with a note of encouragement, and then there ensues an e-silence for a couple of weeks. When I finally send her a second email telling her how eager we all are for her to come next year, a couple of days pass and then a terse and somewhat starchy reply arrives, saying that she's sorry but she's decided to go to another university for graduate school. "But I hope we'll have a chance to interact in the future," she adds politely at the end.

Well, this little episode is all fresh to me. Nicole is a unique individual, our lively conversations with her were all *sui generis*, and the complex configuration of symbols activated in my brain by the whole event is, by definition, unprecedented. And yet on another level that's not true at all.

In my many decades' worth of episodic memory, there are precedents galore for this episode, if I just "hold it loosely in the mind". In fact, without making the slightest effort, I find quite a few old memories bubbling up for the first time in many years, such as that time nearly thirty years ago when a very promising young candidate for our faculty seemed so interested but then, to our great surprise, he turned down our exceedingly generous offer. And that time a few years later when an extremely bright grad student of mine got all excited about accompanying me out to California for my sabbatical year but then changed his mind and soon dropped entirely out of sight, never to be heard from again. And then there's that sad time I was terribly infatuated with that young woman from a far-off land, whose signals to me at first seemed so tinglingly filled with promise, but who then inexplicably drew back a bit, and a week or so later wound up telling me she was involved with someone else (actually, *that* event happened far more than just once, to my chagrin…).

And so, one by one, all these dusty old "books" are pulled off the shelves of dormancy by the current episode, because this "unprecedented" situation, when it is perceived at an abstract level, when its crust is discarded and its core is distilled, points straight at certain other past sagas stored on the shelves of my "library", and one after another of them gets pulled out and placed in the limelight of activation. These old sagas, long ago wrapped up in nice neat mental packages, had been idly sitting around on the shelves of my brain, waiting to be triggered if and when "the same thing" should ever happen, in new guise. And, sad to say, it did!

When all this activity has flowed around for a while, with memories triggering memories triggering memories, something slowly settles out — some kind of "precipitate", to borrow a term from chemistry. In this case, it finally boils down to just one word: "jilted". Yes, I feel *jilted*. My research group has been *jilted*.

What a phenomenal reduction in complexity! We began with an encounter that lasted for hours in two different venues and that involved many people and many thousands of words exchanged and uncountable visual impressions and then some follow-up emails, but in the end the whole thing funneled down to (or should I rather say "fizzled out in"?) just one single very disappointing six-letter word. To be sure, that's not the only idea I retain from the saga, but "jilt" becomes one of the dominant mental categories with which Nicole's visit will forever be associated. And of course, the Nicole saga itself gets neatly bound and stored on the shelves of my episodic memory for potential retrieval by this "I" of mine, somewhere further down the line, who knows when or where.

The Central Loop of Cognition

The machinery that underwrites this wonderfully fluid sort of abstract perception and memory retrieval is at least a little bit like what the skeptics above were clamoring for — it is a kind of perception of internal symbol-patterns, rather than the perception of outside events. Someone seems to be looking at configurations of activated symbols and perceiving their essence, thereby triggering the retrieval of other dormant symbols (which, as we have just seen, can be very large structures — memory packages that store entire romantic sagas, for instance), and round and round it all goes, giving rise to a lively cycle of symbolic activity — a smooth but completely improvised symbolic dance.

The stages constituting this cycle of symbol-triggerings may at first strike you as being wildly different from the act of recognizing, say, a magnolia tree in a flood of visual input, since that involves an *outside* scene being processed, whereas here, by contrast, I'm looking at my own activated symbols dancing and trying to pinpoint the dance's essence, rather than pinpointing the essence of some external scene. But I would submit that the gap is far smaller than one might at first suppose.

My brain (and yours, too, dear reader) is constantly seeking to label, to categorize, to find precedents and analogues — in other words, *to simplify while not letting essence slip away*. It carries on this activity relentlessly, not only in response to freshly arriving sensory input but also in response to its own internal dance, and there really is not much of a difference between these two cases, for once sensory input has gotten beyond the retina or the tympani or the skin, it enters the realm of the *internal*, and from that point on, perception is solely an internal affair.

In short, and this should please the skeptics, there *is* a kind of perceiver of the symbols' activity — but what will not please them is that this "perceiver" is itself just further symbolic activity. There is not some special "consciousness locus" where something magic happens, something *other* than just more of the same, some locus where the dancing symbols make contact with... well, with what? What would please the skeptics? If the "consciousness locus" turned out to be just a physical part of the brain, how would that satisfy them? They would still protest that if *that's* all I claim consciousness is, then it's just insensate physical activity, no different from and no better than the mindless careening of simms in the inanimate arena of the careenium, and has nothing to do with consciousness!

I think it may be helpful at this point to allow my various inner skeptical voices to merge into a single paper persona (hopefully not a paper tiger!), and for that persona to lock horns in an extended dialogue with another persona who essentially represents the ideas of this book. I'll call the voice of this book "Strange Loop #641" and the voice of the skeptics "Strange Loop #642".

It may strike some readers that I am unfairly prejudicing the case by labeling not only myself (or rather, my proxy) a "strange loop", but also my worthy opponent, for that might be seen as suggesting that the game is over before it's begun. But these are nothing more than labels. What counts in the dialogue is what the characters say, not what I call them. And so, if you prefer to give Strange Loops #641 and #642 the alternative names "Inner Light #7" and "Inner Light #8", or perhaps even "Socrates" and "Plato", that's fine by me.

And now, without further ado, we tune in as our two strange loops (or inner lights) begin their amiable debate. Oops! I guess I've been rambling on a bit too long here, and we seem unfortunately to have missed a bit of the two friends' opening repartee. Oh, well, that's life. I expect you and I can jump in at this point without feeling too lost. Let's give it a try…

CHAPTER 20

A Courteous Crossing of Words

ॐ ॐ ॐ

Dramatis personæ:

Strange Loop #641: a believer in the ideas of *I Am a Strange Loop*
Strange Loop #642: a doubter of the ideas of *I Am a Strange Loop*

•   •   •

SL #642:   Dreary, oh so dreary. In fact, your picture of the soul is not just dreary; it's completely empty. Vacuous. There's nothing spiritual there at all. It's just physical activity and nothing more.

SL #641:   What else did you expect? What else *could* you expect? Unless you're a dualist, that is, and you think souls are ghostly, nonphysical things that don't belong to the physical universe, and yet that can push pieces of it around.

SL #642:   No, I don't go for that. It's just that there has to be something extremely special that accounts for the existence of spiritual, mental, feeling, perceiving beings in this physical world — something that explains our inner light, our awareness, our *consciousness*.

SL #641:   I couldn't agree with you more. An explanation of such elusive phenomena surely calls for something special. Building a soul out of physical nuts and bolts is a tall order. But bear in mind that in my view, consciousness is a very unusual sort of intricately organized material pattern, not just any old physical activity. It's not the swinging of a chain, the plopping of a stone in a pond, the splashing of a waterfall, the swirling of a hurricane, the refilling of a flush toilet, the self-regulation of the temperature in a house, the flow of electrons in a program that plays chess, the wiggling of an ovum-seeking sperm, the neural firings in a hungry mosquito's brain… but we are getting ever closer as this list progresses. An "inner light" *starts* to turn on as we rise in this hierarchy. The light is still incredibly dim even at the list's end, but if we extend the list further and sweep upwards through the brains of bees, goldfish, bunnies, dogs, and toddlers, it grows far brighter. It gets very bright when we arrive at human adolescents and adults, and it stays bright for decades. What we know as our own consciousness is, yes, *nothing but* the physical activity inside a human brain that has lived in the world for a number of years.

SL #642:   No, the essence of consciousness is missing from your picture. You've described a complex set of brain activities involving symbols triggering each other, and I'm prepared to believe that something like that does take place inside brains. But that isn't the whole story, because *I* am nowhere in this story. There is no room for an *I*. You've proposed myriads of unconscious particles bouncing around, or perhaps big clouds of activity made of particles — but if the universe were only that, then there would be no me, no you, no points of view. It would be the way the earth was before life evolved — millions of sunrises and sunsets, winds blowing hither and thither, clouds forming and scattering, thunderstorms swooping along valleys, boulders tumbling down mountains and gouging out gulleys, water flowing in riverbeds and carving deep canyons, waves breaking on sandy beaches, tides flowing in and out, volcanoes spewing out red-hot seas of lava, mountain chains bursting up out of plains, continents drifting and breaking apart, and so on. All very scenic, but there would be no inner life, no mind, no inner light, no I — no one to enjoy the great scenery.

SL #641:   I sympathize with your sense of the barrenness of a universe made of physical phenomena only, but some kinds of physical systems can mirror what's on their outside and can launch actions that depend upon their perceptions. That's the thin edge of the wedge. When perception grows sophisticated enough, it can lead to phenomena that have no counterparts in systems that perceive only in a primitive manner. By "primitive" perceiving systems, I mean entities like, for instance, thermostats, knees, sperms, and tadpoles. These are too rudimentary to merit the term "consciousness", but when perception takes place in a system endowed with a truly rich, fluidly extensible set of symbols, then an "I" will arise just as inevitably as strange loops arise in the barren fortress of *Principia Mathematica*.

SL #642: Perception?! Who's doing the perceiving? No one! Your universe is still just a vacuous system of physical objects and their intricate, intertwined, enmeshed movements — galaxies, stars, planets, winds, rocks, water, landslides, ripples, sound waves, fire, radioactivity, and so forth. Even proteins and RNA and DNA. Even your beloved feedback loops — heat-seeking missiles, thermostats, refilling toilets, video feedback, domino chains, pool tables flooded with hordes of microscopic magnetic balls. But something crucial is missing from this bleak scene, and that's *me*-ness. *I* am in a specific *place*. I'm *here*! What would pick out a *here* in a world consisting of water and float-balls in thousands of tanks, or in a world having zillions of different domino chains? There's no *here* there.

SL #641: I really do understand that this matter would nag at you; it should nag at any thinking person. My reply is this: In the vast universe of diverse physical events that you just evoked so vividly, there are certain rare spots of localized activity in which a special kind of abstractly swirling pattern can be found. Those special loci — at least the ones that we have run into so far — are human brains, and "I"'s are restricted to those loci. Such loci are hard to find in the vast universe; they are few and far between. Wherever this special, rare kind of physical phenomenon arises, there's an *I* and a *here*.

SL #642: Your phrase "abstractly swirling pattern" makes me think of a physical vortex, like a hurricane or a whirlpool or a spiral galaxy — but I suppose those aren't abstract enough for you.

SL #642: No, they really aren't. Whirlpools and hurricanes are merely spinning vortices — fluid cousins to tops and gyroscopes. To make an "I" you need *meanings*, and to make meanings you need perception and categories — in fact, a repertoire of categories that keeps on building on itself, growing and growing and growing. Such things are nowhere to be found in the physical vortices you mentioned. That's why a far better metaphor for an "I" is the structure of the self-referring formulas that Gödel found in the barren-seeming universe of *PM*. His formulas, like human "I"'s, are extremely intricately and delicately structured, and are hardly a dime a dozen. "Ordinary" formulas of *PM*, like "0+0=0", say, or a formula that states that every integer is the sum of at most four squares, are the analogues to inert, "I"-less physical objects, like grains of sand or bowling balls. Those simple kinds of formulas don't have wraparound high-level meanings in the way that Gödel's special strings do. It takes a great deal of number-theoretical machinery to build up from ordinary assertions about numbers to the complexity of Gödelian strange loops, and likewise it takes a great deal of evolution to build up from very simple feedback loops to the complexity of strange loops in brains.

SL #642:   Suppose I granted you that there are lots of abstract "strange loops" floating around the universe, which somehow coalesced over the course of billions of years of evolution — strange loops residing in crania, a bit like audio feedback loops residing in auditoriums. They can be as complex as you like; the complexity of their physical activity doesn't matter one whit to me. The knotty issue that simply will not go away is: What would make one of those strange loops *me*? Which one? You can't answer that.

SL #641:   I can, although you won't like my answer. What makes one of them *you* is that it is resident in a particular brain that went through all the experiences that made you you.

SL #642:   That's just a tautology!

SL #641:   Not really. It's a subtle idea whose crux is that what you call "I" is an *outcome*, not a starting point. You coalesced in an unplanned fashion, coming only slowly into existence, not in a flash. At the beginning, when the brain that would later house your soul was taking form, there was no you. But that brain slowly grew, and its experiences slowly accumulated. Somewhere along the way, as more and more things happened to it, were registered by it, and became internalized in it, it started imitating the cultural and linguistic conventions in which it was immersed, and thus it tentatively said "I" about itself (even though the referent of that word was still very blurry). That's roughly when it noticed it was somewhere — and not surprisingly, it was where a certain brain was! At that point, though, it didn't know anything about its brain. What it knew instead was its brain's *container*, which was a certain body. But even though it didn't know anything about its brain, that nascent "I" faithfully followed its brain around just as a shadow always tags along after a moving object.

SL #642:   You're not dealing with my question, which is about how to pick *me* out in a world of indistinguishable physical structures.

SL #641:   All right, let me turn straight to that. To you, all the brains housing strange loops seem no different from thousands of sewing machines scattered hither and yon, all clicking away. You would ask, "Which sewing machine is *me*?" Well, of course, none of them is you — and that's because none of them *perceives* anything. You see brains that house strange loops as being just as inert and identity-lacking as sewing machines, pinwheels, or merry-go-rounds. But the funny thing is that the beings whose brains house those strange loops don't agree with you that they have no identity. One of them insists, "I'm the one right *here*, looking at this purple flower, not the one over *there*, drinking a milkshake!" Another one insists, "I'm the one drinking this chocolate shake, not the one looking at that flower!" Each one of them is convinced of being somewhere and of seeing things and hearing things and having experiences. What makes you reject their claims?

SL #642: I don't reject their claims. Those claims are perfectly valid — it's just that their validity has nothing to do with brains housing strange loops. You're focusing on the wrong thing. Any claims of "being here" and "being conscious" are valid because there is something extra, something over and above strange loops, that makes a brain be the locus of a soul. I can't tell you just what it is, but I know this is true, because *I* am not just physical stuff happening somewhere in the universe. I *experience* things, such as that purple flower in the garden and that loud motorcycle a couple of blocks away. And my experience is the primary data on which everything else that I say is based, so you cannot deny my claim.

SL #641: How is that any different from what I've described? A sufficiently complex brain not only can perceive and categorize but it can verbalize what it has categorized. Like you, it can talk about flowers and gardens and motorcycle roars, and it can talk about itself, saying where it is and where it is not, it can describe its present and past experiences and its goals and beliefs and confusions... What more could you want? Why is that not what you call "experience"?

SL #642: Words, words, words! The point is that experience involves *more* than mere words — it involves *feelings.* Any experiencer worthy of the term has to see that brilliant purple color of the flower and *feel* it as such, not merely drone the sound "purple" like an automated voice in a telephone menu tree. Seeing a vivid purple takes place below the level of words or ideas or symbols — it is more primordial. It's an experience directly felt by an experiencer. That's the difference between true consciousness and mere "artificial signaling" as in a mechanical-sounding telephone menu tree.

SL #641: Would you say nonverbal animals enjoy such "primordial" experiences? Do cows savor the deep purple of a flower just as intensely as you do? And do mosquitoes? If you say "yes", doesn't that come dangerously close to suggesting that cows and mosquitoes have just as much consciousness as you do?

SL #642: Mosquito brains are far less complex than mine, so they can't have the same kinds of rich experiences as I do.

SL #641: Now wait a minute. You can't have it both ways. A moment ago, you were insisting that brain complexity doesn't make any difference — that if a brain lacks that special *je ne sais quoi* that separates things that feel from things that don't feel, then it's not a locus of consciousness. But now you're saying that the complexity of the brain in question *does* make a difference.

SL #642: Well, I guess it has to, to some extent. A mosquito doesn't have the equipment to appreciate a purple flower in the way I do. But maybe a cow does, or at least it comes closer. But complexity alone does not account for the presence of feeling and experience in brains.

SL #641: Let's consider a bit more deeply this notion of experiencing and feeling the world outside. If you were to stare at a big broad sheet of pure, uniform purple, your favorite shade ever, entirely filling your visual field, would you experience the same rush as when you see that color in the petals of a flower blooming in a garden?

SL #642: I doubt it. Part of what makes my experience of a purple flower so intense is all the subtle shades I see on each petal, the delicate way each petal is curved, and the way the petals all swirl together around a glowing center made of dozens of tiny dots…

SL #641: Not to mention the way the flower is poised on a branch, and the branch is part of a bush, and the bush is just one of many in a brightly colored garden…

SL #642: Are you intimating that I don't enjoy the purple for its own sake, but only because of the way it's embedded in a vast scene? This goes too far. The surroundings may *enhance* my experience, but I love that rich velvety purple purely for itself, independently of anything else.

SL #641: Why then do you describe it with the word "velvety"? Do flies or dogs experience purple flowers as "velvety"? Isn't that word a reference to velvet? Doesn't it mean that your visual experience calls up deeply buried memories, perhaps tactile memories from childhood, of running your fingers along a purple cushion made of velvet? Or maybe you're unconsciously reminded of a dark-colored wine you once drank whose label described it as "velvety". How can you claim your experience of purple is "independent of anything else in the world"?

SL #642:　　All I'm trying to say is that there are basic, primordial experiences out of which larger experiences are built, and that even the primordial ones are radically, qualitatively different from what goes on in simple physical systems like ropes dangling in breezes and floats bobbing in toilets.　A dangling rope doesn't feel anything when a breeze impinges on it.　There's no feeling in there, there's no *here* there. But when I see purple or taste chocolate, that's a sensual experience I'm having, and it's from millions of such sensual experiences that my mental life is built up.　There's a big mystery here, in this breach.

SL #641:　　It sounds attractive, but unfortunately I think you've got it all backwards.　Those little sensual experiences are to the grand pattern of your mental life as the letters in a novel are to the novel's plot and characters — irrelevant, arbitrary tokens, rather than carriers of meaning.　There is no meaning to the letter "b", and yet out of it and the other letters of the alphabet, put together in complex sequences, comes all the richness and humanity in a novel or a story.

SL #642:　　That's the wrong level to talk about a story.　Writers choose *words,* not letters, and words are of course imbued with meaning.　Put together a lot of those tiny meanings and you get one big meaning-rich thing.　Similarly, life is made out of many tiny sensual experiences, chained together to make one huge sensuo-emotional experience.

SL #641:　　Hold on a minute.　No isolated word has depth and power. When embedded in a complex context, a word may have great power, but in isolation it does not.　It's an illusion to attribute power to the word itself, and it's a greater illusion to attribute power to the letters constituting the word.

SL #642:　　I agree that letters have no power or meaning.　But words, yes! They are the atoms of meaning out of which larger structures of meaning are built.　You can't get big meanings from atoms that are meaningless!

SL #641:　　Oh, really?　I thought you just conceded that exactly this happens in the case of words and letters.　But all right — let's move on from that example.　Would you say that music has meaning?

SL #642:　　Music is among the most meaningful things I know.

SL #641:　　And yet, are individual notes meaningful to you?　For instance, do you feel attraction or repulsion, beauty or ugliness, when you hear middle C?

SL #642:　　I hope not!　No more than when I see the isolated letter "C".

SL #641:   Is there *any* isolated note that on its own attracts or repels you?

SL #642:   No. An isolated note doesn't carry musical meaning. Anyone who claimed to be moved by a single note would be putting on airs.

SL #641:   Yet when you hear a piece of music you like or hate, you certainly are attracted or repelled. Where does that feeling come from, given that no *note* in it has any intrinsic attraction or repulsion for you?

SL #642:   It depends on how they are arranged in larger structures. A melody is attractive because of some kind of "logic" it possesses. Some other melody could be repulsive because it lacks logic, or because its logic is too simplistic or childish.

SL #641:   That certainly sounds like a response to *pattern*, not like raw sensation. A piece of music can have great emotional meaning despite being made of tiny atoms of sound that have no emotional meaning. What matters, therefore, is the pattern of organization, not the nature of the constituents. This brings us back to your puzzlement about the difference between experiencers such as you and me, and non-experiencers such as dangling ropes and plastic floats. To you, this crucial difference must originate in some special ingredient, some tangible *thing* or *substance*, which experiencers have in their makeup, and which non-experiencers lack. Is that right?

SL #642:   Something like that has to be the case.

SL #641:   Then let's call this special ingredient that allows experiencers to come into existence "feelium". Unfortunately, no one has ever found a single atom or molecule of feelium, and I suspect that even if we did find a mysterious substance present in all higher animals but not in lower ones, let alone in mere machines, you would start wondering how it could be that any mere *substance*, inanimate and insensate on its own, could give rise to sensation.

SL #642:   Feelium, if it existed, would probably be more like electricity than like atoms or molecules. Or maybe it would be like fire or radioactivity — in any case, something that seems living, something that by its very nature dances in crazy ways — not just inert *stuff*.

SL #641:   When you painted a picture of the earth before life evolved, it had volcanoes, thunder and lightning, electricity, fire, light, and sound — even the sun, that great big ball of nuclear fusion. And yet you weren't willing to imagine that the presence of such phenomena, in any combination or permutation, could ever give rise to an experiencer. Yet just now, in talking about the mysterious soul-creating essence I called "feelium", you used the word "dance", as in the phrase "dancing symbols". Are you perhaps unwittingly changing your tune?

SL #642: Well, I can imagine a sparkling, firelike "dance" as being what distinguishes experiencers from non-experiencers. It's even somehow appealing to me to think that the dancing of feelium, if it turned out to exist, might be able to explain the difference between experiencers and non-experiencers. But even if we came to understand the physics of how feelium produces experience, something crucial would *still* be missing. Suppose that the world were populated by experiencers defined by some kind of pattern involving feelium. Let's even suppose that the pattern at the core of each experiencer were a strange loop, as you postulate. So now, because of this elusive but wonderful physical pattern executed at least partially in feelium, there are lots of "lights on" scattered around in special spots here and there in the universe. The sticking point remains: Which one of them is *me*? What makes *one* of them different from all of the others? What is the source of "I"-ness?

SL #641: Why do you say you would be different from the others? Each one would cry out that *it* was different. You'd all be mouthing just the same thoughts. In that sense, you would all be indistinguishable!

SL #642: I think you're teasing me. You know perfectly well that I'm *not* the same as anyone else. My inner fire is *here*, not anywhere else. I want to know what singles out this particular fire from all the others.

SL #641: It's as I said before: you're a satellite to your brain. Like a fireplace, a particular brain is in a particular spot. And wherever it happens to be, its resident strange loop calls that place "here". What's so mysterious about that?

SL #642: You're not answering my question. I don't think you're even *hearing* my question.

SL #641: Oh, sure — I hear you. I here, you there!

SL #642: Ouch. Now just listen for a moment. My question is very straightforward. Anybody can understand it (except maybe you). Why am I in *this* brain? Why didn't I wind up in some *other* brain? Why didn't I wind up in *your* brain, for instance?

SL #641: Because your "I" was not an *a priori* well-defined thing that was predestined to jump, full-fledged and sharp, into some just-created empty physical vessel at some particular instant. Nor did your "I" suddenly spring into existence, wholly unanticipated but in full bloom. Rather, your "I" was the slowly emerging outcome of a million unpredictable events that befell a particular body and the brain housed in it. Your "I" is the self-reinforcing structure that gradually came to exist not only *in* that brain, but *thanks to* that brain. It couldn't have come to exist in *this* brain, because *this* brain went through different experiences that led to a different human being.

SL #642:   But why couldn't *I* have had those experiences as easily as you?

SL #641:   Careful now! Each "I" is defined as a *result* of its experiences, and not vice versa! To think the reverse is a very tempting, seductive trap to fall into. You keep on revealing your tacit assumption that any "I", despite having grown up inside one particular brain, isn't deeply rooted in that brain — that the same "I" could just as easily have grown up in and been attached to any other brain; that there is no deeper a connection between a given "I" and a given brain than the connection between a given canary and a given cage. You can just swap them arbitrarily.

SL #642:   You're still missing my point. Instead of asking why I *ended up* in this brain, I'm asking why I *started out* in that random brain, and not in some other one. There's no reason that it had to be *that* one.

SL #641:   No, *you're* the one who's missing the point. The key point, uncomfortable for you though it will be, is that *no one* started out in that brain — no one at all. It was just as uninhabited as a swinging rope or a whirlpool. But unlike those physical systems, it could perceive and evolve in sophistication, and so, as weeks, months, and years passed, there gradually came to be *someone* in there. But that personal identity didn't suddenly appear full-blown; rather, it slowly coalesced and came into focus, like a cloud in the sky or condensation on a windowpane.

SL #642:   But who was that person destined to be? Why couldn't it have been someone else?

SL #641:   I'm coming to that. What slowly came to pervade that brain was a complicated set of mental tendencies and verbal habits that are now insistently repeating this question, "Why am I *here* and not *there*?" As you may notice, this brain *here* (mine, that is) doesn't make its mouth ask that question over and over again. *My* brain is very different from *your* brain.

SL #642:   Are you telling me that it doesn't make sense to ask the question, "Why am I here and not there?"

SL #641:   Yes, I'm saying that, among other things. What makes all of this so counterintuitive — verging on the incomprehensible, at times —

is that your brain (like mine, like everyone's) has told itself a million times a self-reinforcing story whose central player is called "I", and one of the most crucial aspects of this "I", an aspect that is truly a *sine qua non* for "I"-ness, is that it fluently flits into other brains, at least partially. Out of intimacy, out of empathy, out of friendship, and out of relatedness (as well as for other reasons), *your* brain's "I" continually makes darting little forays into *other* brains, seeing things to some extent from their point of view, and thus convincing itself that it could easily be housed in them. And then, quite naturally, it starts wondering why it *isn't* housed in them.

SL #642:   Well, of course it would ask itself that. What more natural thing to wonder about?

SL #641:   And one piece of the answer is that to a small extent, your "I" *is* housed in other brains. Yes, your "I" is housed a little bit in my frustratingly dense and pigheaded brain, and vice versa. But despite that blurry spillover that turns the strict city-limits version of You into Greater Metropolitan You, your "I" is still very localized. Your "I" is certainly not uniformly spread out among all the brains on the surface of the earth — no more so than the great metropolitan sprawl of Mexico City possesses suburbs in Madagascar! But there is another piece of the answer to your question "Why am I here and not there?", and it is going to trouble you. It is that your "I" isn't housed anywhere.

SL #642:   Come again? This doesn't sound like your usual line.

SL #641:   Well, it's just another way of looking at these things. Earlier, I described your "I" as a self-reinforcing structure and a self-reinforcing story, but now I'll risk annoying you by calling it a self-reinforcing *myth*.

SL #642:   A *myth*?! I'm certainly not a myth, and I'm here to tell you so.

SL #641:   Hold your horses for a moment. Think of the illusion of the solid marble in the box of envelopes. Were I to insist that that box of envelopes had a *genuine* marble in it, you'd say I had fallen hook, line, and sinker for a tactile illusion, wouldn't you?

SL #642:   I would indeed, although the *feeling* that something solid is in there is not an illusion.

SL #641:   Agreed. So my claim is that your brain (like mine and like everyone else's) has, out of absolute necessity, invented something it calls an "I", but that that thing is as real (or rather, as unreal) as is that "marble" in that box of envelopes. In that sense, your brain has tricked itself. The "I" — yours, mine, everyone's — is a tremendously effective illusion, and falling for it has fantastic survival value. Our "I"'s are self-reinforcing illusions that are an inevitable by-product of strange loops, which are themselves an inevitable by-product of symbol-possessing brains that guide bodies through the dangerous straits and treacherous waters of life.

SL #642:   You're telling me there is not *really* any "I". Yet my brain tells me just as assuredly that there *is* an "I". Then you tell me that this is just my brain pulling a trick on me. But excuse me — pulling a trick on *whom*? You've just told me that this *me* doesn't exist, so who is my brain pulling a trick on? And — pardon me once again — how can I even call it "*my* brain" if there is no *me* for it to belong to?

SL #641:   The problem is that in a sense, an "I" is something created out of nothing. And since making something out of nothing is never possible, the alleged something turns out to be an illusion, in the end, but a very powerful one, like the marble among the envelopes. However, the "I" is an illusion far more entrenched and recalcitrant than the marble illusion, because in the case of "I", there is no simple revelatory act corresponding to turning the box upside down and shaking it, then peering in between the envelopes and finding nothing solid and spherical in there. We don't have access to the inner workings of our brains. And so the only perspective we have on our "I"-ness marble comes from the counterpart to squeezing all the envelopes at once, and *that* perspective says it's real!

SL #642:   If that's the only possible perspective, then what would ever give us even the slightest sense that we might be lending credence to a myth?

SL #641:   One thing that gives many people a sneaking suspicion that something about this "I" notion might be mythical is precisely what you've been troubled about all through our discussion — namely, there seems to be something incompatible between the hard laws of physics and the existence of vague, shadowy things called "I"'s. How could experiencers come to exist in a world where there are just inanimate things moving around? It seems as if perception, sensation, and experience are something *extra*, above and beyond physics.

SL #642:   Unless, of course, there's feelium, but that's not by any means clear. In any case, I agree that conflicts with physics give a hint that this "I" notion is very elusive and cries out for an explanation.

SL #641:   A second hint that something needs revision has to do with what we perceive as causing what. In our everyday life, we take it for granted that an "I" can cause things, can push things around. If I decide to drive to the grocery store, my one-ton automobile winds up taking me there and bringing me back. Now that seems pretty peculiar in the world of physics, where everything comes about solely as a result of how particles interact. How does the particle story leave room for a shadowy, ethereal "I" to cause a heavy car to move somewhere? This, too, casts a bit of doubt on the reality of the notion of "I".

SL #642:   Perhaps — but if so, it's very very slight.

SL #641:   No matter. That extremely slight doubt flies in the face of what we all take for granted ever since our earliest childhood, which is that "I"'s *do* exist — and in most people, the latter belief simply wins out, hands down. The battle is never even engaged, in most people's minds. On the other hand, for a few people the battle starts to rage: physics versus "I". And various escape hatches have been proposed, including the notion that consciousness is a novel kind of quantum phenomenon, or the idea that consciousness resides uniformly in all matter, and so on. My proposal for a truce to end this battle is to see the "I" as a hallucination perceived by a hallucination, which sounds pretty strange, or perhaps even stranger: the "I" as a hallucination *hallucinated* by a hallucination.

SL #642:   That sounds way beyond strange. That sounds crazy.

SL #641:   Perhaps, but like many strange fruits of modern science, it can sound crazy yet be right. At one time it sounded crazy to say that the earth moved and the sun was still, since it was patently obvious that it was the other way around. Today we can see it either way, depending on circumstances. When we're in an everyday frame of mind, we say, "The sun is setting", and when we're in a scientific frame of mind we remember that the earth is merely turning. We are flexible creatures, able to shift point of view according to circumstance.

SL #642:   And so, in your view, should we also be able to shift points of view concerning the existence of an "I"?

SL #641:   Definitely. My claim that an "I" is a hallucination perceived by a hallucination is somewhat like the heliocentric viewpoint — it can yield new insights but it's very counterintuitive, and it's hardly conducive to easy communication with other human beings, who all believe in their "I"'s with indomitable fervor. We explain our own behavior, and that of others, through the positing of our own "I" and its analogues in other people. This naïve viewpoint allows us to talk about the world of people in terms that make perfect sense to people.

SL #642:   *Naïve*?! I notice that *you* haven't stopped saying "I"! You've probably said it a hundred times in the last five minutes!

SL #641:   To be sure. You're absolutely right. This "I" is a necessary, indispensable concept to all of us, even if it's an illusion, like thinking that the sun is circling the earth because it rises, moves across the sky, and sets. It's only when our naïve viewpoint about "I" bangs up against the world of physics that it runs into all sorts of difficulties. It's at that point that those of us who are scientifically inclined realize that there has to be some other story to be told about it. But believing in the easy story about "I" is a million times more important to most of us than figuring out a scientific explanation for "I", so the upshot is that there's no contest. The "I" myth wins hands down, without a debate ever taking place — even in the minds of the majority of scientifically inclined people!

SL #642:   How can that be?

SL #641:   I surmise it's for two reasons. One is that the "I" myth is infinitely more central to our belief systems than is the "sun circling the earth" myth, and the other is that any scientific alternative to it is far subtler and more disorienting than the shift to heliocentrism was. And so the "I" myth is much harder to dislodge from our minds than the "sun circling the earth" myth. Deconstructing the "I" holds about as much appeal for a typical adult as deconstructing Santa Claus would hold for a typical toddler. Actually, giving up Santa Claus is trivial compared to giving up "I". Ceasing to believe altogether in the "I" is in fact impossible, because it is indispensable for survival. Like it or not, we humans are stuck for good with this myth.

SL #642:   Why do you keep on saying the "I" is just a myth or a hallucination or an illusion, just like that blasted non-marble? I'm tired of your trotting out your tired old marble metaphor. I want to know what's hallucinated.

SL #641:   All right, let's put the marble metaphor to bed for a while. The basic idea is that the dance of symbols in a brain is itself perceived by symbols, and that step extends the dance, and so round and round it goes. That, in a nutshell, is what consciousness is. But if you recall, symbols are simply large phenomena made out of nonsymbolic neural activity, so you can shift viewpoint and get rid of the language of symbols entirely, in which case the "I" disintegrates. It just poofs out of existence, so there's no room left for downward causality.

SL #642:   What does that mean, more specifically?

SL #641:   It means that in the new picture there are no desires, beliefs, character traits, senses of humor, ideas, memories, or anything mentalistic; just itty-bitty physical events (particle collisions, in essence) are left.   One can do likewise in the careenium, where you can shift points of view, either looking at things at the level of simmballs or looking at things at the level of simms.   At the former level, the simms are totally unseen, and at the latter level, the simmballs are totally unseen.   These rival viewpoints really are extreme opposites, like the heliocentric and geocentric views.

SL #642:   All of this I see, but why do you keep implying that one of these views is an illusion, and the other one is the truth?   You always give primacy to the *particle* viewpoint, the lower-level microscopic viewpoint.   Why are you so prejudiced?   Why don't you simply see two equally good rival views that we can oscillate between as we find appropriate, in somewhat the way that physicists can oscillate between thermodynamics and statistical mechanics when they deal with gases?

SL #641:   Because, most unfortunately, the non-particle view involves several types of magical thinking.   It entails making a division of the world into two radically different kinds of entities (experiencers and non-experiencers), it involves two radically different kinds of causality (downward and upward), it involves immaterial souls that pop into being out of nowhere and at some point are suddenly extinguished, and on and on.

SL #642:   You are so bloody inconsistent!   You *liked* the explanation of the falling domino that invoked the primeness of 641!   You *preferred* it!   You kept on saying it was the *real* reason the domino didn't fall, and that the other explanation was myopic and hopelessly useless.

SL #641:   Touché!   I admit that my stance has a definite ironic tinge to it. Sometimes the strict scientific viewpoint *is* hopelessly useless, even if it's correct.   That's a dilemma.   As I said, the human condition is, by its very nature, one of believing in a myth.   And we're permanently trapped in that condition, which makes life rather interesting.

SL #642:   Taoism and Zen long ago sensed this paradoxical state of affairs and made it a point to try to dismantle or deconstruct or simply get rid of the "I".

SL #641:   That sounds like a noble goal, but it's doomed to failure.   Just as we need our eyes in order to *see*, we need our "I"'s in order to *be*! We humans are beings whose fate it is to be able to perceive abstractions, and to be driven to do so.   We are beings that spend their lives sorting the world into an ever-growing hierarchy of patterns, all represented by symbols in our brains. We constantly come up with new symbols by putting together previous symbols in new kinds of structures, nearly *ad infinitum*. Moreover, being macroscopic, we can't see way down to the level where physical causality happens, so in compensation, we find all sorts of marvelously efficient shorthand ways of describing what goes on, because the world, though it's pretty crazy and chaotic, is nonetheless filled with regularities that can be counted on most of the time.

SL #642:   What kinds of regularities are you talking about?

SL #641:   Oh, for example, swings on a playground will swing in a very predictable way when you push them, even though the detailed motions of their chains and seats are way beyond our ability to predict. But we don't care in the least about that level of detail. We feel we know extremely well how swings move. Similarly, shopping carts go pretty much where you want them to when you push them, even if their wobbly wheels, rather predictably, lend them an amusing trace of unpredictability. And someone ambling down the sidewalk in your direction may make some slightly unpredictable motions, but you can count on them not turning into a giant and gobbling you up. These sorts of regularity are what we all know intimately and take for granted, and they are amazingly remote from the level of particle collisions. The most efficient and irresistible shorthand of all is that of imputing abstract desires and beliefs to certain "privileged" entities (those with minds — animals and people), and of wrapping all of those things together in one single, supposedly indivisible unity that represents the "central essence" of such an entity.

SL #642:   You mean that entity's "soul"?

SL #641:   Pretty much. Or if you don't want to use that word, then it's the way that you presume that that thing feels inside — its inner viewpoint, let's say. And then, to cap it all off, since each perceiver is always swimming in its own activities and their countless consequences, it can't keep itself from fabricating a particularly intricate tale about its *own* soul, its *own* central essence. That tale is no different in kind from the tales it makes up for the other mind-owning entities that it sees — it's just far more detailed. Moreover, the story of an "I" is a tale about a central essence that never disappears from view (in contrast to "you"'s and "she"'s and "he"'s, which tend to come into view for a scene or two and then go off stage).

SL #642:   So it's the fact that the system can watch itself that dooms it to this illusion.

SL #641:   Not just that it *can* watch itself, but that it *does* watch itself, and does so all the time. That, plus the crucial fact that it has no choice but to radically simplify everything. Our categories are vast simplifications of patterns in the world, but the well-chosen categories are enormously efficient in allowing us to fathom and anticipate the behavior of the world around us.

SL #642:   And why can't we get rid of our hallucinations? Why can't we attain that pure and selfless "I"-less state that the Zen people would aim for?

SL #641:   We can try all we want, and it is an interesting exercise for a short while, but we can't turn off our perception machinery and still survive in the world. We can't make ourselves *not* perceive things like trees, flowers, dogs, and other people. We can play the game, can tell ourselves we've succeeded, can claim that we have "unperceived" them, but that's just plain self-fooling. The fact is, we are macroscopic creatures, and so our perception and our categories are enormously coarse-grained relative to the fabric at which the true causality of the universe resides. We're stuck at the level of radical simplification, for better or for worse.

SL #642:   Is that a tragedy? You make it sound like a sad fate.

SL #641:   Not at all — it's our glory! It's only those who take Zen and the Tao very seriously who consider this to be a condition to be fought against tooth and nail. They resent words, they resent breaking the world up into discrete chunks and giving them names. And so they give you recipes — such as their droll koans — to try to combat this universal built-in drive to use words. I myself have no desire to fight against the use of words in understanding the world's mysteries — quite the reverse! But I admit that using words has one very major drawback.

SL #642:   What is that?

SL #641:   It is that we have to live with paradox, and live with it in the most intimate fashion. And the word "I" epitomizes all of that.

SL #642:   I don't see anything in the least paradoxical about the word "I". In fact, I see no analogy at all between the commonplace, straightforward, down-to-earth notion of "I" and the esoteric, almost ungraspably elusive notion of a Gödelian strange loop.

SL #641:   Well, consider this. On the one hand, "I" is an expression denoting a set of very high abstractions: a life story, a set of tastes, a bundle of hopes and fears, some talents and lacunas, a certain degree of wittiness, some other degree of absent-mindedness, and on and on. And yet on the other hand, "I" is an expression denoting a physical object made of trillions of cells, each of which is doing its own thing without the slightest regard for the supposed "whole" of which it is but an infinitesimal part. Put another way, "I" refers at one and the same time to a highly tangible and palpable biological substrate and also to a highly intangible and abstract psychological pattern. When you say "I am hungry", which one of these levels are you referring to? And to which one are you referring when you declare, "I am happy"? And when you confess, "I can't remember our old phone number"? And when you exult, "I love skiing"? And when you yawn, "I am sleepy"?

SL #642:   Yes, now that you mention it, I do agree that what "I" stands for is a little hard to pin down. Sometimes its referent is concrete and physical, sometimes it's abstract and mental. And yet when you come down to it, "I" is always both concrete and abstract at the same time.

SL #641:   It is just one thing described in two phenomenally different ways, and that's just the same as Gödel's sentence. That's why it is valid to say that it is both about numbers and about itself. Likewise, "I" is both about a myriad of separate physical objects and also about one abstract pattern — the very pattern causing the word to be said!

SL #642:   It seems that this little pronoun is the nexus of all that makes our human existence mysterious and mystical. It's so different from anything else around. The intrinsically self-pointing loop that the pronoun "I" involves — its indexicality, as philosophers would call it — is quintessentially different from all other structures in the universe.

SL #641:   I don't quite agree with that. Or rather, I don't agree with it at all. The pronoun "I" doesn't involve a stronger or deeper or more mysterious self-reference than the self-reference at the core of Gödel's construction. Quite the contrary. It's just that Gödel spelled out what "I" really means. He revealed that behind the scenes of so-called "indexicals", there are merely codes and correspondences depending on stable, reliable systems of analogies. The thing we call "I" comes from that referential stability, and that's all. There's nothing more mystical about "I" than about any other word that refers. If anything, it is *language* that is so different from other structures in the universe.

SL #642:   So for you, "I" is not mystical? Being is not mysterious?

SL #641:   I didn't say that. Being feels *very* mysterious to me, because, like everyone else, I'm finite and don't have the ability to see deeply enough into my substrate to make my "I" poof out of existence. If I did, I guess life would be very uninteresting.

SL #642:   I should think so!

SL #641:   When we *do* look down at our fine-grained substrates through scientific experiments, we find small miracles just as Gödelian as is "I".

SL #642:   Ah, yes, to be sure — little microgödelinos! But… such as?

SL #641:   I mean the self-reproduction of the double helix of DNA. The mechanism behind it all involves just the same abstract ideas as are implicated in Gödel's type of self-reference. This is what John von Neumann unwittingly revealed when he designed a self-reproducing machine in the early 1950's, and it had exactly the same abstract structure as Gödel's self-referential trick did.

SL #642:   Are you saying microgödelinos are self-replicating machines?

SL #641:   Yes! It's a subtle but beautiful analogy. The analogue of the Gödel number k is a particular blueprint. The "parent" machine examines this blueprint and follows its instructions exactly — that is, it builds what the blueprint depicts. To do that, it has to know what icons stand for what objects — a Gödelian kind of code, or mapping. The newly-built object is a machine that lacks one crucial part. To fill this lacuna, the parent machine then *copies* the blueprint and sticks the copy (which is the key missing part) into the new machine, and *voilà!* — the new composite object is a "child" machine, identical to its parent.

SL #642:   This reminds me of the Morton Salt logo. Would the "child machine" lacking the crucial part be like the "umbrella girl" standing there empty-handed? And the blueprint would be a little blue salt box?

SL #641:   Right! Hand her the little blue box, and she's off to the races! Infinity, ho! And amazingly, only a few years later molecular biologists found that von Neumann's Gödelian mechanism was the same trick Nature had discovered for making self-reproducing physical entities. DNA is the blueprint, of course. It all hinges on the existence of stable mappings (in this case, the mapping called the "genetic code") and the meanings that come from them. And look where that led — to all of life, as far as it has come, and wherever it's still heading! Infinity, ho!

SL #642:   So you claim that the sense of being a unique living thing, reflected in the magical indexicality of the elusive word "I", is not a profound phenomenon, but just a mundane consequence of mappings?

SL #641:   I don't think I said that! The sense of being alive and being a unique link in the infinite chain certainly *is* profound. It's just that it doesn't transcend physical law. To the contrary, it is a profound exploitation of physical law — hardly mundane! On the other hand, the all-too-common desire to mystify the pronoun "I", as if it concealed a deeper mystery than other words do, truly muddies up the picture. The sole root of all these strange phenomena is *perception*, bringing symbols and meanings into physical systems. To perceive is to make a fantastic jump from William James' "blooming, buzzing confusion" to an abstract, symbolic level. And then, when perception twists back and focuses on itself, as it inevitably will, you get rich, magical-seeming consequences. Magical-*seeming*, mind you, but not *truly* magical. You get a level-crossing feedback loop whose apparent solidity dominates the reality of everything else in the world. This "I", this unreal but unutterably stubborn marble in the mind, this "Epi" phenomenon, simply takes over, anointing itself as Reality Number One, and from there on out it won't go away, no matter what words are spoken.

SL #642:   So the "I" is all too marbelous — too marbelous for words?

SL #641:   What?! I thought you thought my "I" idea was for the birds.

SL #642:   It's true, I did, but I think I'm catching your drift. Perhaps I'm coming around a little bit. Your strange-loop view of an "I" is close to paradoxical, and yet not quite. It's like Escher's *Drawing Hands* — paradoxical when you're sucked *into* the drawing by its wondrous realism, yet the paradox dissolves when you step back and see it from *outside*. Then it's just another drawing! Most intriguing. It's all too much, and just too very Berry… to ever be in Russell's Dictionary.

SL #641:   Ah, music to my ears! I'm so delighted you find a bit of merit in my ideas. As you know, they are only metaphors, but they help me to make some sense of the great puzzle of being alive and, as you kept on stressing, the great puzzle of being *here*. I thank you for the splendid opportunity of exchanging views on such subtle matters.

SL #642:   The pleasure, I assure you, was all mine. And I shall await our next meeting with alacrity, celerity, assiduity, vim, vigor, vitality, savoir-faire, and undue velocity. Adieu till then, and cheerio!

[Exeunt.]

❧ ❧ ❧

CHAPTER 21

A Brief Brush with Cartesian Egos

ᐧᐧᐧ ᐧᐧᐧ ᐧᐧᐧ

Well-told Stories Pluck Powerful Chords

IN THE preceding dialogue, the query most insistently posed by Strange Loop #642 was, "What makes me housed in this particular brain, rather than in any other one?" However, even though Strange Loop #641 tried to provide an answer to this enigma in several different fashions, Strange Loop #642 always had the nagging feeling that Strange Loop #641 hadn't really gotten the question, and hadn't understood how profoundly central it is to human existence. Could it be that there is a fundamental breach of communication here, and that some people simply never will get the question because it is too subtle and elusive?

Well, if one is not averse to using a science-fiction scenario, this same question can be posed so vividly and starkly that hopefully no one could fail to understand and feel deeply troubled by the enigma. One way of doing this appears in the path-breaking book *Reasons and Persons* by the Oxford philosopher Derek Parfit. Here is how Parfit poses the riddle:

> I enter the Teletransporter. I have been to Mars before, but only by the old method, a space-ship journey taking several weeks. This machine will send me at the speed of light. I merely have to press the green button. Like others, I am nervous. Will it work? I remind myself what I have been told to expect. When I press the button, I shall lose consciousness, and then wake up at what seems a moment later. In fact I shall have been unconscious for about an hour. The Scanner here on Earth will destroy my brain and body, while recording the exact states of all my cells. It will then transmit this information by radio.

Travelling at the speed of light, the message will take three
minutes to reach the Replicator on Mars. This will then create,
out of new matter, a brain and body exactly like mine. It will be
in this body that I shall wake up.

Though I believe that this is what will happen, I still hesitate.
But then I remember seeing my wife grin when, at breakfast
today, I revealed my nervousness. As she reminded me, she has
been often teletransported, and there is nothing wrong with *her*.
I press the button. As predicted, I lose and seem at once to
regain consciousness, but in a different cubicle. Examining my
new body, I find no change at all. Even the cut on my upper lip,
from this morning's shave, is still there.

*

Several years pass, during which I am often Teletransported.
I am now back in the cubicle, ready for another trip to Mars.
But this time, when I press the green button, I do not lose
consciousness. There is a whirring sound, then silence. I leave
the cubicle, and say to the attendant, "It's not working. What
did I do wrong?"

"It's working," he replies, handing me a printed card. This
reads: "The New Scanner records your blueprint without
destroying your brain and body. We hope that you will welcome
the opportunities which this technical advance offers."

The attendant tells me that I am one of the first people to use
the New Scanner. He adds that, if I stay an hour, I can use the
Intercom to see and talk to myself on Mars.

"Wait a minute," I reply, "If I'm here I can't *also* be on
Mars."

Someone politely coughs, a white-coated man who asks to
speak to me in private. We go to his office, where he tells me to
sit down, and pauses. Then he says: "I'm afraid that we're
having problems with the New Scanner. It records your
blueprint just as accurately, as you will see when you talk to
yourself on Mars. But it seems to be damaging the cardiac
systems which it scans. Judging from the results so far, though
you will be quite healthy on Mars, here on Earth you must
expect cardiac failure within the next few days."

The attendant later calls me to the Intercom. On the screen
I see myself just as I do in the mirror every morning. But there
are two differences. On the screen I am not left–right reversed.
And, while I stand here speechless, I can see and hear myself, in
the studio on Mars, starting to speak.

Since my Replica knows that I am about to die, he tries to console me with the same thoughts with which I recently tried to console a dying friend. It is sad to learn, on the receiving end, how unconsoling these thoughts are. My Replica assures me that he will take up my life where I leave off. He loves my wife, and together they will care for my children. And he will finish the book that I am writing. Besides having all of my drafts, he has all of my intentions. I must admit that he can finish my book as well as I could. All these facts console me a little. Dying when I know that I shall have a Replica is not quite as bad as simply dying. Even so, I shall soon lose consciousness, forever.

What Pushovers We Are!

The concerns around which Parfit's two-part story revolves are clearly those that haunted Strange Loop #642. In the first part, we worry along with Parfit whether he will truly exist again after he is atomized on Earth and the signals carrying his ultra-detailed blueprint have reached Mars and directed the construction of a new body; we fear that the newly built person will merely be someone who looks precisely like and thinks precisely like Parfit, but is not Parfit. Soon, however, we are relieved to find out that our worries are unfounded: Parfit himself made it, down to the last tiny scratch. Great! And how do we know that he did? Because he told us so! But which "he" is it that gives us this good news? Is this Derek Parfit the philosopher–author, or is it Derek Parfit the intrepid space voyager?

It is Parfit the space voyager. As it happens, Parfit the philosopher is just spinning a good yarn, doing his best to make it sound teddibly realistic, but we soon find out that, in fact, he doesn't believe in several parts of his own story. The second episode in his fantasy starts out by contradicting the first one. When we find out that the New Scanner, in contrast to the old one, *doesn't* destroy the "original", we go right along with the tacit idea that Parfit the intrepid space voyager has not voyaged anywhere. We don't question his stepping out of the cubicle on Earth, because *he's still here.*

Oh, but what mindless pushovers we are! Whereas we bought right into the "teleportation equals travel" theme of Episode I, falling for it hook, line, and sinker, we seem in Episode II to have unthinkingly taken the path of least resistance, which runs something like this: "If there are two different things that look like, think like, and quack like Derek Parfit, and if one of those things is located where we last saw Parfit and the other one of them is farther away, then, by God, the close one is obviously the *real* one, and the far one is just a *copy* — a clone, a counterfeit, an impostor, a fake."

This already is plenty of food for thought. If the copy on Mars is a fake in Episode II, why wasn't it a fake in Episode I? Why were we such suckers when we read Episode I? We naïvely bought into his wife's reassuring smile at breakfast, and then, when he stepped out of the Martian cubicle, that telltale nick on his face convinced us beyond all doubt. We took his word for it that it was indeed *he* who was stepping out of the cubicle. But what else could we have expected? Was the newborn body going to step out of the cubicle and proclaim, "Oh, horrors, I'm not me! I'm someone else who merely *looks* like me, and who has all of my memories stretching all the way back to childhood, and even my memory of breakfast only a few moments ago with my wife! I'm just a sham, but oh, such a good one!"

Of course the newly built Martian is not going to utter something incoherent like that, because he would have no way of knowing that he is a fake. He would believe for all the world that he *is* the original Derek Parfit, only moments ago disintegrated in the scanner on Earth. After all, that's what his brain would tell him, since it's identical to Derek Parfit's brain! This shows that we have to treat claims of personal identity, even ones coming straight from the first person's mouth, with extreme caution.

Well then, given our new no-nonsense attitude, what should we think about Episode II? We have been told that Parfit the would-be space voyager instead stepped out of the cubicle *on Earth,* and with heart damage. But how do we know that *that one* is Parfit? Why didn't Parfit the storyteller tell us the story from the vantage point of the new Martian who also calls himself "Derek Parfit"? Suppose the story had been told this way: "The moment I stepped out of the Martian cubicle, I was told the terrible news that the *other* Parfit — that poor fellow way down on Earth — had suffered cardiac damage in beaming me up here. I was devastated to hear it. Soon he and I were talking on the phone, and I found myself in the odd position of trying to console him just as I had recently consoled a dying friend..."

If it had been recounted sufficiently smoothly, we might not have been able to resist the thought that *this* body, the Mars-borne one, is really Derek Parfit. Indeed, Derek Parfit the skilled philosopher–storyteller might even have gotten us to imagine that the earthbound body with the damaged heart was merely a pretender to the Unique Soul linked by birth and by divine decree to the name "Derek Parfit".

Teleportation of a Thought Experiment across the Atlantic

It seems that the way in which a science-fiction scenario is related is crucial in determining our intuitions about its credibility. This is a point that my old colleague and friend Dan Dennett has made many times in his discussions of philosophers' crafty thought experiments. Indeed, Dan calls such carefully crafted fables *intuition pumps,* and he knows very well whereof he speaks, since he has dreamt up some of the most insight-providing intuition pumps in the field of philosophy of mind.

And I have to say that as I was typing Parfit's story from his 1984 book into this chapter, a little voice murmured softly to me, "Say, doesn't this remind you of Dan's foreword to *The Mind's I,* his ingenious teleportation fantasy that drew so many readers to our book when it came out in 1981?" And so after the Parfit story had been all typed in, I pulled a copy of *The Mind's I* off my shelf and reread its first few pages. I have to say that my jaw fairly dropped. It was exactly the same fantasy, only with planets reversed and sexes reversed, and told in a more American style. There was exactly the same bipartite structure, the first part featuring a "Teleclone Mark IV" that destroyed the original, and the second part featuring a new-and-improved version ("Mark V") that preserved the original.

What can I say? I love both of these stories, one from each side of the Atlantic, whether one is a "clone" of the other or their pedigrees are independent (though that seems unlikely, since *The Mind's I* is in Parfit's bibliography). In any case, now that I've got this little matter off my chest, I'll continue with my commentary on Parfit's provocative tale (and also, of course, on Dan's, thanks to the referential power of analogy).

The Murky Whereabouts of Cartesian Egos

The key question raised by Parfit's tale is this: "Where is space voyager Derek Parfit *really,* after the teletransportation has taken place in Episode II?" Put otherwise, which of the two claimants to being Parfit really *is* Parfit? In Episode I, Parfit the storyteller plants a most plausible-seeming answer, but then in Episode II he just as plausibly undermines that answer. At this point, you can probably almost hear Strange Loop #642 intensely identifying with the space traveler and screaming out, "Which of the two would I be?"

To my mind, one cannot claim to have said anything significant about the riddle of consciousness if one cannot propose (and defend) some sort of answer to this extremely natural-seeming and burning question. I think that by now you know my answer to the question, but maybe not. In any case, I'll let you ponder the issue for a moment, and meanwhile, I'll go on to tell you more or less how Parfit sees the matter.

This issue lies at the very core of Parfit's book, and the explanation of his position occupies about a hundred pages. The key notion to which he is opposed is what he dubs "Cartesian Pure Ego", or "Cartesian Ego", for short. To put it in my words, a Cartesian Ego constitutes one exact quantum of pure soul (also known as "personal identity"), and it is 100 percent indivisible and undilutable. In short, it is what makes you be *you* and me be *me*. My Cartesian Ego is mine and no one else's, has been from birth and will be to death, and that's that. It's my very own, completely private, unshared and unsharable, first-person world. It's the subject of my experiences. It's my totally unique inner light. You know what I mean!

I have to admit, parenthetically, that every time I see the phrase "Cartesian Ego", although my eyes perceive only one "g" there, some part of me invariably hallucinates another "g", and the image of an egg bubbles up in my brain — a "Cartesian Eggo", if you'll permit — a beautifully formed egg with a pristine white shell protecting a perfectly spherical and infinitely precious yolk at its core. In my strange distorted imagery, that yolk is the secret of human identity — and alas, Parfit's central mission in his book is to mercilessly crush the whole egg, and with it, the sacred yolk!

There are two questions that Parfit does his best to answer. The first one is: When Parfit is teleported to Mars in Episode I, is his Cartesian Ego teleported along with him, or is it destroyed along with his body? The second question, seemingly even more urgent and confusing, is this: When Parfit is teleported to Mars in Episode II, where does his Cartesian Ego go? Could it possibly go to Mars, abandoning him on Earth? In that case, who is it that remains on Earth? Or conversely, does Parfit's Cartesian Ego simply stay put on Earth? In that case, who, if anyone, is it that debarks from the cubicle on Mars? (Note that we are conflating the word "who" or the phrase "who it is" with the notion of a specific, uniquely identifiable Cartesian Ego.) The temptation to ask such questions (and to believe that these questions have objectively correct answers) is nearly irresistible, but nonetheless, the nearly universal intuitions that give rise to this temptation are what Parfit is out to crush in his book.

To be more specific, Parfit staunchly resists the idea that the concept of "personal identity" makes sense. To be sure, it makes sense in the everyday world that we inhabit — a world without telecloning or fanciful cut-and-paste operations on brains and minds. The fact is, we all more or less take for granted this notion of "Cartesian Ego" in our daily lives; it is built into our common sense, into our languages, and into our cultural backgrounds as profoundly, as tacitly, as seamlessly, and as invisibly as is the notion that time passes or the notion that things that move preserve their identity. But Parfit is concerned with investigating how well this primordial notion of Cartesian Egos stands up under extreme and unprecedented pressures. As a careful thinker, he is doing something analogous to what Einstein did when he imagined himself moving at or near the speed of light — he is pushing the limits of classical notions — and, like Einstein, he finds that classical worldviews do not always work in worlds that are very different from those in which they were born and grew.

Am I on Venus, or Am I on Mars?

In his hundred or so pages of musings on this issue, Parfit analyzes many thought experiments, some dreamt up by himself and some by other contemporary philosophers, and his analysis is always keen and clear. I have no intention to reproduce here those thought experiments or his analyses, but I will summarize what his conclusions are. The essence of his position is that when pushed to its limits, personal identity becomes an indeterminate notion. In extreme circumstances such as Episode II, the question "Which one of them am I?" has no valid answer.

This will be extremely unsatisfying and unsettling to many readers of Parfit's book, and to many readers of this book, as well. Our intuitions as we grew up on planet Earth have not prepared us for anything in the least like a nondestructive teleportation scenario, and so we clamor for a simple, straightforward answer, yet somehow we also intuit that none will be forthcoming. After all, we could invent Episode III, featuring a *destructive* teleportation scenario as in Episode I, but with signals simultaneously sent out to receiving stations on Venus and on Mars. In this scenario, shortly after the destruction of the original Parfit body and brain, two brand-new Parfits (both complete with shaving nick) would be assembled more or less simultaneously on the two planets, and now there really doesn't seem to be any valid claim of primacy for either one above the other (unless you argue that the *first one finished* should get to claim the honor of the Cartesian Ego, but in that case, we can simply posit that they are assembled in synchrony, thus barring that easy escape route).

To our everyday, downhome, SL #642–style minds, it's very stark and very simple: one of the Parfits is a fake. We cannot imagine being in two places at once, so we think (identifying ourselves with the intrepid voyager), "Either I've got to be the *Venus* one, or the *Mars* one, or *neither* one." And yet none of these answers is in the least satisfying to our classical intuitions.

Parfit's own answer is actually closer to the thought that I brusquely dismissed in the previous paragraph: that we are in two places at once! I say it's *closer* to that answer rather than saying that it *is* that answer, because Parfit's view, like mine in this book, is that these things that seem so black-and white to us actually come in shades of gray — it's just that in ordinary circumstances, things are always so close to being *pure* black or white that any hints of grayness remain hidden from view, not only thanks to the obvious external fact that we all have separate physical brains housed in separate skulls, but also thanks to an extensive web of linguistic and cultural conventions that collectively and subliminally insist that we each are exactly one person (this is the "caged-bird metaphor" of Chapter 18, and it's also the Cartesian Ego notion), and which implicitly discourage us from imagining any kind of blending, overlapping, or sharing of souls.

There is also, I cannot deny it, an absolute certainty, deep down in each one of us, that *I cannot be in two places at once.* In earlier chapters, I went to great lengths to give counterexamples of many sorts to this idea, and Parfit, too, takes great pains to give other kinds of evidence about the possibility of spread-out identity. In fact, he eschews the term "personal identity", preferring to replace it by a different term, one less likely to conjure up images of indivisible "soul quanta" (analogous to unique factory-issued serial numbers or government-issued identity cards). The term Parfit prefers is "psychological continuity", by which he means what I would tend to call "psychological similarity". In other words, although he doesn't propose anything that would smack of mathematics, Parfit essentially proposes an abstract "distance function" (what mathematicians would call a "metric") between personalities in "personality space" (or between brains, although at what structural level brains would have to be described in order for this "distance calculation" to take place is never specified, and it is hard to imagine what that level might be).

Using such a mind-to-mind metric, I would be very "close" to the person I was yesterday, slightly less close to the person I was two days ago, and so forth. In other words, although there is a great degree of overlap between the individuals Douglas Hofstadter today and Douglas Hofstadter yesterday, they are *not identical.* We nonetheless standardly (and reflexively) choose to consider them identical because it is so convenient, so natural, and so easy. It makes life much simpler. This convention allows us to give things (both animate and inanimate) fixed names and to talk about them from one day to the next without constantly having to update our lexicon. Moreover, this convention is ingrained in us when we are infants — at about the same Piagetian developmental stage as that in which we learn that when a ball rolls behind a box, it still exists even though it's not visible, and may even reappear on the other side of the box in a second or two!

The Radical Nature of Parfit's Views

To dismantle unconscious beliefs that are so deeply rooted and that have such a degree of primacy in our worldview is an extremely daunting and bold undertaking, comparable in subtlety and difficulty to what Einstein accomplished in creating special relativity (undermining, through sheer logic, our deepest and most unquestioned intuitions about the nature of time), and what a whole generation of brilliant physicists, with Einstein at their core, collectively accomplished in creating quantum mechanics (undermining our deepest and most unquestioned intuitions about the nature of causality and continuity). The new view that Parfit proposes is a radical reperception of what it is to *be*, and in certain ways it is extremely disturbing. In other ways, it is extremely liberating! Parfit even devotes a page or two to explaining how this radical new view of human existence has freed him up and profoundly changed his attitudes towards his life, his death, his loved ones, and other people in general.

In Chapter 12 of *Reasons and Persons*, boldly entitled "Why Our Identity Is Not What Matters", there is a series of penetrating musings, all of which have wonderfully provocative titles. Since I so much admire this book and its style, I will simply quote those section titles for you here, hoping thereby to whet your appetite to read it. Here they are: "Divided Minds"; "What Explains the Unity of Consciousness?"; "What Happens When I Divide?"; "What Matters When I Divide?"; "Why There is No Criterion of Identity that Can Meet Two Plausible Requirements"; "Wittgenstein and Buddha"; "Am I Essentially My Brain?"; and finally, "Is the True View Believable?"

Even though all eight of these sections are rife with insight, it is the last section that I admire the most, because in the end, Parfit asks himself if he really believes in the edifice he has just built. It is as if Albert Einstein had just realized that his own ideas would bring Newtonian mechanics crashing down in rubble, and then paused to ask himself, "Do I really have such deep faith in my own mind's pathways that I can believe in the bizarre, intuition-defying conclusions I have reached? Am I not being enormously arrogant in rejecting a whole self-consistent web of interlocked ideas that were carefully worked out by two or three centuries' worth of extraordinary physicists who came before me?"

And although Einstein was exceedingly modest throughout his lifetime, his answer to himself (though to my knowledge he never wrote any such introspective essay) was, in effect, "Yes, I do have this strange faith in my own mind's correctness. Nature *has* to be this way, no matter what other people have said before me. I have somehow been given the opportunity to glimpse the inner logic of nature more deeply and more accurately than anyone else before me has. I am unaccountably lucky in this fact, and though I take no personal credit for it, I do wish to publish it so that I may share this valuable vision with others."

Self-confidence, Humility, and Self-doubt

Parfit is far more prudent than this. His conclusions, to my mind, are just as radical as those of Einstein (although I find it a bit of a stretch to imagine radical ideas about the ineffability of personal identity leading to any marvelous technological consequences, whereas Einstein's ideas of course did), but he is not quite as convinced of them as Einstein must have been. He feels confident, but not absolutely confident, of his edifice of thought. He doesn't think it will start to shake and soon tumble down if he stands on it, but then again he admits that it just might do so. Let us hear him express himself on this topic in his own words:

> [The philosopher of mind Thomas Nagel] once claimed that, even if the Reductionist View is true, it is psychologically impossible for us to believe this. I shall therefore briefly review my arguments given above. I shall then ask whether *I* can honestly claim to believe my conclusions. If I can, I shall assume that I am not unique. There would be at least some other people who can believe the truth.
>
> [A few pages later] ....I have now reviewed the main arguments for the Reductionist View. Do I find it impossible to believe this View?
>
> What I find is this. I can believe this view at the intellectual or reflective level. I am convinced by the arguments in favour of this view. But I think it likely that, at some other level, I shall always have doubts....
>
> I suspect that reviewing my arguments would never wholly remove my doubts. At the reflective or intellectual level, I would remain convinced that the Reductionist View is true. But at some lower level I would still be inclined to believe that there must always be a real difference between some future person's being me, and his being someone else. Something similar is true when I look through a window at the top of a sky-scraper. I know that I am in no danger. But, looking down from this dizzying height, I am afraid. I would have a similar irrational fear if I was about to press the green button.
>
> ....It is hard to be serenely confident in my Reductionist conclusions. It is hard to believe that personal identity is not what matters. If tomorrow someone will be in agony, it is hard to believe that it could be an empty question whether this agony will be felt by *me*. And it is hard to believe that, if I am about to lose consciousness, there may be no answer to the question "Am I about to die?"

I must say, I find Parfit's willingness to face and to share his self-doubts with his readers to be extremely rare and wonderfully refreshing.

Morphing Parfit into Bonaparte

In the last paragraph quoted above, Parfit alludes to a thought experiment invented partly by philosopher Bernard Williams and partly by himself (in other words, invented by a Williams–Parfit hybrid who might be called "Bernek Willfits"), in which he is about to undergo a special type of neurosurgery whose exact nature is determined by a numerical parameter — namely, how many switches will be thrown. What do the individual switches do? Each one of them converts one of Parfit's personality traits into a different personality trait belonging to none other than Napoleon Bonaparte (and I literally mean "none other than", as I will shortly explain). For example, one switch makes Parfit far more irascible, another switch removes his repugnance at the idea of seeing people killed, and so forth. Note that in the previous sentence I used the proper noun "Parfit" and the pronoun "his", which presumably is an unambiguous reference to Parfit. However, the whole question here is whether or not such usages are legitimate. If switch after switch were thrown, converting Parfit more and more into Napoleon, at what stage would he — or rather, at what stage would *this slowly morphing person* — simply *be* Napoleon?

As I have already made clear, asking exactly where along the line the switchover would take place makes no sense from Parfit's point of view, for what matters is psychological continuity (*i.e.*, proximity in that quasi-mathematical space of personalities or brains that I suggested a little while ago), and that is a feature that comes in all shades of gray. It is not a 0/1 matter, not all-or-nothing. A person can be *partly* Derek Parfit and *partly* Napoleon Bonaparte, and drifting from the one to the other as the switches are thrown. And this doesn't merely mean that this person is becoming more and more *like* Napoleon Bonaparte — it means that this person really is slowly becoming Bonaparte himself.

In Parfit's view, the Cartesian Ego of Napoleon is not indivisible, nor is that of Derek Parfit. Rather, it is as if there were a slider on a wire, and the two individuals (who are *not* really "individuals" in the etymological sense, since the word means "undividable") can be merged or morphed arbitrarily by sliding that slider to any desired position on the wire. The result is a hybrid person, a tenth or a third or halfway or three-quarters of the way between the two ends — whatever proportions one wishes, ranging from Derek Parfit to Deren Parfite to Dereon Parpite to Deleon Parapite to Doleon Paraparte to Daoleon Panaparte to Dapoleon Ponaparte to Napoleon Bonaparte.

Most people, unlike Parfit, want there to be and are convinced that there *must* be, at each point along the spectrum of cases, a sharp yes–no answer to the question, "Is this person Derek Parfit?" This is the classical view, of course — the view that takes for granted the notion of Parfit's own Cartesian Ego. And so most people are put into the awkward position of having to say that there would be a particular spot along the wire at which all of a sudden, without warning, at the instant when the slider passes it, the Cartesian Ego of Parfit would poof out of existence, to be replaced by that of Napoleon Bonaparte. Where only a moment ago we had been dealing with a somewhat personality-modified Derek Parfit, but still and all a Derek Parfit who genuinely felt Derek Parfit's feelings, now we suddenly have a modified Napoleon Bonaparte, and he feels *Napoleon's* feelings, and not Parfit's whatsoever!

The Radical Redesign of Douglas R. Hofstadter

The intuitions being pushed here are very emotional and run very deep in our culture and our whole view of life. It gets particularly intense for me when I insert myself into this scenario and start imagining the personality-trait substitutions that a neurosurgeon might carry out by throwing one switch after another.

For example, I begin by imagining that, upon the throwing of Switch #1, my love for Chopin and Bach is replaced by a visceral loathing of their music and that instead, a sudden yet powerful veneration for Beethoven, Bartók, Elvis, and Eminem flowers in "my" brain.

Next, I imagine that Switch #2 causes me every single weekend (and every other spare moment as well) to elect, instead of designing ambigrams or working hard on my book about being a strange loop, to spend hours on end watching professional football games on a huge-screen television and delightedly ogling all the busty babes in the beer ads.

And then (Switch #3) I imagine my political leanings being turned on their head, including my decades of crusading against sexist language. Now, I come out with "you guys" every other sentence, and anyone who objects to it I chortlingly deride as "a politically correct monkey" (as you might imagine, that's just one of the milder epithets I use).

With the next switch, I jettison my lifelong inclination towards vegetarianism and trade it in for a passion for shooting deer and other wild animals — and of course the larger they are, the better. Thus, after Switch #4 has been thrown, I just *adore* toppling elephants and rhinos with my trusty rifle! The most fun thing in the world! And each time one of the noble beasts bows humbly down to my triumphant bullets, I give one of those "I'm great" jerks with my arm, which one so often sees when a football player scores a touchdown.

And lastly, needless to say, after Switch #5 has been thrown, I totally agree with John Searle's Chinese Room experiment, and I think that Derek Parfit's ideas about personal identity are a complete crock. Oh, I forgot — can't do that, since I never think about philosophical issues at all!

You may have noticed that when I discussed Switch #1, I put quotes around the word "my" when talking about the brain in which a veneration for Ludwig, Béla, Elvis, and Eminem flowers. From there on out, though, I didn't bother with the quote marks, but I probably should have. After all, everything I suggested in the paragraphs above is the diametric opposite of what I consider *core me-ness*. Letting go of even one of these traits is enough to make me think, "That person wouldn't be me any more. That *couldn't* be me. That is incompatible with the deepest fiber of my being."

Of course we can imagine milder changes, such as an alternate life in which I somehow never ran into Prokofiev's violin concerto #1. That would be another version of me, and surely a more impoverished one, but it would still feel like me, to this me. Or we can imagine that I still eat hamburgers on occasion but feel guilty about it, or that once in a blue moon I voluntarily turn on a football game on TV. These are shades of gray that create a halo of "possible Dougs" around the Doug that I happen to have become, thanks to a million accidental events that have befallen me over the decades, and thanks to hundreds of particular individuals who happen to have entered my life (and millions of others who never did, not to mention an infinite number of counterfactual individuals who never entered my life!). We don't normally think of "who/what/how I am" in such shades of gray, but there they are, spelled out a bit, in my case.

On "Who" and on "How"

I might add, by the way, that I think the word "who" is sometimes granted a bit too much subliminal power, in much the way as are the personal pronouns "he" and "she" (you may recall my brief interchange with Kellie about pronouns applied to animals, in Chapter 1). In the 1980's, Pamela McCorduck wrote a history of artificial intelligence with the provocative and ingenious title "Machines Who Think". The word "who" in the title conjures up an image radically different from our knee-jerk associations with standard machines such as can-openers, refrigerators, typewriters, and even computers; it suggests that with at least certain machines, there is someone "in there", or as Thomas Nagel would say, "there is something it is like to be that machine" (a hard phrase to translate into other languages, by the way). It implicitly suggests, once again, a sharp, black-and-white dichotomy between a set of hypothetical "machines *that* think" (such machines would *merely* think but would have no inner life) and a different set of hypothetical "machines *who* think" (these machines *would* have an inner life, and each one would be a *particular someone*).

It has often seemed to me that ultimately, when I am thinking about *who* my closest friends are, it all comes down to *how* they are — how they smile, how they talk, how they laugh, how they listen, how they suffer, how they share, and so on. I think to myself that the innermost essence of each friend is made up of thousands of such "how"'s, and that that collection of "how"'s is the answer — the *full* answer — to "Who is this person?"

It may seem that this is purely a third-person, external perspective, and that it takes away, or even denies, the whole first-person perspective. It may seem to short-change or even to casually dismiss the "I". I don't think so, however, for I think that even *to itself,* that is all an "I" is. The rub is, an "I" is very good at convincing itself that it is a lot more than that — in fact, that is the entire business that the word "I" is in! "I" has a vested interest in continuing this scam (even if it is its own victim)!

Double or Nothing

At long last, we return to the Venus-versus-Mars enigma of Episode III. I have already told you that Parfit somewhat sidesteps the question by simply denying the existence of Cartesian Egos, and thus saying that the question has no meaningful answer. But in his book he also refers quite often to what he terms "double survival", which means essentially that he is simultaneously in two places at once. More than once, he writes that double survival is hardly equivalent to death (which would be *no* survival), and that the number two should not be conflated with the number zero! So what is he really saying? Is he saying that there is no answer to the question, or is he saying that in fact he has been doubled, and there are now two Derek Parfits?

It's hard for me to figure this out since I think he says both things often enough that one could argue it either way. But where do *I* come down on this issue? I think I come down on the "two me's" side. At first, this almost sounds as if I am embracing the Cartesian Ego theory, just imagining that the egg is cloned and two identical Cartesian Egos come to exist, one on Venus and one on Mars. But then SL #642 would start screaming, "Which one is me?" It sounds as if I haven't answered the question at all, or as if I want to have my egg on Mars and eat it too, on Venus.

In order to regain some semblance of consistency, I have to return to SL #641's theme in the dialogue, which is that the "I" notion is, fundamentally and in the end, a hallucination. Let's let Episode III, my teleportation scenario with fresh copies on Venus and Mars and no copy left on Earth, apply to me instead of to Parfit. In that case, each of the new brains — the one on Mars and the one on Venus — is convinced that it is *me*. It feels just like it always felt to be me. The same old urge to say, "I am *here* and not *there*" zooms up in both brains as automatically as when someone taps my knee and my leg jerks upwards. But knee-jerk reflex or not, the truth of the matter is that there is no *thing* called "I" — no hard marble, no precious yolk protected by a Cartesian eggshell — there are just tendencies and inclinations and habits, including verbal ones. In the end, we have to believe both Douglas Hofstadters as they say, "This one *here* is me," at least to the extent that we believe the Douglas Hofstadter who is right now sitting in his study typing these words and saying to you in print, "This one here is me." Saying this and insisting on its truth is just a tendency, an inclination, a habit — in fact, a knee-jerk reflex — and it is no more than that, even though it seems to be a great deal more than that.

Ultimately, the "I" is a hallucination, and yet, paradoxically, it is the most precious thing we own. As Dan Dennett points out in *Consciousness Explained*, an "I" is a little like a bill of paper money — it *feels* as if it is worth a great deal, but ultimately, it is just a social convention, a kind of illusion that we all tacitly agree on without ever having been asked, and which, despite being illusory, supports our entire economy. And yet the bill is just a piece of paper with no intrinsic worth at all.

Trains Who Roll

In Chapters 15 through 18, I argued that each of us is spread out and that, despite our usual intuitions, each of us is housed at least partially in different brains that may be scattered far and wide across this planet. This viewpoint amounts to the idea that one *can* be in two places at once, despite our initial knee-jerk rejection of such a crazy-sounding thought. If being in two or more places at once seems to make no sense, think about reversing the roles of space and time. That is, consider that you have no trouble imagining that you will exist tomorrow and also the next day. Which one of those future people will *really* be you? How can two *different* you's exist, both claiming your name? "Ah," you reply, "but I will shortly be getting there, like a train pulling through different stations." But that just begs the question. Why is it the *same* train, if in the meantime it has dropped some passengers off and picked others up, perhaps changed a car or two, maybe even its locomotive? It is simply *called* "Train 641", and *that's* why it is "the same train". It's a linguistic convention, and a very good one, too. It is a very natural convention in the classical world in which we exist.

If Train 641, heading east from Milano, always were to split up in Verona into two pieces, one that headed north to Bolzano and one that continued eastwards to Venice, then we would probably not call either half "Train 641" any longer, but would give them separate numbers. But we could also call them "Train 641a" and "Train 641b", or even just leave them both as "Train 641". It might happen, after all, that upon reaching Bolzano, the northern half always veers suddenly eastwards, and likewise that upon reaching Venice, the eastern half always veers suddenly northwards, and the two halves always rejoin and fuse together in Belluno, on their way — or rather, on *its* way — to Udine!

You may object that trains have no *inner* perspective on the matter — that "641" is just a third-person label rather than a first-person point of view. All I can say is, this is a very tempting viewpoint, but it is to be resisted. Trains *who* roll and trains *that* roll are the same thing, at least if they have sufficiently rich representational systems that allow them to wrap around and self-represent. Most trains today don't (in fact none of them do), so we don't usually give them the benefit of the "who" pronoun. But maybe someday they will, and then we will. However, the transition from one pronoun to the other won't be sharp and sudden; it will be gradual, like the fading of the belief in Cartesian Egos as people grow in sophistication.

The Glow of the Soular Corona

It may strike you that this whole chapter has been predicated upon such weird science-fiction scenarios that it has no bearing at all upon how we think about the real world of real human beings, and their real lives and deaths. But I believe that that is mistaken.

I have a close friend whose aging father Jim has Alzheimer's Disease. For some years my friend has been sadly watching his father lose contact, bit by bit, with one aspect after another of the reality that only a few years ago constituted the absolute bedrock, the completely reliable *terra firma*, of his inner life. He no longer knows his address, he has lost his former understanding of such mundane things as credit cards, and he isn't quite sure who his children are, though they look vaguely familiar. And it is all getting dimmer, never brighter.

Perhaps Jim will forget his own name, where he grew up, what he likes to eat, and much more. He is heading into the same terrible, thick, all-enveloping fog that former President Reagan lived in during the closing few low-huneker years of his life. And yet, something of Jim is surviving strongly — surviving in *other* brains, thanks to human love. His easy-going sense of humor, his boundless joy at driving the wide open spaces of the

prairies, his ideals, his generosity, his simplicity, his hopes and dreams —
and (for what it's worth) his understanding of credit cards. All of these
things survive at different levels in many people who, thanks to having
interacted with him intimately over many years or decades, constitute his
"soular corona" — his wife, his three children, and his many, many friends.

Even before Jim's body physically dies, his soul will have become so
foggy and dim that it might as well not exist at all — the soular eclipse will
be in full force — and yet despite the eclipse, his soul will *still* exist, in
partial, low-resolution copies, scattered about the globe. Jim's first-person
perspective will flicker in and out of existence in other brains, from time to
time. *He* will exist, albeit in an extremely diluted fashion, now here, now
there. *Where will Jim be?* Not very much anywhere, admittedly, but to some
extent he will be in many places at once, and to different degrees. Though
terribly reduced, he will be wherever his soular corona is.

It is very sad, but it is also beautiful. In any case, it is our only
consolation.

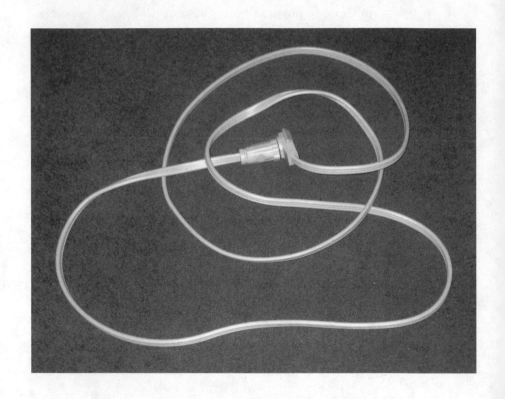

CHAPTER 22

A Tango with Zombies and Dualism

ବ ବ ବ

Pedantic Semantics?

TO ARGUE over whether the appropriate relative pronoun to apply to some hypothetical thinking machine one day in the future will be "who" or merely "which" would doubtless strike certain people as the quintessence of pedantic semantic quibbling, yet there are other people for whom the question would raise issues of life-or-death importance. Indeed, this is a quintessentially semantic issue, in that it involves deciding what verbal label to apply to something never seen before, but since category assignments go right to the core of thinking, they are determinant of our attitude toward each thing in the world, including such matters as life and death. For that reason, I feel that this pronoun issue, even if it is "merely semantics", is of great importance to our sense of who or what we are.

The well-known Australian philosopher of mind David Chalmers, which not only is a cherished friend but also is my former doctoral student, has devoted many years to arguing for the provocative idea that there could be both "machines *that* think" and also "machines *who* think". For me, the notion of both types of machine coexisting makes no sense, because, as I declared in Chapter 19, the word "thinking" stands for the dancing of symbols in a cranium or careenium (or some such arena), and this is also what is denoted by the word "consciousness". Since being conscious merits the use of the pronoun "who" (and also, of course, the pronouns "I", "me", and so on), so does thinking — and that settles the question for me. In other words, "machine that thinks" is an incoherent phrase because of its relative pronoun, and if some day there really *are* machines that think, then by definition they will be machines *who* think.

Two Machines

Dave Chalmers explores these issues in an unprecedented new fashion. He paints a picture of a world that has two machines identical down to the last nail, transistor, atom, and quark, and these two machines, sitting side by side on an old oaken table in Room 641 of the Center for Research into Consciousness and Cognetics at Pakistania University, are carrying out exactly the same task. For concreteness' sake, let's say both machines are struggling to prove, using informal geometrical insights rather than formal algebraic manipulations, the simple but surprising "chord–angle theorem" of Euclidean geometry, which states that if a point (*A* in the figure below) moves along an arc of a circle, then the angle (*α*) subtended by a fixed chord (*BC*) that the point is "looking at" as it moves along will be constant.

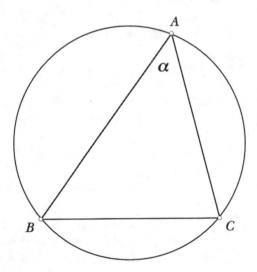

I chose this elementary but elegant theorem because it is one that Dave and I discussed together with great pleasure many years ago, and some of his comments on it gave me insights that literally changed my life. In fact, that fateful fork in the road way back when allows me to imagine Switch #6, the throwing of which would subtract from my brain all knowledge of this theorem and all the subsequent passion for geometry that was sparked by my thinking carefully about it…

As I was saying, these two exactly identical machines are launched on this task in the exact same terasecond by an atomic clock, and they proceed in exact lockstep synchrony towards its solution, simulating, let us say, the exact processes that took place in Dave Chalmers' own brain when he first found an insight-yielding visual proof. The details of the program running in both machines are of no import to us here; what does matter is that Machine Q (it stands for "qualia") is actually *feeling* something, whereas Machine Z (it rhymes with "dead") is feeling nothing. This is where Dave's ideas grow incomprehensible to me.

Now I have to admit that in order to make it a bit easier to envision, I have slightly altered the story that Dave tells. I placed these two machines side by side on the old oaken table in Room 641 of CRCC, while Dave never does that. In fact, he would protest, saying something such as, "It's bloody incoherent to postulate two identical machines running identical processes on the very same oaken table with one of them feeling something and the other one not. That violates the laws of the universe!"

I fully accept this objection and plead guilty to having distorted Dave's tale. To atone for my sin and to turn my story back into his, I first remove one of the machines from the old oaken table in Room 641. Let's call the machine who remains, no matter what we'd called it before, "Machine Q". Now (following Dave), we take a rather unexpected step: we imagine a different but isomorphic (*i.e.*, "separate but indistinguishable") universe. We'll call the first one "Universe Q" and the new one "Universe Z". Both universes have exactly the same laws of physics, and in each universe the laws of physics are all one needs to know in order to predict what will happen, given any initial configuration of particles.

When I say these two universes are indistinguishable, one of the myriad consequences is that Universe Z, just like Universe Q, has a Milky Way galaxy, a star therein called "Sol" with a nine-planet solar system whose third planet is called "Earth", and on Universe Z's Earth there is a Pakistania University with a Center for Research into Consciousness and Cognetics, and in it, good old Room 641. There is even "the same" old oaken table, and there, lo and behold, is "the same machine" sitting on it. Surely you see it, do you not? But since this machine is in Universe Z, we will call it "Machine Z", just so that we have different names for these indistinguishable machines located in indistinguishable surroundings.

Now of course we can't launch Machines Q and Z at "the same instant", because they belong to different universes with independent timelines, but luckily these two universes have exactly the same laws of physics, so synchronization isn't necessary. We just start them up and let them do their things. As before, they do *exactly the same thing,* since they are both following the same laws of physics, and physics suffices to determine all behavior down to the finest detail. And yet, what do you suppose turns out to be the case? Oddly enough, although both machines do exactly the same thing down to the quark level and far beyond, Machine Q enjoys *feelings* about what it is doing while Machine Z does not. Machine Q is in fact ecstatic, whereas Machine Z feels nothing. That is, Zilch. Zero.

"How is that possible?", you might ask. I too, no less bewildered, ask the same question. But Dave most cheerfully explains: "Oh, it's because the universe in which Machine Q exists has something extra, on top of the laws of physics, that allows *feelings* to accompany certain types of physical processes. Even though these feelings don't have and *can't* have any effect on anything physical, they are nonetheless real, and they are really there."

In other words, although physics is identical in Universes Q and Z, there are no feelings anywhere in Universe Z — just empty motions. Thus Machine Z mouths all the same words as Machine Q does. It *claims* to be ecstatic about its proof (exactly as does Machine Q), and it goes on and on about the beauty it sees in it (exactly as does Machine Q) — but in fact it is feeling nothing. Its words are all hollow.

Two Daves

What is this extra ingredient that makes Universes Q and Z so vitally different? Dave doesn't say, but he tells us that it is the very stuff of consciousness — I'll dub it *élan mental* — and if you're born in a universe *with* it, then lucky you, whereas if you're born in a universe without it, well, tough luck, because there's no you-ness, no I-ness, no who-ness, no me-ness (or he-ness or she-ness) in you — there's just *it*-ness. Despite this enormous difference, all the objective phenomena in both universes are identical. Thus there are Marx Brothers movies in both of these universes, and when Z-people in Universe Z look at *A Night at the Opera,* they laugh exactly the same as when Q-people in Universe Q look at *A Night at the Opera.*

Most deliciously ironically of all, just as there is a Dave Chalmers in Universe Q (the one in which *we* live), there is also a Dave Chalmers in Universe Z, and it goes around the world giving lectures on why there *is* feeling in the universe in which it was born but *no* feeling in the isomorphic universe into which its unfortunate "zombie twin" was born. The irony, of course, is that Universe Z's Dave Chalmers is lying through its teeth, yet without having the foggiest idea it's lying. Although it *believes* it is conscious, in truth it is not. Sadly, this Dave is an innocent victim of the *illusion* of consciousness, which is nothing but a trivial by-product of having a deeply entrenched strange loop in its brain, whereas its isomorphic counterpart in Universe Q, using the same words and intonations, is telling the truth, for *he* truly *is* conscious! Why? Because he not only has a strange loop in his brain but also — lucky fellow! — lives in a universe with *élan mental.*

Now please don't think I am poking fun at my friend Dave Chalmers, for Dave truly *does* go around the world visiting philosophy departments, giving colloquia in which he most gleefully describes his "zombie twin" and chortles merrily over that twin's helpless deludedness, since the zombie twin gives word for word and chortle for chortle the very same lecture, believing every word of it but not feeling a thing. Dave is a very insightful thinker, and he is every bit as aware as I am of the *seeming* craziness of his distinction between Universes Q and Z, between Machines Q and Z, and between himself and his alleged zombie twin, but whereas I find all of this unacceptably silly, Dave is convinced that, outrageous though such a distinction seems at first to be, Universe Q's mysterious, nonphysical, and causality-lacking extra ingredient *élan mental* — a close kin to the notion of "feelium" discussed by Strange Loops #641 and #642 — is the missing key to the otherwise inexplicable nature of consciousness.

The Nagging Worry that One Might Be a Zombie

Of late, not a few philosophers of mind have, like Dave, been caught in a tidal wave of fascination with this notion called "zombies". (Actually, it's more like "the notion we love to hate".) It seems to have originated in voodoo rites in the Caribbean and to have spread from there to horror films and then to the world of literature. A Web search will quickly give you all the information you want, and most of it is pretty funny.

Basically, a zombie is an unconscious humanoid who acts — oops, I mean "that acts" — as if it were conscious. There's no one home inside a zombie, though from the outside one might think so. I have to admit, once in a blue moon I've run into someone whose glazed eyes give me the eerie sense that there's no one home behind them. Of course, I don't take such impressions seriously. Yet for many philosophers, the hollow, glazed-eyes image has turned into a paradigmatic fear, and today there is no paucity of philosophers of mind who find the notion of a zombie not just painfully abhorrent but in fact perplexingly coherent. These philosophers are so troubled by the specter of zombies that they have taken as their sacred mission to show that our world is not the cold and empty Universe Z, but the warm and fuzzy Universe Q.

Now you might say that this whole book buys into the cold, glazed-eyes, zombie vision of human beings, since it posits that the "I" is, when all is said and done, an illusion, a sleight of mind, a trick that a brain plays on itself, a hallucination hallucinated by a hallucination. That would mean that we all *are* unconscious but we all *believe* we are conscious and we all *act* conscious. All right, fine. I agree that that's a fair characterization of my views. But the swarm of zombie-fearing philosophers all want our inner existence to be richer than that. They claim that they can easily conceive of a cold, icy universe populated solely by nightmarishly hollow zombies, yet not distinguishable in any objective way from our own universe; at the same time, they insist that such is not the universe we live in. According to them, we humans don't just *act* conscious or *claim* to be conscious; we truly *are* conscious, and that's another matter entirely. Therefore Hofstadter and Parfit are wrong, and David Chalmers is right.

Well, I think Dan Dennett's criticism of such philosophers hits the nail on the head. Dan asserts that these thinkers, despite their solemn promises, are *not* conceiving of a world identical to ours but populated by zombies. They don't even seem to try very hard to do so. They are like SL #642, who, when imagining what a strange loop would say on looking at a brilliant purple flower, chose the dehumanizing verb "drone" to describe how it would talk, and likened its voice to a mechanical-sounding recorded voice in a hated phone menu tree. SL #642 has a stereotype of a strange loop as soul-less, and that prejudice rides roughshod over the image of perfectly natural, normal human behavior. Likewise, philosophers who fear zombies fear them because they fear the mechanical drone, the glazed eyes, and the frigid inhumanity that would surely pervade a world of mere zombies — even if, only a moment before, they signed off on the idea that such a world would be *indistinguishable* from our world.

Consciousness Is Not a Power Moonroof

In debates about consciousness, one of the most frequently asked questions goes something like this: "What is it about consciousness that helps us *survive*? Why couldn't we have had all this cognitive apparatus but simply been machines that don't feel anything or have any experience?" As I hear it, this question is basically asking, "Why did consciousness get *added on* to brains that reached a certain level of complexity? Why was consciousness thrown into the bargain as a kind of *bonus*? What *extra* evolutionary good does the possession of consciousness contribute, if any?"

To ask this question is to make the tacit assumption that there could be brains of any desired level of complexity that are *not* conscious. It is to buy into the distinction between Machines Q and Z sitting side by side on the old oaken table in Room 641, carrying out identical operations but one of them doing so *with* feeling and the other doing so *without* feeling. It assumes that consciousness is some kind of orderable "extra feature" that some models, even the fanciest ones, might or might not have, much as a fancy car can be ordered with or without a DVD player or a power moonroof.

But consciousness is not a power moonroof (you can quote me on that). Consciousness is not an optional feature that one can order independently of how the brain is built. You cannot order a car with a two-cylinder motor and then tell the dealer, "Also, please throw in *Racecar Power*® for me." (To be sure, nothing will keep you from placing such an order, but don't hold your breath for it to arrive.) Nor does it make sense to order a car with a hot sixteen-cylinder motor and then to ask, "Excuse me, but how much more would I have to throw in if I also want to get *Racecar Power*®?"

Like my fatuous notion of optional "*Racecar Power*®", which in reality is nothing but the upper end of a continuous spectrum of horsepower levels that engines automatically possess as a result of their design, consciousness is nothing but the upper end of a spectrum of self-perception levels that brains automatically possess as a result of their design. Fancy 100-huneker-and-higher racecar brains like yours and mine have a lot of self-perception and hence a lot of consciousness, while very primitive wind-up rubber-band brains like those of mosquitoes have essentially none of it, and lastly, middle-level brains, with just a handful of hunekers (like that of a two-year-old, or a pet cat or dog) come with a modicum of it.

Consciousness is not an add-on option when one has a 100-huneker brain; it is an inevitable emergent consequence of the fact that the system has a sufficiently sophisticated repertoire of categories. Like Gödel's strange loop, which arises *automatically* in any sufficiently powerful formal system of number theory, the strange loop of selfhood will automatically arise in any sufficiently sophisticated repertoire of categories, and once you've got self, you've got consciousness. *Élan mental* is not needed.

Liphosophy

Philosophers who believe that consciousness comes from something over and above physical law are dualists. They believe we inhabit a world like that of magical realism, in which there are two types of entities: magical entities, which possess *élan mental*, and ordinary entities, which lack it. More specifically, a magical entity has a nonphysical soul, which is to say, it is imbued with exactly one "dollop of consciousness" (a dollop being the standard unit of *élan mental*), while ordinary entities have no such dollop. (Dave Chalmers believes in two types of universe rather than two types of entity in a single universe, but to me it's a similar dichotomy, since we can consider various universes to be entities inside a greater "meta-verse".) Now I should like to be very sure, dear reader, that you and I are on the same page about this dichotomy between magical and ordinary entities, so to make it maximally clear, I shall now parody it, albeit ever so gently.

Imagine a philosophical school called "liphosophy" whose disciples, known as "liphosophers", believe in an elusive — in fact, undetectable — and yet terribly important nonphysical quality called *Leafpilishness* (always with a capital "L") and who also believe that there are certain special entities in our universe that are imbued with this happy quality. Now, not too surprisingly, the entities thus blessed are what you and I would tend to call "leaf piles" (with all the blurriness that any such phrase entails). If you or I caught a glimpse of such a thing and were in the right mood, we might exclaim, "Well, what do you know — a leaf pile!" Such an enthusiastic outburst would more than suffice for you and me, I suspect. We would not be likely to dwell much further on the situation.

But for a liphosopher, it would lead to the further thought, "Aha! So there's another one of those rare entities imbued with one dollop of Leafpilishness, that mystical, nonphysical, other-worldly, but very real aura that doesn't ever attach itself to haystacks, reams of paper, or portions of French fries, but only to piles of leaves! If it weren't for Leafpilishness, a leaf pile would be nothing but a motley heap of tree debris, but thanks to Leafpilishness, all such motley heaps become Leafpilish! And since each dollop of Leafpilishness is different from every other one, that means that each leaf pile on Earth is imbued with a totally unique identity! What an amazing and profound phenomenon is Leafpilishness!"

No matter what your opinion is on consciousness, reader, I suspect you would scratch your head at the tenets of liphosophy. It would be unnatural if you didn't wonder, "What is this nutty Capitalized Essence all about? What follows from having this invisible, undetectable aura?" You would also be likely to wonder, "Who or what agent in nature decides which entities in the physical world will receive dollops of Leafpilishness?"

Such musings might lead you to posing other hard questions, such as: What exactly constitutes a leaf pile? How many leaves, and of what size, does it take to make a leaf pile? Which leaves belong to it, and which ones do not? Is "belonging" to a given leaf pile always a black-and-white matter? What about the air between the leaves? What about the dirt on a leaf? What if the leaves are dry, and a few (or half, or most) of them have been crushed into tiny pieces? What if there are two neighboring leaf piles that share a few leaves between them? Is it 100 percent clear at all times where the borders of a leaf pile are? In short, how does Mother Nature figure out in a perfectly black-and-white fashion what things are worthy recipients of dollops of Leafpilishness?

If you were in a yet more philosophical mood, you might ask yourself questions such as: What would happen if, through some freak accident or bizarre mistake, a dollop of Leafpilishness got attached to, say, a leaf pile with an ant crawling in it (that is, to the *compound* entity consisting of leaf pile plus ant)? Or to just the upper two-thirds of a leaf pile? Or to a pile of seaweed? Or to a child's crumbly sand castle on the beach? Or to the San Francisco Zoo? Or to Andromeda galaxy? Or to my dentist appointment next week? What would happen if *two* dollops of Leafpilishness accidentally got attached to just *one* leaf pile? (Or zero dollops, yielding a "zombie" leaf pile?) What dreadful or marvelous consequences would ensue?

I suspect, reader, that you would not take seriously a liphosopher who argued that Leafpilishness was a central and mystical aspect of the cosmos, that it transcended physical law, that items possessing Leafpilishness were inherently different from all other items in the universe, and that each and every leaf pile had a unique identity — thanks not to its unique internal composition but rather to the particular dollop of Leafpilishness that had been doled out to it from who knows where. I hope you would join me in saying, "Liphosophy is a motley belief pile!" and in paying it no heed.

Consciousness: A Capitalized Essence

So much for liphosophers. Now let's turn to philosophers who see consciousness as an elusive — in fact, undetectable — and yet terribly important nonphysical aspect of the universe. In order to distinguish *this* notion of consciousness from the one I've been talking about all through this book, I'm going to capitalize it: "Consciousness". Whenever you see this word capitalized, just think of the nonphysical essence called *élan mental,* or else make an analogy to *Racecar Power*® or Leafpilishness; either way, you won't be far off.

At this point, I have to admit that I have a rather feeble imagination for Capitalized Essences. In trying to picture in my mind a physical object imbued with a nonphysical essence (such as Leafpilishness or *élan mental*), I inadvertently fall back on imagery derived from the purely physical world. Thus for me, the attempt to imagine a "dollop of Consciousness" or a "nonphysical soul" inevitably brings to mind a translucent, glowing swirl of haze floating within and perhaps a little bit around the physical object that it inhabits. Mind you, I know all too well that this is most wrong, since the phenomenon is, by definition, *not* a physical one. But as I said, my imagination is feeble, and I need this kind of physical crutch to help it out.

In any case, the idea of a sharp dichotomy between objects imbued with dollops of Consciousness and those deprived of such leads to all sorts of puzzling riddles, such as the following:

Which physical entities possess Consciousness, and which ones do not? Does a whole human body possess Consciousness? Or is it just the human's *brain* that is Conscious? Or could it be that only a certain *part* of the brain is Conscious? What are the exact boundaries of a Conscious physical entity? What organizational or chemical property of a physical structure is it that graces it with the right to be invaded by a dollop of Consciousness?

What mechanism in nature makes the elusive elixir of Consciousness glom onto some physical entities and spurn others? What wondrous pattern-recognition algorithm does Consciousness possess so as to infallibly recognize just the proper kinds of physical objects that deserve it, so it can then bestow itself onto them?

How does Consciousness know to do this? Does it somehow go around the physical world in search of candidate objects to glom onto? Or does it shine a metaphorical flashlight metaphorically down at the world and examine it piece by piece, occasionally saying to itself, "Aha! So *there's* an entity that deserves one standard-size dollop of me!"

How does Consciousness get attached to some specific physical structure and not accidentally onto nearby pieces of matter? What kind of "glue" is used to make this attachment? Can the "glue" possibly wear out and the Consciousness accidentally fall off or transfer onto something else?

How is *your* Consciousness different from *my* Consciousness? Did our respective dollops come with different serial numbers or "flavors", thus establishing the watertight breach between us? If your dollop of Consciousness had been attached to my brain and vice versa, would you be writing this and I reading it?

How does Consciousness coexist with physical law? That is, how does a dollop of Consciousness push material stuff around without coming into sharp conflict with the fact that physical law *alone* would suffice to determine the behavior of those things?

A Sliding Scale of Élan Mental

Now some readers might say that I am not giving *élan mental* (a.k.a. Consciousness) enough respect. They might say that there are gradations in the dispensation of this essence, so that some entities receive a good deal of it while others get rather little or none of it. It's not just all-or-nothing;

rather, the amount of Consciousness attached to any given physical structure is not precisely one dollop but can be any number of dollops (including fractional amounts). That's progress!

And yet, for such readers, I would still have numerous questions, such as the following:

How is it determined exactly how many dollops (or fractional dollops) of Consciousness get attached to a given physical entity? Where are these dollops stored in the meantime? In other words, where is the Central Consciousness Bank?

Once a certain portion of Consciousness has been dished out to a recipient entity (Ronald Reagan, a chess-playing computer, a cockroach, a sperm, a sunflower, a thermostat, a leaf pile, a stone, the city of Cairo), is it a permanent allotment, or is the size of the allotment variable, depending on what physical events take place involving the recipient? If the recipient is in some way altered, does its allotment (or part of it) revert to the Central Consciousness Bank, or does it just float around forevermore, no longer attached to a physical anchor? And if it floats around unattached, does it retain traces of the recipient to which it was once attached?

What about people with Alzheimer's disease and other forms of dementia — are they still "just as Conscious" as they always were, until the moment of their death? What makes something be "the same entity" over long periods of time, anyway? Who or what decreed that the changing pattern that over several decades was variously known as "Ronnie Reagan", "Ronald Reagan", "Governor Reagan", "President Reagan", and "Ex-President Reagan" was "one single entity"? And if it truly, objectively, indisputably *was* one single entity no matter how ephemeral and wispy it became, then mightn't that entity *still* exist?

And what about Consciousness for fetuses (or for their growing brains, even when they consist of just two neurons)? What about for cows (or their brains)? What about for goldfish (or their brains)? What about for viruses?

As I hope these lists of enigmas make clear, the questions entailed by a Capitalized Essence called "Consciousness" or *élan mental* abound and multiply with out end. Belief in dualism leads to a hopelessly vast and murky pit of mysteries.

Semantic Quibbling in Universe Z

There is one last matter I wish to deal with, and that has to do with Dave Chalmers' famous zombie twin in Universe Z. Recall that this Dave sincerely *believes* what it is saying when it claims that it enjoys ice cream and purple flowers, but it is in fact telling falsities, since it enjoys nothing at all, since it feels nothing at all — no more than the gears in a Ferris wheel feel something as they mesh and churn. Well, what bothers me here is the uncritical willingness to say that this utterly feelingless Dave *believes* certain things, and that it even believes them *sincerely*. Isn't sincere belief a variety of feeling? Do the gears in a Ferris wheel sincerely believe anything? I would hope you would say no. Does the float-ball in a flush toilet sincerely believe anything? Once again, I would hope you would say no.

So suppose we backed off on the sincerity bit, and merely said that Universe Z's Dave *believes* the falsities that it is uttering about its enjoyment of this and that. Well, once again, could it not be argued that *belief* is a kind of feeling? I'm not going to make the argument here, because that's not my point. My point is that, like so many distinctions in this complex world of ours, the apparent distinction between phenomena that *do* involve feelings and phenomena that do *not* is anything but black and white.

If I asked you to write down a list of terms that slide gradually from fully emotional and sentient to fully emotionless and unsentient, I think you could probably quite easily do so. In fact, let's give it a quick try right here. Here are a few verbs that come to my mind, listed roughly in descending order of emotionality and sentience: *agonize, exult, suffer, enjoy, desire, listen, hear, taste, perceive, notice, consider, reason, argue, claim, believe, remember, forget, know, calculate, utter, register, react, bounce, turn, move, stop.* I won't claim that my extremely short list of verbs is impeccably ordered; I simply threw it together in an attempt to show that there is unquestionably a spectrum, a set of shades of gray, concerning words that do and that do not suggest the presence of feelings behind the scenes. The tricky question then is: Which of these verbs (and comparable adjectives, adverbs, nouns, pronouns, etc.) would we be willing to apply to Dave's zombie twin in Universe Z? Is there some precise cutoff line beyond which certain words are disallowed? Who would determine that cutoff line?

To put this in perspective, consider the criteria that we effortlessly apply (I first wrote "unconsciously", but then I thought that that was a strange word choice, in these circumstances!) when we watch the antics of the humanoid robots R2-D2 and C-3PO in *Star Wars*. When one of them acts fearful and tries to flee in what strike us as appropriate circumstances, are we not justified in applying the adjective "frightened"? Or would we need to have obtained some kind of word-usage permit in advance, granted only when the universe that forms the backdrop to the actions in question is a universe imbued with *élan mental*? And how is this "scientific" fact about a universe to be determined?

If viewers of a space-adventure movie were "scientifically" informed at the movie's start that the saga to follow takes place in a universe completely unlike ours — namely, in a universe without a drop of *élan mental* — would they then watch with utter indifference as some cute-looking robot, rather like R2-D2 or C-3PO (take your pick), got hacked into little tiny pieces by a larger robot? Would parents tell their sobbing children, "Hush now, don't you bawl! That silly robot wasn't *alive*! The makers of the movie told us at the start that the universe where it *lived* doesn't have creatures with feelings! Not one!" What's the difference between *being alive* and *living*? And more importantly, what merits being sobbed over?

Quibbling in Universe Q

At chapter's end, we are thus brought back full circle to the "pedantic semantic" pronoun issues with which we began. Should we use different pronouns to refer to Universe Q's Dave Chalmers (which is clearly a "he") and to its indistinguishable zombie twin in Universe Z (who is just as clearly an "it")? Of course such semantic quibbles aren't limited to humans and their zombie twins. If a mosquito in our universe — our warm and fuzzy Universe Q overflowing with *élan mental* — is unquestionably a swattable "it", then what about a turkey? And if a turkey is unquestionably just a Thanksgiving dinner, then what about a chinchilla? And if a chinchilla is just a fur coat, then what about a bunny and a cat and a dog? And then what about a human fetus? And what about a newborn baby? Where lies the "who"/"which" cutoff line?

As I said at the chapter's outset, I see these as important questions — questions that in the end have everything to do with matters of life and death. They may not be easy to answer, but they are important to ponder. Semantics is not always just pedantic quibbling.

❧ ❧ ❧

CHAPTER 23

Killing a Couple of Sacred Cows

ॐ ॐ ॐ

A Cerulean Sardine

T HERE'S an idea in the philosophical literature on consciousness that makes me sea-blue, and that is the so-called "problem of the inverted spectrum". After describing this sacred cow as accurately I can, I shall try to slaughter it as quickly as I can. (It suffers from mad sacred cow disease.)

It all comes from the idea that you are supposedly so different from me that there is no way to cross the gap between our interiorities — no way for you to know what I am like inside, or vice versa. In particular, when you look at a bunch of red roses and I look at the same bunch of red roses, we both externalize what we are seeing by making roughly the same noise ("red roses"), but maybe, for all you know, what *I* am experiencing as redness inside my private, inaccessible cranium is what *you,* if only you could "step inside" my subjectivity for a moment or two, would actually call "blue". (By the way, advocates of the inverted-spectrum riddle would spurn any suggestion that you and I actually *are* already inside each other, even the littlest bit. Their riddle is predicated upon the existence of an Unbridgeable You–Me Chasm — that is, the absolute inaccessibility by one person of any other person's interiority. In other words, belief in the inverted spectrum is a close cousin to belief in Cartesian Egos — the idea that we are all disjoint islands and that "you can't get there from here".)

Bleu Blanc Rouge = Red, White, and Blue

Let's consider this idea. Maybe, just maybe, when all fifty million French people look at blood and declare that its color is "rouge", they are actually experiencing an inner sensation of blueness; in other words, blood looks to them just the way melted blueberry ice cream looks to Americans. And when they gaze up at a beautiful cloudless summer sky and voice the word "bleu", they are actually having the visual experience of melted raspberry ice cream. Sacrebleu! There is a systematic deception being pulled on them, and simultaneously a systematic linguistic coverup is going on, preventing anyone, including themselves, from ever knowing it.

We'd be convinced of this reversal if only we could get inside their skulls and experience colors in their uniquely *bleu-blanc-rouge* way, but alas, we'll never do that. Nor will they ever see colors in our red-white-and-blue way. And by the way, it's *not* the case that some wires have been crossed inside those French skulls — their brains look no different from ours, on every scale, from neurotransmitters to neurons to visual cortex. It's not something fixable by rewiring, or by any other physical operation. It's just a question of, well, ineffable *feelings*. And what's worse is that although it's true, nobody will ever know that it's true, since nobody can ever flit from one interiority to another — we're all trapped inside our own cranium.

Now this scenario sounds downright silly, doesn't it? How could it ever come about that the fifty million people living inside the rather arbitrary frontiers of a certain hexagonally shaped country would all mistakenly take redness for blueness and blueness for redness (though never revealing it linguistically, since they had all been taught to call that blue sensation "red" and that red sensation "blue")?

Even the most diehard of inverted-spectrum proponents would find this scenario preposterous. And yet it's just the same as the standard inverted spectrum; it's simply been promoted to the level of entire cultures, which makes it sound as it should sound — like a naïve fairy tale.

Inverting the Sonic Spectrum

Let's explore the inverted spectrum a little further by twisting some other knobs. What if all the chirpy high notes on the piano (we do agree they are chirpy, dear reader, don't we?) sounded very deep and low to, say, Diana Krall (though she always *called* them "high"), and all the deep low notes sounded chirpy and high to her (though she always called them "low")? This, too, would be the "inverted spectrum" problem, merely involving a sonic spectrum instead of the visual one. Now this scenario strikes me as much less plausible than the original one involving colors, and I hope strikes you that way, too. But why would there be any fundamental difference between an auditory inverted spectrum and a visual one?

Well, it's pretty clear that as musical notes sink lower and lower, the individual vibrations constituting them grow more and more perceptible. If you strike the leftmost key on a piano, you will feel very rapid pulsations at the same time as you (sort of) sense what pitch it is. Such a note is so low that we reach the boundary line between hearing it as a unitary *pitch* and hearing it — or rather, feeling it — as a rapid sequence of individual oscillations. The low "note" floats somewhere between singularity and plurality, somewhere between being auditory and being tactile. And if we had a piano that had fifteen or twenty extra keys further to the left (some Bösendorfers have a handful, but this piano would go quite a ways further down than they do), the superlow notes would start to feel even more like vibrations of our skin and bones rather than like pitches of sound. Two neighboring keys, when struck, wouldn't produce distinguishable tones, but just low, gruff rumbles that felt like long, low, claps of thunder or distant explosions, or perhaps cars passing by with subwoofers blasting out their amazing primordial shaking rather than a singable sequence of pitches.

In general, low notes, as they sink ever lower, glide imperceptibly into bodily shakings as opposed to being pitches in a spectrum, whereas high notes, as they grow higher, do not do so. This establishes a simple and obvious objective difference between the two ends of the audible spectrum. For this reason, it is inconceivable that Diana Krall could have an inverted-spectrum experience — that is, could experience what you or I would call a very *high* sound when the lowest piano note is struck. After all, there are no objective bodily shakings produced by a high note!

Glebbing and Knurking

Well, all right. If the idea of a *sonic* inverted spectrum is incoherent, then why should the *visual* inverted spectrum seem any more plausible? The two ends of the visible range of the electromagnetic spectrum are just as physically different from each other as are the two ends of the audible sonic spectrum. One end has light of lower frequencies, which makes certain pigments absorb it, while the other end has light of higher frequencies, which makes *other* pigments absorb it. Unlike rumbles, though, those cell-borne pigments are just intellectual abstractions to us, and this gives some philosophers the impression that our *experiences* of redness and blueness are totally disconnected from physics. The *feeling* of a color, they have concluded, is just some kind of personal invention, and two different people could "invent" it differently and never be the wiser for it.

To spell this idea out a little more clearly, let's posit that knurking and glebbing (two words I just concocted) are two vastly different sensations that any human brain can enjoy. All humans are created in the womb with these experiences as part of their built-in repertoire. You and I were born with knurking and glebbing as standard features, and ever since our cradle days, we've enjoyed these two sensations countless thousands of times. In some folks, though, it's red light that makes them knurk and blue that makes them gleb, while in others it's the reverse. When you were tiny, one of the colors red and blue happened to trigger knurking more often, while the other one triggered glebbing more often. By age five or so, this initial tendency had settled in for good. No science could predict which way it would go, nor tell which way it wound up — but it happened anyway. And thus you and I, dear reader, may have wound up on opposite sides of the gleb/knurk fence — but who knows? Who could *ever* know?

I must stress that, in the inverted-spectrum scenario, the association of red light (or blue light) with knurking is not any kind of postnatal *wiring pattern* that gets launched in a baby's brain and reinforced as it grows. In fact, although I stated above that to knurk and to gleb are *experiences* that all babies' brains come innately equipped with, they are not distinguishable brain *processes*. It's not possible to determine, no matter how fancy are the brain-scanning gadgets that one has access to, whether my brain (or yours) is knurking or glebbing. In short, we are not talking about objectively observable or measurable facts about the brain.

If *objectively observable* facts were all the inverted-spectrum riddle was about, it would be as easy as pie to tell the difference between ourselves and the fifty million French people whose inner sensations are all wrong! We would just examine their gray matter and pinpoint the telltale spot where certain key connections were flipped with respect to ours. Then we could watch their French brains engage in glebbing when the identical retinal stimulus would provoke knurking in our brains. But that's not in the least the meaning of the inverted-spectrum idea. The meaning is that, despite having *identical* brain wirings, two people looking at the same object experience completely different color sensations.

The Inverted Political Spectrum

This hypothetical notion makes our inner experiences of the colors in the rainbow sound like a set of floating pre-existent pure abstractions that are not intimately (in fact, not at all) related to the physics outside our skull, or even to any physics inside it; rather, these inner experiences are *arbitrarily* mappable onto outside phenomena. As we grow up, the rainbow colors get mapped onto the spectrum of prefabricated feelings with which our brains all come equipped "from the factory", but this mapping is not mediated by neural wiring; after all, neural wiring is observable from a detached third-person perspective, such as that of a neurosurgeon, so that rules it out.

Let's now ponder the implications of this notion of the independence of subjective feelings and external stimuli. Maybe, just to pick a random example, the abstraction of "liberty" feels to me like what the abstraction of "imprisonment" feels like to you — it's just that we both use the same word "liberty" for it, and so we are deluded into thinking that it is the same experience for both of us. This sounds pretty unlikely, doesn't it? After all, liberty is pleasant whereas imprisonment is unpleasant. But then again, who can say for sure? Maybe experiences that I feel are pleasurable are unpleasurable for you, and vice versa.

Or maybe that churning feeling that I feel inside me when I run into right-wing flag-wavers and pro-lifers (those who dominated in the "red" states in the 2004 election) is identical to that churning feeling that you feel inside you when you encounter left-wing flag-burners and pro-choicers (those who dominated in the "blue" states in the 2004 election), and vice versa! This would be the inverted political spectrum! Are you getting a bit dizzy at this point? (Perhaps what you experience as dizziness I experience as clarity, and vice versa. But let's not go there.)

The philosophers who take the inverted visual spectrum with total seriousness would not take the inverted political spectrum in the least seriously. But why not? Presumably because they don't think our brains come from the factory with prefabricated political "feelings" inside them, feelings that can be arbitrarily attached to right-wing or left-wing politics as we grow up. And yet they truly *do* think that we come with knurking and glebbing built in (although they don't use my words).

I once again wish to remind you that knurking is not an identifiable physical phenomenon in a brain (nor is glebbing). Knurking is that inherently incommunicable *sensation* that you supposedly have when red light (or blue light, if you're French, reader) hits your eyes. French people have all the same internal physical events happen in their brains as we do, but they don't have the same *experiences* as ours. French people experience glebbing when red light hits their retina, and knurking when blue light hits it. So just what *is* this knurking "experience", then, if it isn't anything physically identifiable in a brain?

The inverted-spectrists say it is pure *feeling*. Since this distinction is completely independent of physics, it amounts to dualism (something we already knew, in effect, since belief in Cartesian Egos is a kind of dualism).

Violets Are Red, Roses Are Blue

Why is it that those who postulate the inverted spectrum always do so only for experiences that lie along a one-dimensional numerical scale? It seems like a great paucity of imagination to limit oneself to swapping red and blue. If you think it's coherent to say to someone else, "Maybe *your* private inner experience of red is the same as *my* private inner experience of blue", then why would it not be just as coherent to say, "Maybe *your* private inner experience of looking at a red rose is the same as *my* private inner experience of looking at a blue violet"?

What is sacrosanct about the idea of shuffling colors inside a spectrum? Why not shuffle all sorts of experiences arbitrarily? Maybe *your* private inner experience of redness is the same as *my* private inner experience of hearing very low notes on a piano. Or maybe *your* private inner experience of going to a baseball game is the same as *my* private inner experience of going to a football game. Then again, maybe your private inner experience of going to a baseball game is the same as my private inner experience of going on a roller-coaster ride. Or maybe it's the same as my private inner experience of wrapping Christmas presents.

I hope that these sound ridiculously incoherent to you, and that you can move step by step backwards from these variations on the inverted-spectrum theme to the original inverted-spectrum riddle without losing the sense of ridiculousness. That would be most gratifying to me, because I see no fundamental difference between the original riddle and the patently silly caricatures of it just offered.

A Scarlet Sardine

The inverted-spectrum riddle depends on the idea that we are all born with a range of certain "pure experiences" that have no physical basis but that can get attached, as we grow, to certain external stimuli, and thus specific experiences and specific stimuli get married and from then on they are intimately tied together for a lifetime. But these "pure experiences" are supposedly not physical states of the brain. They are, rather, subjective *feelings* that one simply "has", without there being any physical explanation for them. Your brain state and mine could look as identical as anyone could ever imagine (using ultra-fine-grained brain-scanning devices), but whereas I would be feeling blueness, you would be feeling redness.

The inverted-spectrum fairy tale is a feeble mixture of bravado and timidity. While it boldly denies the physical world's relevance to what we feel inside, it meekly limits itself to a one-dimensional spectrum, and to the electromagnetic one, to boot. The sonic spectrum is too tied to objective physical events like shaking and vibrating for us to imagine it as being inverted, and if one tries to carry the idea beyond the realm of one-dimensional spectra, it becomes far too absurd to give any credence to.

Yes, People Want Things

There's something else in the philosophical literature on consciousness that gives me the willies, and that is the so-called "problem of free will". Let me describe this second sacred cow, and then try to dispatch it, too, as quickly as possible. (It, too, suffers from sacred mad cow disease.)

When people decide to do something, they often say, "I did it of my own free will." I think what they mean by this is usually, in essence, "I did it because I wanted to, not because someone else forced me to do it." Although I am uncomfortable with the phrase "I did it of my own free will", the paraphrase I've suggested sounds completely unobjectionable to me. We do indeed have wants, and our wants do indeed cause us to do things (at least to the extent that 641's primeness can cause a domino in a domino chain to fall).

The Hedge Maze of Life

Sometimes our desires bang up against obstacles. Somebody else drank that last soft drink in the refrigerator; the formerly all-night grocery store now closes at midnight; my friend's car has a flat tire; the dog ate my homework; the plane just pulled out of the gate thirty seconds ago; the flight has been canceled because of a snowstorm in Saskatoon; we're having computer troubles and we can't seem to make PowerPoint work in here; I left my wallet in my other pair of pants; you misread the final deadline; the reviewer was someone who hates us; she didn't hear about the job until too late; the runner in the next lane is faster than I am; and so on and so forth.

In such cases our will alone, though it pushes us, does not get us what we want. It pushes us in a certain direction, but we are maneuvering inside a hedge maze whose available paths were dictated by the rest of the world, not by our wants. And so we move willy-nilly, but not freewilly-nilly, inside the maze. A combination of pressures, some internal and some external, collectively dictates our pathway in this crazy hedge maze called "life".

There's nothing too puzzling about this. And I repeat, there is nothing puzzling about the idea that some of the pressures are our *wants*. What makes no sense is to maintain, over and above that, that our wants are somehow "free" or that our decisions are somehow "free". They are the outcomes of physical events inside our heads! How is that free?

There's No Such Thing as a Free Will

When a male dog gets a whiff of a female dog in heat, it has certain extremely intense desires, which it will try extremely hard to satisfy. We see the intensity only too clearly, and when the desire is thwarted (for instance, by a fence or a leash), it pains us to identify with that poor animal, trapped by its innate drives, pushed by an abstract force that it doesn't in the least understand. This poignant sight clearly exemplifies *will*, but is it *free* will?

How do we humans have anything that transcends that dog-like kind of yearning? We have intense yearnings, too — some of them in the sexual arena, some in more exalted arenas of life — and when our yearnings are satisfied, we attain some kind of happy state, but when they are thwarted, we are forlorn, like that dog on a tight leash.

What, then, is all the fuss about "free will" about? Why do so many people insist on the grandiose adjective, often even finding in it humanity's crowning glory? What does it gain us, or rather, what *would* it gain us, if the word "free" were accurate? I honestly do not know. I don't see any room in this complex world for my will to be "free".

I am pleased to have a will, or at least I'm pleased to have one when it is not too terribly frustrated by the hedge maze I am constrained by, but I don't know what it would feel like if my will were *free*. What on earth would that mean? That I didn't follow my will sometimes? Well, why would I do that? In order to frustrate myself? I guess that if I wanted to frustrate myself, I might make such a choice — but then it would be because I *wanted* to frustrate myself, and because my meta-level desire was stronger than my plain-old desire. Thus I might choose not to take a second helping of noodles even though I — or rather, part of me — would still like some, because there's *another* part of me that wants me not to gain weight, and the weight-watching part happens (this evening) to have more votes than the gluttonous part does. If it didn't, then it would lose and my inner glutton would win, and that would be fine — but in either case, my non-free will would win out and I'd follow the dominant desire in my brain.

Yes, certainly, I'll make a decision, and I'll do so by conducting a kind of inner vote. The count of votes will yield a result, and by George, one side will come out the winner. But where's any "freeness" in all this?

Speaking of George, the analogy to our electoral process is such a blatant elephant in the room that I should spell it out. It's not as if, in a brain, there is some kind of "neural suffrage" ("one neuron, one vote"); however, on a higher level of organization, there is some kind of "desire-level suffrage" in the brain. Since our understanding of brains is not at the state where I can pinpoint this suffrage physically, I'll just say that it's essentially "one desire, n votes", where n is some weight associated with the given desire. Not all values of n are identical, which is to say, not all desires are born equal; the brain is not an egalitarian society!

In sum, our decisions are made by an analogue to a voting process in a democracy. Our various desires chime in, taking into account the many external factors that act as constraints, or more metaphorically, that play the role of hedges in the vast maze of life in which we are trapped. Much of life is incredibly random, and we have no control over it. We can will away all we want, but much of the time our will is frustrated.

Our will, quite the opposite of being free, is steady and stable, like an inner gyroscope, and it is the stability and constancy of our non-free will that makes me me and you you, and that also keeps me me and you you. Free Willie is just another blue humpback.

CHAPTER 24

On Magnanimity and Friendship

ॐ ॐ ॐ

Are There Small and Large Souls?

HERE and there in this book, alluding to James Huneker's droll warning to "small-souled men" quoted in Chapter 1, I have somewhat light-heartedly referred to the number of "hunekers" comprising various human souls, but I have never been specific about the kinds of traits a high-huneker or low-huneker soul would tend to exhibit. Indeed, any hint at such a distinction risks becoming inflammatory, because in our culture there is a dogma that states, roughly, that all human lives are worth exactly the same amount.

And yet we violate that dogma routinely. The most obvious case is that of a declared war, where as a society we officially slip into an alternate collective mode in which the value of the lives of a huge subset of humanity is suddenly reduced to zero. I needn't spell this out because it is so blatant. Another clear violation of our dogma is capital punishment, where society collectively chooses to terminate a human life. Basically, society has judged that a certain soul merits no respect at all. Short of capital punishment, there is incarceration, where society treats people with many different levels of dignity or lack thereof, implicitly showing different levels of respect for different-sized souls. Consider also the phenomenal differences in the measures taken by physicians in attempting to save lives. A head of state (or the head of any large corporation) who has a heart attack will receive far better care than a random citizen, not to mention an illegal alien.

Why do I see such unequal treatments by society as tacit distinctions between the values of *souls*? Because I think that wittingly or unwittingly, we all equate the size of a living being's soul with the "objective" value of that being's life, which is to say, the degree of respect that we outsiders pay to that being's interiority. And we certainly do not place equal values on all beings' lives! We don't hesitate for a moment to draw a huge distinction between the values of a human life and an animal life, and between the values of the lives of different "levels" of animals.

Thus most humans willingly participate, directly or indirectly, in the killing of animals of many different species and the eating of their flesh (sometimes even mixing together fragments of the bodies of pigs, cows, and lambs in a single dish). We also nonchalantly feed our pets with pieces of the bodies of animals we have killed. Such actions establish in our minds, obviously, a hierarchy within the realm of animal souls (unless someone were to argue in good old black-and-white style that the word "soul" does not even apply to animals, but such absolutism seems to me more like received dogma than like considered reflection).

Most people I know would rate (either explicitly, in words, or implicitly, through choices made) cat souls as higher than cow souls, cow souls higher than rat souls, rat souls higher than snail souls, snail souls higher than flea souls, and so forth. And so I ask myself, if soul-size distinctions *between* species are such a commonplace and non-threatening notion, why should we not also be willing to consider some kind of explicit (not just tacit) spectrum of soul-sizes *within* a single species, and in particular within our own?

From the Depths to the Heights

Having painted myself into a corner in the preceding section, I'll go out on a limb and make a very crude stab at such a distinction. To do so I will merely cite two ends of a wide spectrum, with yourself and myself, dear reader, presumably falling somewhere in the mid-range (but hopefully closer to the "high" end than to the "low" one).

At the low end, then, I would place uncontrollably violent psychopaths — adults essentially incapable of internalizing other people's (or animals') mental states, and who because of this incapacity routinely commit violent acts against other beings. It may simply be these people's misfortune to have been born this way, but whatever the reason, I class them at the low end of the spectrum. To put it bluntly, these are people who are *not as conscious* as normal adults are, which is to say, they have *smaller souls*.

I won't suggest a numerical huneker count, because that would place us in the domain of the ludicrous. I simply hope that you see my general point and don't find it an immoral view. It's not much different, after all, from saying that such people should be kept behind bars, and no one I know considers prisons to be immoral institutions *per se* (it's another matter how they are run, of course).

What about the high end of the spectrum? I suspect it will come as no surprise that I would point to individuals whose behavior is essentially the opposite of that of violent psychopaths. This means gentle people such as Mohandas Gandhi, Eleanor Roosevelt, Raoul Wallenberg, Jean Moulin, Mother Teresa, Martin Luther King, and César Chávez — extraordinary individuals whose deep empathy for those who suffer leads them to devote a large part of their lives to helping others, and to doing so in nonviolent fashions. Such people, I propose, are *more conscious* than normal adults are, which is to say, they have *greater souls*.

Although I seldom attach much weight to the etymologies of words, I was delighted to notice, when preparing a lecture on these ideas a few years ago, that the word "magnanimity", which for us is essentially a synonym of "generosity", originally meant, in Latin, "having a great soul" (*animus* meaning "soul"). It gave me much pleasure to see this familiar word in a new light, thanks to this X-ray. (And then, to my surprise, in preparing this book's rather fanatical index, I discovered that "Mahatma" — the title of respect usually given to Gandhi — also means "great soul".) Another appealing etymology is that of "compassion", which comes from Latin roots meaning "suffering along with". These hidden messages echoing down the millennia stimulated me to explore this further.

The Magnanimity of Albert Schweitzer

My personal paragon for great-souledness is the theologian, musician, writer, and humanitarian Albert Schweitzer, who was born in 1875 in the tiny village of Kaysersberg in Alsace (which was then part of Germany, even though my beloved old French encyclopedia *Le Petit Robert 2*, dating from exactly one century later, claims him as *français!*), and who became world-famous for the hospital that he founded in 1913 in Lambaréné, Gabon, and where he worked for over fifty years.

Already at a very young age, Schweitzer identified with others, felt pity and compassion for them, and wanted to spare them pain. Where did this empathic generosity come from? Who can say? For example, on his very first day at school, six-year-old Albert noticed that he had been decked out by his parents in fancier clothes than his schoolmates, and this disparity disturbed him greatly. From that day onward, he insisted on dressing just like his poorer schoolmates.

A vivid excerpt from Schweitzer's autobiographical opus *Aus meiner Kindheit und Jugendzeit* portrays the compassion that pervaded his life:

As far as I can peer back into my childhood, I suffered from all
the misery that I saw in the world around me. I truly never knew a
simple, youthful *joie de vivre*, and I believe that this is the case for many
children, even if from the outside they give the appearance of being
completely happy and carefree.

In particular, I was tormented by the fact that poor animals had
to endure such great pain and need. The sight of an old, limping
horse being dragged along by one man while another man beat it
with a stick as it was being driven to the Colmar slaughterhouse
haunted me for weeks. Even before I entered school, I found it
incomprehensible that in my evening prayer I was supposed to pray
only for the sake of human beings. And so I secretly spoke the words
to a prayer that I had made up myself. It ran this way: "Dear God,
protect and bless everything that breathes, keep it from all evil, and
let it softly sleep."

Schweitzer's compassion for animals was not limited to mammals but
extended all the way down the spectrum to such lowly creatures as worms
and ants. (I say "all the way down" and "lowly" not to indicate disdain, but
only to suggest that Schweitzer, like nearly all humans, must have had a
"consciousness cone", vaguely like mine on page 19. Such a mental
hierarchy can just as easily give rise to a sense of concern and responsibility
as to a sense of disdain.) He once remarked to a ten-year-old boy who was
about to step on an ant, "That's my personal ant. You're liable to break its
legs!" He would routinely pick up a worm he saw in the middle of a street
or an insect flailing in a pond and place it in a field or on a plant so that it
could try to survive. Indeed, he commented rather bitterly, "Whenever I
help an insect in distress, I do so in an attempt to atone for some of the guilt
contracted by humanity for its crimes against animals."

As is well known, Schweitzer's simple but profound guiding principle
was what he termed "reverence for life". In the address delivered when he
was awarded the Nobel Prize for Peace in 1953, Schweitzer declared:

> The human spirit is not dead. It lives on in secret... It has come
> to believe that compassion, in which all ethics must take root, can
> only attain its full breadth and depth if it embraces all living creatures
> and does not limit itself to human beings.

The following anecdote, also from *Aus meiner Kindheit und Jugendzeit*, is
particularly revealing. In the springtime, with Easter approaching, little
Albert, seven or eight years old, had been invited by a comrade — a

comrade-in-arms, quite literally! — to go on an adventure of killing birds with slingshots that they had just made together. Looking back at this turning point in his life from the perspective of many decades later, Schweitzer recalls:

> This was an abhorrent proposal, but I dared not refuse out of fear that he would mock me. Soon we found ourselves standing near a leafless tree whose branches were filled with birds singing out gaily in the morning, without any fear of us. My companion, crouching low like an Indian on a hunt, placed a pebble in the leather pouch of his slingshot and stretched it tightly. Obeying the imperious glance he threw at me, I did the same, while fighting sharp pangs of conscience and at the same time vowing firmly to myself that I would shoot when he did.
>
> Just at that moment, church bells began to ring out, mingling with the song of the birds in the sunshine. These were the early bells that preceded the main bells by half an hour. For me, though, they were a voice from Heaven. I threw my slingshot down, startling the birds so that they flew off to a spot safe from my companion's slingshot, and I fled home.
>
> Ever since that day, whenever the bells of Holy Week ring out amidst the leafless trees of spring, I have remembered with deep gratitude how on that fateful day they rang into my heart the commandment: "Thou shalt not kill." From that day on, I swore that I would liberate myself from the fear of other people. Whenever my inner convictions were at stake, I gave less weight to the opinions of other people than I once had. And I did my best to overcome the fear of being mocked by my peers.

Here we have a classic conflict between peer pressure and one's own inner voices, or as we usually phrase it (and as Schweitzer himself put it), one's *conscience*. In this case, fortunately, conscience was the clear winner. And indeed, this was a decision that lasted a lifetime.

Does Conscience Constitute Consciousness?

In this region of semantic space there is one further linguistic observation that strikes me as most provocative. That is the fact that in the Romance languages, the words for "conscience" and "consciousness", which strike us English speakers as very distinct concepts, are one and the same (for example, the French word *conscience* has both meanings, a fact that I learned when, as a teen-ager, I bought a book entitled *Le cerveau et la conscience*). This may merely be a lexical gap or a confusing semantic blur in these languages (the meaning on a literal level is "co-knowledge"), but even if that's the case, I nonetheless think of it as offering us an insight that might otherwise never occur to us: that the partial internalization of other creatures' interiority (conscience) is what most clearly marks off creatures who have large souls (much consciousness) from creatures that have small souls, and from yet others that have none or next to none.

I think it's obvious, or nearly so, that mosquitoes have no conscience and likewise no consciousness, hence nothing meriting the word "soul". These flying, buzzing, blood-sucking automata are more like miniature heat-seeking missiles than like soulful beings. Can you imagine a mosquito experiencing mercy or pity or friendship? 'Nough said. Next!

What about, say, lions — the very prototype of the notion of carnivore? Lions stalk, pounce on, rip into, and devour giraffes and zebras that are still kicking and braying, and they do so without the slightest mercy or pity, which suggests a complete lack of compassion, and yet they seem to care a great deal about their own young, nuzzling them, nurturing them, protecting them, teaching them. This is quite unmosquito-like behavior! Moreover, I suspect that lions can easily come to care for certain beasts of other species (such as humans). In this sense, a lion can and will internalize certain limited aspects of the interiority of at least *some* other creatures (especially those of a few other lions, most particularly its immediate family), even though it may remain utterly oblivious to and indifferent to those of most other creatures (a quality that sounds depressingly like most humans).

I think it's also obvious, or nearly so, that most dogs care about other creatures — particularly humans who belong to their inner circle. Indeed, it's well known that some dogs, displaying incredible magnanimity, will lay down their lives for their owners. I have yet to hear about a lion doing such a thing for any animal of another species, although I suppose some dog-like lion, somewhere or other, may have once fought to the death against another beast in order to save the life of a human companion. It's a bit too much of a stretch, however, to imagine a lion choosing to be a vegetarian.

And yet a quick Web search shows that the idea of a vegetarian lion is not all that rare (usually in fiction, admittedly, but not always). Indeed, such a lion, a female named "Little Tyke", was apparently brought up as a pet near Seattle. For four years (so says the Web site), Little Tyke refused all meat offered her until finally her owners gave up trying and accepted her vegetarian ways and her joy at playing with lambs, chickens, and other beasts. Until her dying day, Little Tyke was a vegetarian lion. Will miracles never cease?

In any case, having a conscience — a sense of morality and of caring about doing "the right thing" towards other sentient beings — strikes me as the most natural and hopefully also the most reliable sign of consciousness in a being. Perhaps this simply boils down to how much one puts into practice the Golden Rule.

Albert Schweitzer and Johann Sebastian Bach

I have to admit that I have always intuitively felt there was another and quite different yardstick for measuring consciousness, although a most blurry and controversial one: musical taste. I certainly cannot explain or defend my own musical taste, and I know I would be getting myself into very deep, hot, and murky waters if I were to try, so I won't even begin. I will, however, have to reveal a little bit of my musical taste in order to talk about Albert Schweitzer and his musical profundity.

For my sixteenth birthday, my mother gave me a record of the first eight preludes and fugues of Book One of J. S. Bach's monumental work, *The Well-Tempered Clavier*, as played on the piano by Glenn Gould. This was my first contact with the notion of "fugue", and it had an electrifying effect on my young mind. For the next several years, every time I went into a record store, I would seek out other parts of *The Well-Tempered Clavier* on piano, for it was a genuine rarity those days (even on harpsichord, but especially on piano, which I preferred). Every time I found a new set of preludes and fugues from either volume, the act of putting the needle down in the grooves of the new record and listening to it for the first time was among the most exciting events in my life.

In my parents' record collection, there was also a recording of several Bach organ works as performed by Albert Schweitzer, but it took me a long time to come around to giving it a try, because I feared it would be too "heavy". But when I finally did, what I heard was incredibly moving and I became as addicted to it as I had ever been to *The Well-Tempered Clavier*. I then naturally expanded my search in record stores to include Bach organ works, but I soon discovered something that troubled me, which was that many performers took them very swiftly and jauntily, as if they were merely virtuoso exercises as opposed to profound statements about the human condition. Schweitzer's playing was humble and simple, and it charmed me that he made mistakes now and then but simply went on unperturbedly (in no other recordings would one hear even a *single* mistake anywhere, which struck me as unnatural and even bizarre). It also happened, although I didn't know it then, that these performances had all been recorded on a simple organ in the very church in the Alsatian village of Günsbach whose bells had pealed one bright spring morning, saving the lives of a bird or two, and transforming young Albert's life, and therewith, the lives of thousands of people.

Dig that Profundity!

Over the years, Bach as played by Schweitzer became a deep part of me. I obtained several more recordings by him, all belonging to the same series, each one revealing new depths of a cosmic wisdom (perhaps that sounds grandiose, but to me it is exactly on the mark) that emanated from both composer and performer.

I was naturally filled with gratification when the popularity of my book *Gödel, Escher, Bach* linked my name in some fashion in the musical community with that of Bach (this was a true honor), and in Bach's 300th birthyear, 1985, I had the pleasure of participating in several tricentennial celebrations, including a tiny one on his exact birthday that I organized in Ann Arbor for the members of a class I was teaching, plus a few friends, the highlight of which was the small firestorm unleashed when we lit all 300 candles on the giant birthday cake I had ordered.

Fifteen years later, I was surprised to be invited to participate in a commemoration in Rovereto, Italy, of the 250th anniversary of Bach's death (which had taken place in July of 1750), but since I was going to be in northern Italy at that time in any case, I gladly accepted. Several memorable talks were delivered in the afternoon, and after a banquet there was to be a treat — a performance of a number of Bach pieces (transcribed for small chorus) by a well-known singing troupe. I remembered their skill and was looking forward to a rewarding evening of moving music.

What I heard, however, was something quite different, although I should perhaps have anticipated it: a nonstop display of unrestrained vocal virtuosity, and nothing but that. It was terribly impressive, but to my mind it was also terribly vapid. The lowlight of the entire performance for me was when the singers came to one of the most profound of all the Bach organ fugues — the G minor fugue often called simply "The Great" (BWV 542), a work that I loved as played by Albert Schweitzer in all his modesty, but with unrivaled depth of feeling. Regrettably, I will never forget how they tackled this meditative fugue at roughly twice the speed it should be taken at, lighting into it as if they were sprinting to catch a train, and struttin' their stuff like nobody's business. They bounced on their toes, as if to try to get the audience to swing along with their snappy rhythm, and they even snapped their fingers to the beat (even the word "beat" sounds ridiculous in this hallowed context). Several of the singers periodically flashed bright grins at the audience, as if to say, "Aren't we fabulous? Ever heard anyone sing so many notes per second in your life? How about those trills! Isn't this music sexy? Hope you're all diggin' it! And don't forget, we have lots of CD's you can buy after the show!"

All of this threw me for a real loop. Of course there is room in this world for many ways to perform any work of music, and of course there was something *interesting* about these singers' speed and slickness, and the way that they executed ultrarapid trills flawlessly — it was impressive in much the same way as the engineering details of a beautiful sports car are impressive — but for me it had nothing to do with the *meaning* of the music. That meaning was contemplative and cosmic, not frilly and show-offy. I am tolerant of many diverse ways of playing pieces of music, but I also have my limits and this went considerably beyond them. It made me long to hear the slightly flawed, very mortal, and reflective profundity of Albert Schweitzer at his little village organ in Günsbach, but that was not in the cards that evening. It was a classic case of sacred versus profane, and it remains vivid in my memory.

Only in preparing this chapter did I come across some writings by Schweitzer himself that strangely echo (if echoes can precede their causes!) my great troubledness that evening in Rovereto. Here is what he wrote, almost one hundred years earlier, about performances of Bach in that era:

> Many performers have been performing Bach for years without experiencing for themselves the deepening that Bach is capable of bringing out in any true artist. Most of our singers are far too caught up in technique to sing Bach correctly. Only a very small number of them can reproduce the spirit of his music; the rest of them are incapable of penetrating into the Master's spiritual world. They do not feel what Bach is trying to say, and therefore cannot transmit it to anyone else. Worst of all, they consider themselves to be outstanding Bach interpreters, and have no awareness of what it is that they lack. Sometimes one has to wonder how listeners who attend such superficial performances are able to detect even the slightest sign of the depth of Bach's music.
>
> Those who understand the situation today will not consider these comments to be exaggeratedly pessimistic. Our enchantment with Bach is undergoing a crisis. The danger is that our love for Bach's music will become superficial and that

too much vanity and smugness will be mixed in with it. Our era's lamentable trend towards imitation comes out also in the way that we take Bach over, which is all too visible these days. We act as if we wanted to praise Bach, but in truth we only praise ourselves. We act as if we had rediscovered him, understood him, and performed him as no one has ever done before. A bit less noise, a bit less "Bach dogmatism", a bit more ability, a bit more humility, a bit more tranquility, a bit more devotion… Only thus will Bach be more honored in spirit and in truth than he has been before.

There is little I can add to this trenchant criticism of superficiality taking itself for depth; I will simply say that running across it comforted me, even though I did so several years after the Rovereto event, as it made me know that I am not alone in my lamentation. Schweitzer was the most humble and self-effacing of people, and so his remarks have to be taken as nothing other than an honest reaction to a deplorable trend that was already clear a century ago and that seems only to be increasing today.

Alle Grashüpfer Müssen Sterben

What, some readers may be asking themselves, does any of this have to do with "I" or consciousness or souls? My response would be, "What could have *more* to do with consciousness or souls than merging oneself with the combined spirituality of Albert Schweitzer and J. S. Bach?"

The other night, in order to refresh my musty memories of Schweitzer playing Bach organ music (which I listened to hundreds of times in my teen-age years and my twenties), I pulled all four of the old vinyl records off my shelf and put them on in succession. I began with the prelude and fugue in A major (BWV 536, nicknamed by Schweitzer the "walking fugue") and went through many others, winding up with my very favorite, the beatific prelude and fugue in G major (BWV 541), and then as a final touch, I listened to the achingly sweet–sad chorale-prelude "Alle Menschen Müssen Sterben" ("We All Must Die" — or perhaps, in order to echo the trochaic meter of the German, "Human Beings All Are Mortal").

While I was sitting silently in my living room, listening intently to the soft notes of these fathomless meditations, I noticed a lone grasshopper sitting on the rug. At first I thought it was dead (after all, all grasshoppers must die, too), but when I approached, it took a big hop, so I quickly grabbed a glass bowl from a nearby table, flipped it over to trap the little jumper, then carefully slid a record cover underneath, so as to form a floor for this glassy room. Then I carried the improvised craft and its diminutive passenger to my front door, opened it, and let the grasshopper leap down onto a bush in the dark night. Only while I was in the middle of this mini-samaritan act did the resonance with Schweitzer's spirit cross my mind — in fact, it happened just as I slid the record cover, which has a drawing by Ben Shahn of Schweitzer at the organ, underneath the glass bowl, so that the grasshopper was sitting on Schweitzer's hand. Something felt just right about this fortuitous conjunction.

An hour or so later, as I got up to stretch, I chanced to notice a carpenter ant crawling under a table, and so once again I made a little transport vehicle for it and escorted my six-legged friend outside. It started to seem rather curious to me that all this mini-samaritanism was happening while I was so immersed in Bach's profound spirituality and Schweitzer's pacifistic mentality of "reverence for life".

Perhaps to break this spell, or perhaps to underline my own personal dividing line, I then saw another little black dot moving in a certain familiar zigzaggy fashion in the air near a lamp, and I went to investigate. The small black dot landed on the table below the lamp, and there was no question what it was: a mosquito, *un moustique, una zanzara, eine Mücke.* One moment later, that *Mücke* was history (I'll spare you the details). By this point, I suspect, my views on the expendability of mosquitoes have probably become an annoyingly familiar refrain to readers of this book, but I have to say that I felt not the tiniest twinge of regret at the late blood-seeking missile's demise.

A little before midnight, I interrupted my music-listening session to call my aging and ailing mother out in California, since I have a routine of phoning her every evening to give her a bit of family news and a bit of cheer. After our brief chat, I returned to my music, and when the Dorian toccata and fugue came on, I found my thoughts turning to a close friend who deeply loves that piece, and to his son, who had just been diagnosed with a worrisome illness. The music went on, and all these thoughts about beloved people and the precious, frightening fragility of human life somehow blended naturally with it.

To cap it all off, at some point after midnight, I heard a knock at the back door (not a standard event at our house, I assure you!), and I went to see who it was. It turned out to be a teen-ager whom I had met once or twice, who had been kicked out of his home a month earlier by his parents and who was sleeping in parks. He said it was a bit nippy that night and asked me if he could sleep in our playroom. I thought about it for a moment and since I knew my daughter trusted him, I said yes.

All at once it seemed an extremely strange coincidence that all of these intensely human things, these events dependent on my mirroring of other beings' interiorities, were taking place right while I was so focused on the concepts of compassion and magnanimity.

Friends

Compassion, magnanimity, reverence for life — all these are qualities epitomized by Albert Schweitzer, who in addition had the remarkable quality of being a reverential Bach organist — but to my mind, this is no accident. Some might say that Schweitzer and people of his rare caliber are *selfless*. I understand this idea and I think there's some truth to it, but on the other hand, oddly enough, I have been arguing, as does etymology, that the more magnanimous one is, the *greater* one's self or soul is, not the smaller! So I would say that those who strike us as self-less are in fact very soul-full — that is, they house many other souls inside their own skulls/brains/minds/souls — and I don't think this sharing of mind-space diminishes their central core but enlarges and enriches it. As Walt Whitman put it in his poem "Song of Myself", "I am large, I contain multitudes." All this richness is a consequence of the fact that at some point in the dim past, the generic human brain surpassed a critical threshold of flexibility and became quasi-universal, able to internalize the abstract essences of other human brains. It is something to marvel at.

One day, as I was trying to figure out where I personally draw the line for applying the word "conscious" (even though of course there's no sharp cutoff), it occurred to me that the most crucial factor was whether or not the entity in question could be said to have some notion, perhaps only very primitive, of "friend", a friend being someone you care about and who cares about you. It seems clear that human babies acquire the rudiments of this notion pretty early on, and it also seems clear that some kinds of animals — mostly but not only mammals — have a pretty well-developed sense of the "friend" concept.

It's clear that dogs feel that certain humans and dogs are their friends, and possibly also a few other animals. I won't try to enumerate which types of animals seem capable of acquiring the "friend" notion because it's blurry and because you can run down a mental list just as easily I can. But the more I think about this, the righter it feels to me. And so I find myself led to the unexpected conclusion that what seems to be the epitome of selfhood — a sense of "I" — is in reality brought into being if and only if along with that self there is a sense of *other* selves with whom one has bonds of affection. In short, only when generosity is born is an ego born.

How different this is from the view held by the majority of philosophers of mind about the nature of consciousness! That view is that consciousness is the consequence of having so-called *qualia,* the supposedly primordial experiences (such as the retinal buzz made by the color purple, the sound of middle C, or the taste of Cabernet Sauvignon) out of which all "higher" experiences are built in bottom-up fashion. My view, in contrast, posits a high abstraction as the threshold at which consciousness starts to emerge from the gloom. Mosquitoes may "experience" the quale of the taste of blood, but they are unconscious of that quale, in just the way that toilets respond to but are totally unconscious of the various qualia of different water levels. Now if mosquitoes only had big enough brains to allow them to have *friends,* then they could be *conscious* of that great taste! Alas, the poor small-brained mosquitoes are constitutionally deprived of that chance.

But our glory as human beings is that, thanks to being beings with brains complicated enough to allow us to have friends and to feel love, we get the bonus of *experiencing* the vast world around us, which is to say, we get consciousness. Not a bad deal at all.

EPILOGUE

The Quandary

ॐ ॐ ॐ

Not a Tall!

IN THE foregoing four-and-twenty chapters, I have given my best shot at saying what an "I" is, which means, perforce, that I have also done my best at saying what a self, a soul, an inner light, a first-person viewpoint, interiority, intentionality, and consciousness are. A tall order, to be sure, but I hope not a tall tale. To some readers, however, my story may still seem to be a tall — a terribly tall — a too-tall — tale. With such readers I sincerely sympathize, for I concede that there still are troubling issues.

The key problem is, it seems to me, that when we try to understand what we are, we humans are doomed, as spiritual creatures in a universe of mere stuff, to eternal puzzlement about our nature. I vividly remember how, as a teen-ager reading about brains, I was forced for the first time in my life to face up to the idea that a human brain, especially my own, must be a physical structure obeying physical law. Although it may seem strange to you, just as it does to me now, this realization threw me for a loop.

In a nutshell, our quandary is this. Either we believe that our consciousness is something *other* than an outcome of physical law, or we believe it *is* an outcome of physical law — but making either choice leads us to disturbing, perhaps even unacceptable, consequences. My purpose in these final pages is to face this dilemma head-on.

The Pull and Pitfalls of Dualism

In Chapter 22, I discussed dualism — the idea that over and above physical entities governed by physical law, there is a Capitalized Essence called "Consciousness", which is an invisible, unmeasurable, undetectable aspect of the universe possessed by certain entities and not others. This notion, very close to the traditional western religious notion of "soul", is appealing because it conforms with our everyday experience that the world is divided up into two kinds of things — animate and inanimate — and it also gives some kind of explanation for the fact that we experience our own interiority or inner light, something of which we are so intimately aware that to deny its existence would seem absurd if not impossible.

Dualism also holds out the hope of explaining the mysterious division of the *animate* world into two types of entity: *myself* and *others*. Otherwise put, this is the seemingly unbridgeable gap between the subjective, first-person view of the world and an impersonal, third-person view of the world. If what we call "I" is a squirt of some unanalyzable Capitalized Essence magically doled out to each human being at the moment in which it is conceived, with each portion imbued with a unique savor permanently defining the recipient's identity, then we need look no further for an explanation of what we are (even if it depends on something inexplicable).

Furthermore, the idea that each of us is intrinsically defined by a unique incorporeal essence suggests that we have immortal souls; belief in dualism may thus remove some of the sting of death. It is not very hard for someone who grows up drenched in the pictorial and verbal imagery of western religion to imagine a wispy, ethereal aura being released from the body of someone who has just died, and sailing up, up, up into some kind of invisible celestial realm, where it will survive eternally. Whether we are believers or skeptics, such imagery is part and parcel of our western heritage, and for that reason it is hard to shuck it entirely, no matter how solidly one's belief system is anchored in science.

Not long after my wife Carol died, I organized a memorial service for her, interleaving reminiscences by a few dear friends and relatives with musical selections that had meant a great deal to her. To close this sad ceremony, I chose the final two-and-a-half minutes of the opening movement of Sergei Prokofiev's first violin concerto, an astonishing work of musical poetry under whose spell Carol had fallen as deeply as I had. The beautiful and moving passage that I selected from this concerto (as well as its twin, at the end of the piece as a whole) might as well have been written to evoke the image of an ascending soul, so tenuous, tremulous, and delicate is it throughout, but most of all in its final upward-drifting tones. Though neither Carol nor I was religious in the least, there was something that to me rang so true in this naïve image of her purest essence leaving her mortal remains and soaring up, up, forever up, even if, in the end, it was not into *the sky* that her soul was flying, but merely into *this guy*...

As this story reveals, this guy, for all his years of scientific training and hardheaded thinking about mind and spirit as rooted in physics, is at times susceptible to the traditional dualistic imagery with which most of us are brought up — if not by our families, then by our wider culture. I can fall for the alluring imagery, even if I reject the ideas. But in my more rational moments, such imagery makes no sense to me, for I know only too well how dualism leads to a long list of unanswerable questions, some of which I wrote out in Chapter 22, showing it to be fraught with such arbitrariness and illogicality that it would seem to collapse under its own weight.

The Lure and Lacunas of Nondualism

If instead one believes that consciousness (now with a small "c") is an outcome of physical law, then no room remains for anything extra "on top". This is appealing to a scientific mind because it is far simpler than dualism. It gets rid of a puzzling dichotomy between ordinary physical entities and extraordinary nonphysical essences, and it cancels the long list of questions about the nature of the nonphysical Capitalized Essence.

On the other hand, throwing dualism out the window is troubling as well, because, at least on first glance, doing so seems to leave us with no distinction between animate and inanimate entities, and no explanation for our unique experience of our own interiority or inner light, no explanation for the gulf between *our* self and *other* selves. A more careful look at this viewpoint, however, shows that there is room in it for such distinctions.

In the Introduction, I wrote of "the miraculous appearance of selves and souls in substrates consisting of inanimate matter", a phrase I suspect made more than one reader bristle. "How can the author refer to a human brain — the most animate of all entities in the universe — as 'inanimate matter'?" Well, one of the leitmotifs of this book has been that the presence or absence of animacy depends on the level at which one views a structure. Seen at its highest, most collective level, a brain is quintessentially animate and conscious. But as one gradually descends, structure by structure, from cerebrum to cortex to column to cell to cytoplasm to protein to peptide to particle, one loses the sense of animacy more and more until, at the lowest levels, it has surely vanished entirely. In one's mind, one can move back and forth between the highest and lowest levels, and in this fashion oscillate at will between seeing the brain as animate and as inanimate.

A nondualistic view of the world can thus include animate entities perfectly easily, as long as different levels of description are recognized as valid. Animate entities are those that, at some level of description, manifest a certain type of loopy pattern, which inevitably starts to take form if a system with the inherent capacity of perceptually filtering the world into discrete categories vigorously expands its repertoire of categories ever more towards the abstract. This pattern reaches full bloom when there comes to be a deeply entrenched self-representation — a story told by the entity to itself — in which the entity's "I" plays the starring role, as a unitary causal agent driven by a set of desires. More precisely, an entity is animate to the *degree* that such a loopy "I" pattern comes into existence, since this pattern's presence is by no means an all-or-nothing affair. Thus to the extent that there is an "I" pattern in a given substrate, there is animacy, and where there is no such pattern, the entity is inanimate.

Rainbows or Rocks?

There still remains a sticky question: What would make a loopy abstract pattern, however fancy it might be, constitute a locus of interiority, an inner light, a site of first-person experience? Otherwise put, where does *me*-ness come from? The notion that such a pattern grows enormously in size and complexity over time, perceives itself, and entrenches itself so deeply as to become all but undislodgeable will constitute a satisfactory answer for some seekers of truth (such as Strange Loop #641). For others, however (such as Strange Loop #642), it will not do at all.

For the latter sort of thinker, there will always remain the kind of riddle posed in Chapter 21 about the two freshly minted atom-for-atom copies of a destroyed body, one on Mars and one on Venus: "Where will I wake up? Which, if either, of the two bodies will house *my* inner light?" Thinkers of this kind cling fiercely to the instinctive notion of a unique Cartesian Ego that constitutes the identity, the "I"-ness, the inner light, the interiority of any sentient being. To such thinkers, it will be totally unacceptable to suggest that their precious notion of *me*-ness is more like a shimmering, elusive rainbow than it is like a solid, mass-possessing rock, and that there is thus no right answer to the perplexing "Which one will I be?" riddle. They will insist that there has to be a genuine marble of "I"-ness in one of the two bodies and not in the other one, as opposed to an elusive rainbow-like entity that first recedes and then disintegrates entirely as one draws ever closer. But to believe in such an indivisible, indissoluble "I" is to believe in nonphysical dualism.

Thrust: The Hard Problem

And this is our central quandary. Either we believe in a nonmaterial soul that lives outside the laws of physics, which amounts to a nonscientific belief in magic, or we reject that idea, in which case the eternally beckoning question "What could ever make a mere physical pattern be *me*?" — the question that philosopher David Chalmers has seductively and successfully nicknamed "The Hard Problem" — seems just as far from having an answer today (or, for that matter, at any time in the future) as it was many centuries ago.

After all, a phrase like "physical system" or "physical substrate" brings to mind for most people, including a substantial proportion of the world's philosophers and neurologists, an intricate structure consisting of vast numbers of interlocked wheels, gears, rods, tubes, balls, pendula, and so forth, even if they are tiny, invisible, perfectly silent, and possibly even probabilistic. Such an array of interacting inanimate stuff seems to most people as unconscious and devoid of inner light as a flush toilet, an automobile transmission, a fancy Swiss watch (mechanical or electronic), a cog railway, an ocean liner, or an oil refinery. Such a system is not just *probably* unconscious, it is *necessarily* so, as they see it. This is the kind of single-level intuition so skillfully exploited by John Searle in his attempts to convince people that computers could never be conscious, no matter what abstract patterns might reside in them, and could never mean anything at all by whatever long chains of lexical items they might string together.

Riposte: A Soft Poem

And yet to you, my faithful reader who has plowed all through this book up to its nearly final page, I would hope that things seem otherwise. Together, you and I have gone through instance after instance of increasingly sophisticated structures having loops, from the ever-darting-off Exploratorium red dot to fine-grained television cameras taking in the screens they fill, then to formulas asserting that they have no *PM* proof, and winding up with the strange loop that comes about inside the ever-growing repertoire of symbols in each human being's brain. (*Élan mental* we have no truck with, for it leads to endless traps.)

If there were ever, in our physics-governed world, a kind of magic, it is surely in these self-reflecting, self-defining patterns. Such strange loops, inspired by Gödel's Trojan horse that sneaked self-consciousness inside the very fortress that was built to keep it out, and recalling Roger Sperry's tower of forces within forces within forces (found inside each teet'ring bulb of dread and dream), give the only explanation I can fancy for how animate, desire-driven beings can arise from just plain matter, and for how, among the swarm of loops that populate our planet, there is one, and only one, that you call "I" (and I call "you").

A Billion Trillion Ants in One's Leg

You and I are mirages who perceive themselves, and the sole magical machinery behind the scenes is perception — the triggering, by huge flows of raw data, of a tiny set of symbols that stand for abstract regularities in the world. When perception at arbitrarily high levels of abstraction enters the world of physics and when feedback loops galore come into play, then "which" eventually turns into "who". What would once have been brusquely labeled "mechanical" and reflexively discarded as a candidate for consciousness has to be reconsidered.

We human beings are macroscopic structures in a universe whose laws reside at a microscopic level. As survival-seeking beings, we are driven to seek efficient explanations that make reference only to entities at our own level. We therefore draw conceptual boundaries around entities that we easily perceive, and in so doing we carve out what seems to us to be reality. The "I" we create for each of us is a quintessential example of such a perceived or invented reality, and it does such a good job of explaining our behavior that it becomes the hub around which the rest of the world seems to rotate. But this "I" notion is just a shorthand for a vast mass of seething and churning of which we are necessarily unaware.

Sometimes, when my leg goes to sleep (as we put it in English) and I feel a thousand pins and needles tingling inside it, I say to myself, "Aha! So *this* is what being alive *really* is! I'm getting a rare glimpse of how complex I truly am!" (In French, one says that one has "ants in one's leg", and the cartoon character Dennis the Menace once remarked that he had "ginger ale in his leg" — two unforgettable metaphors for this odd yet universal sensation.) Of course we can never come close to experiencing the full tingling complexity of what we truly are, since we have, to take just one typical example, six billion trillion (that is, six thousand million million million) copies of the hemoglobin molecule rushing about helter-skelter through our veins at all moments, and in each second of our lives, 400 trillion of them are destroyed while another 400 trillion are created. Numbers like these are way beyond human comprehension.

But our own unfathomability is a lucky thing for us! Just as we might shrivel up and die if we could truly grasp how minuscule we are in comparison to the vast universe we live in, so we might also explode in fear and shock if we were privy to the unimaginably frantic goings-on inside our bodies. We live in a state of blessed ignorance, but it is also a state of marvelous enlightenment, for it involves floating in a universe of mid-level categories of our own creation — categories that function incredibly well as survival enhancers.

I Am a Strange Loop

In the end, we self-perceiving, self-inventing, locked-in mirages are little miracles of self-reference. We believe in marbles that disintegrate when we search for them but that are as real as any genuine marble when we're not looking for them. Our very nature is such as to prevent us from fully understanding its very nature. Poised midway between the unvisualizable cosmic vastness of curved spacetime and the dubious, shadowy flickerings of charged quanta, we human beings, more like rainbows and mirages than like raindrops or boulders, are unpredictable self-writing poems — vague, metaphorical, ambiguous, and sometimes exceedingly beautiful.

To see ourselves this way is probably not as comforting as believing in ineffable other-worldly wisps endowed with eternal existence, but it has its compensations. What one gives up on is a childlike sense that things are exactly as they appear, and that our solid-seeming, marble-like "I" is the realest thing in the world; what one acquires is an appreciation of how tenuous we are at our cores, and how wildly different we are from what we seem to be. As Kurt Gödel with his unexpected strange loops gave us a deeper and subtler vision of what mathematics is all about, so the strange-loop characterization of our essences gives us a deeper and subtler vision of what it is to be human. And to my mind, the loss is worth the gain.

NOTES

᪥ ᪥ ᪥

Page *xi* *gave me the impetus to read a couple of lay-level books about the human brain...* These were [Pfeiffer] and [Penfield and Roberts]. Another early key influence was [Wooldridge].

Page *xi* *the physical basis...of being...un "I", which...* Placing commas and periods outside quotation marks when they are not part of what is being quoted exhibits greater logic than does American usage, which puts them inside regardless of circumstance. In this book, the logical convention (also the standard in British English) is adopted.

Page *xiv* *Hofstadter's Law...* This comes from Chapter V of [Hofstadter 1979].

Page *xiv* *"What is it like to be a bat?"...* See Chapter 24 in [Hofstadter and Dennett].

Page *xv* *I have spent nearly thirty years...* See, for instance, [Hofstadter and Moser], [Hofstadter and FARG], [Hofstadter 1997], and [Hofstadter 2001].

Page *xviii* *virtually every thought in this book...is an analogy...* See [Hofstadter 2001].

Page *xviii* *not indulging in Pushkinian digressions...* See James Falen's sparkling anglicization of Pushkin's classic novel-in-verse *Eugene Onegin* [Pushkin 1995], or see my own translation [Pushkin 1999]. There is no sublimer marriage of form to content than *Eugene Onegin*.

Page *xviii* *typeset it down to the finest level of detail...* In this book, one of my chief esthetic concerns has been where page breaks fall. A cardinal rule has been that no paragraph (or section) should ever break in such a way that only one line of it occurs at the top or bottom of a page. Another guiding principle has been that the interword spacing in each line should look pleasing, and, in particular, not too loose (which is a frequent and annoying eyesore in computer-set text). In order to avoid such blemishes, I have done touch-up rewriting, often quite extensive, of just about every paragraph in the book. Page *xviii* itself is a typical example of the end result. And of course the page you are right now reading (and that I am right now touching up so that it will please your eye) is another such example.

The foregoing esthetic constraints (along with a number of others that I won't describe here) amount to random darts being thrown at every page in the book, with each dart saying to me, in effect, "Here — don't you think you could rewrite this sentence so that it not only *looks* better but also makes its *point* even more clearly and elegantly?" Some authors might find this tiresome, but I freely confess that I love these random darts and the two-sided challenges that they offer me, and I have worked extremely hard to meet those challenges throughout. There is not a shadow of a doubt that form–content pressures — relentless, intense, and unpredictable — have greatly improved the quality of this book, not only visually but also intellectually.

For a more explicit spelling-out of my views on the magical power of form–content interplay, see [Hofstadter 1997], especially its Introduction and Chapter 5.

Page 5 *no machine can know what words are, or mean...* This ancient idea is the rallying cry of many philosophers, such as John Searle. See Chapter 20 of [Hofstadter and Dennett].

Page 5   *the laws of whose operation are arithmetical...*   This is an allusion to the idea that a
"Giant Electronic Brain", whose very fiber is arithmetical, could act indistinguishably
from a human or animal brain by modeling the arithmetical behavior of all of its
neurons. This would give rise to a kind of artificial intelligence, but very different from
models in which the basic entities are words or concepts governed by rules that reflect
the abstract flow of ideas in a mind rather than the microscopic flow of currents and
chemicals in biological hardware. Chapter XVII of [Hofstadter 1979], Chapter 26 of
[Hofstadter and Dennett], and Chapter 26 of [Hofstadter 1985] all represent
elaborations of this subtle distinction, which I was beginning to explore in my teens.

Page 10   *I don't know what effect it had on her feelings about the picture...*   With some trepidation,
I recently read aloud this opening section of my book to my mother, who, at almost 87,
can only move around her old Stanford house in a wheelchair, but who remains sharp
as a tack and intensely interested in the world around her. She listened with care and
then remarked, "I must have changed a lot since then, because now, those pictures
mean *everything* to me. I couldn't live without them." I doubt that what I said to her
that gloomy day nearly sixteen years ago played much of a role in this evolution of her
feelings, but I was glad in any case to hear that she had come to feel that way.

Page 10   *a tomato is a desireless, soulless, nonconscious entity...*   On the other hand, [Rucker]
proposes that tomatoes, potatoes, cabbages, quarks, and sealing-wax are all conscious.

Page 11   *a short story called "Pig"...*   Found in [Dahl].

Page 16   *In his preface to the volume of Chopin's études...*   All the prefaces that Huneker wrote
in the Schirmer editions can be found in [Huneker].

Page 18   *What gives us word-users the right to make...*   See [Singer and Mason].

Page 20   *it is made of 'the wrong stuff'...*   That brains but not computers are made of "the
right stuff" is a slogan of John Searle. See Chapter 20 in [Hofstadter and Dennett].

Page 23   *Philosophers of mind often use the terms...*   See, for example, [Dennett 1987].

Page 25   *"What do I mean...by 'brain research'?"...*   See [Churchland], [Dennett 1978],
[Damasio], [Flanagan], [Hart], [Harth], [Penfield], [Pfeiffer], and [Sperry].

Page 26   *these are all legitimate and important objects of neurological study...*   See [Damasio],
[Kuffler and Nicholls], [Wooldridge], and [Penfield and Roberts].

Page 26   *abstractions are central...in the study of the brain...*   See [Treisman], [Minsky 1986],
[Schank], [Hofstadter and FARG], [Kanerva], [Fauconnier], [Dawkins], [Blackmore],
and [Wheelis] for spellings-out of these abstract ideas.

Page 27   *Just as the notion of "gene" as an invisible entity that enabled...*   See [Judson].

Page 27   *and just as the notion of "atoms" as the building blocks...*   See [Pais 1986], [Pais 1991],
[Hoffmann], and [Pullman].

Page 28   *Turing machines are...idealized computers...*   See [Hennie] and [Boolos and Jeffrey].

Page 29   *In his vivid writings, Searle gives...*   See Chapter 22 of [Hofstadter and Dennett].

Page 29   *one particular can that would "pop up"...*   In his smugly dismissive review [Searle] of
[Hofstadter and Dennett], Searle states: "So let us imagine our thirst-simulating
program running on a computer made entirely of old beer cans, millions (or billions) of
old beer cans that are rigged up to levers and powered by windmills. We can imagine
that the program simulates the neuron firings at the synapses by having beer cans bang
into each other, thus achieving a strict correspondence between neuron firings and
beer-can bangings. And at the end of the sequence a beer can pops up on which is
written 'I am thirsty.' Now, to repeat the question, does anyone suppose that this Rube
Goldberg apparatus is literally thirsty in the sense in which you and I are?"

Page 30   *Dealing with brains as multi-level systems...*   See [Simon], [Pattee], [Atlan], [Dennett 1987], [Sperry], [Andersen], [Harth], [Holland 1995], [Holland 1997], and the dialogue "Prelude... Ant Fugue" in [Hofstadter 1979] or in [Hofstadter and Dennett].

Page 31   *such as a column in the cerebral cortex...*   See [Kuffler and Nicholls].

Page 31   *I once saw a book whose title was "Molecular Gods..."*   This was [Applewhite].

Page 31   *to quote here a short passage from Sperry's essay...*   Taken from [Sperry].

Page 32   *taken from "The Floor"...*   See [Edson], which is a thin, remarkably vivid, highly surrealistic, often hilarious, and yet profoundly depressing collection of prose poems.

Page 33   *such macroscopic phenomena as friction...*   A beautiful and accessible account of the emergence of everyday phenomena (such as how paper tears) out of the surrealistically weird quantum-mechanical substrate of our world is given in [Chandrasekhar].

Page 34   *quarks, gluons, W and Z bosons...*   See [Pais 1986] and [Weinberg 1992].

Page 35   *Drastic simplification is what allows us to...discover abstract essences...*   See [Kanerva], [Kahneman and Miller], [Margolis], [Sander], [Schank], [Hofstadter and FARG], [Minsky 1986], and [Gentner *et al.*].

Page 38   *641, say...*   I chose the oddball integer 641 because it plays a famous role in the history of mathematics. Fermat conjectured that all integers of the form $2^{2^n} + 1$ are prime, but Euler discovered that 641 (itself a prime) divides $2^{2^5} + 1$, thus refuting Fermat's conjecture. See [Wells 1986], [Wells 2005], and [Hardy and Wright].

Page 41   *Deep understanding of causality...*   See [Pattee], [Holland 1995], [Holland 1997], [Andersen], [Simon], and Chapter 26 of [Hofstadter 1985].

Page 45   *The Careenium...*   Chapter 25 of [Hofstadter 1985] is a lengthy Achilles–Tortoise dialogue spelling out the careenium metaphor in detail.

Page 49   *The effect...was explained...by Albert Einstein...*   See [Hoffmann] and [Pais 1986].

Page 49   *From this perspective, there are no simmballs, no symbols...*   This view approaches the extreme reductionist philosophy expressed in [Unger 1979] and also in [Unger 1979].

Page 52   *Why does this move to a goal-oriented — that is, teleological — shorthand...*   See [Monod], [Cordeschi], [Haugeland 1981], and [Dupuy 2000].

Page 53   *In the video called "Virtual Creatures" by Karl Sims...*   This is found easily on the Web.

Page 53   *a strong pressure to shift...to the goal-oriented level of cybernetics...*   See [Dupuy 2000], [Monod], [Cordeschi], [Simon], [Andersen], and Chapter 11 in [Hofstadter and Dennett], which discusses a trio of related "isms" — holism, goalism, and soulism.

Page 54   *the story of a sultan who commanded...*   Found in the charming old book [Gamow].

Page 55   *contains the seeds of its own destruction...*   Compare this scenario of self-breaking to the story recounted in the dialogue "Contracrostipunctus" in [Hofstadter 1979].

Page 57   *I stumbled upon...a little paperback...*   Of course this was [Nagel and Newman].

Page 57   *I'm sure I didn't think "he or she"...*   See Chapters 7 and 8 of [Hofstadter 1985].

Page 60   *pushed my luck and invented the more threeful phrase...*   Although I didn't know it, I was dimly sensing the infinite hierarchy of arithmetical operations and what I would later come to know as "Ackermann's function". See [Boolos and Jeffrey] and [Hennie].

Page 61   *a pathological retreat from common sense...*   I cannot resist pointing out that *Principia Mathematica* opens with a grand flourish of self-reference, its first sentence unabashedly declaring: "The mathematical treatment of the principles of mathematics, which is the subject of the present work, has arisen from the conjunction of two different studies, both in the main very modern." *Principia Mathematica* thus points at itself through the proud phrase "the present work" — exactly the kind of self-pointer that, in more formal contexts, its authors were at such pains to forbid categorically. Perhaps more weirdly, the chapter in which the self-reference–banning theory of types is presented also opens self-referentially: "The theory of logical types, to be explained in the present Chapter, recommended itself to us in the first instance by its ability to solve certain contradictions…" Note finally that the pronoun "us" is yet another self-pointer that Russell and Whitehead have no qualms using. Were they not aware of these ironies?

Page 62   *the topic of self-reference in language…*   See Chapters 1–4 of [Hofstadter 1985].

Page 62   *This pangram tallies…*   This perfectly self-tallying or self-inventorying "pangram" was discovered by Lee Sallows using an elaborate analog computer that he built.

I have often mused about a large community of sentences somewhat like Sallows', each one inventorying not only *itself* (*i.e.*, giving 26 letter-counts as above), but in addition some or perhaps all of the others. Thus each sentence would be far, far longer than Sallows' pangram. However, in my fantasy, these "individuals", unlike Sallows' remarkable sentence, do not all give accurate reports. Some of what they say is dead wrong. In the *self*-inventorying department, I imagine most of them as being fairly accurate (most of their 26 "first-person" counts would be precisely right, with just a few perhaps being a little bit off). On the other hand, each sentence's inventory of *other* sentences would vary in accuracy, from being somewhat close to being wildly far off.

Needless to say, this is a metaphor for a society of interacting human beings, each of whom has a fairly accurate self-image and less accurate images of others, often based on very quick and inaccurate glances. Two sentences that "know each other well" (*i.e.*, that have reasonably accurate though imperfect inventories of each other) would be the analogue of good friends, whereas two sentences that have rough, partial, or vacuous representations of each other would be the analogue of strangers.

A more complex variation on this theme involves a population of Sallows-type sentences varying in time. At the outset, they would all be filled with random numbers, but then they would all get updated in parallel. Specifically, each one would replace its wrong inventories by counting letters inside itself and in a few other sentences, and replacing the wrong values by the values just found. Of course, since everything is a moving target, the letter-counts would still be wrong, but hopefully over the course of a long series of such parallel iterations, each sentence would tend, at least on average, to gain greater accuracy, especially concerning itself, and simultaneously to form a small clique of "friends" (sentences that it inventories fairly fully and well), while remaining remote from most members of the population (*i.e.*, representing them at best sparsely and with many errors, or perhaps not even at all). This is a kind of caricature of my ideas about people "living inside each other", proposed in Chapters 15 through 18.

Page 63   *Perhaps there is no harm…*   Quoted from [Skinner] in George Brabner's letter.

Page 63   *I wrote a lengthy reply to it…*   This is found in Chapter 1 of [Hofstadter 1985].

Page 68   *If dogs were a bit more like robots…*   As I was putting the finishing touches on these notes, my children and I flew out to California for Christmas break. We were gliding low, approaching the San Jose airport at night, when Danny, who was peering out the window, said to me, "You know what I just saw?" "What?" I replied, having not the foggiest idea. He said, "A parking lot packed with cars whose headlights and taillights were all flashing on and off at random!" "Why were they all doing *that*?" I asked, a bit densely. Danny instantly supplied the answer: "Their alarm systems were all triggering each other. I know that's what it was, because I've seen fireworks set car alarms off." Seeing this in my mind's eye, I grinned from ear to ear with delight and amazement, all the more so since Danny hadn't read any of my manuscript and had no idea how relevant his sighting of reverberant honking and flashing was to my book — in fact to the chapter that I was writing notes for just then (Chapter 5). Danny's reverberant parking lot truly put reverberant barking to shame, and what an infernal racket it must have been for people down on the ground! And yet, as observed from above by chance voyeurs in the plane, it was a totally silent, surrealistic vision of robots who had gotten one another all excited, and who certainly weren't about to calm down, as dogs will. What a stupendous last-minute addition to my book!

Page 69   *the amazing visual universe discovered around 1980...*   See [Peitgen and Richter].

Page 76   *winds up triggering a small set...*   See [Kanerva] and [Hofstadter and FARG].

Page 77   *Suppose we begin with a humble mosquito...*   See [Griffin] and [Wynne]. The latter contains a remarkable account of analogy-making by bees, of all creatures!

Page 80   *cars that drive themselves down...highways or across rocky deserts...*   See [Davis 2006].

Page 82   *structure that represents itself (i.e., the dog itself, not the symbol itself!)...*   This sounds like a joke, but not entirely. When it comes to the self-symbols of humans — their "I"'s — much of the structure of the "I" involves pointers that point right back at the abstraction "I", and not just at the body. This is discussed in Chapters 13 and 16.

Page 83   *their category systems became arbitrarily extensible...*   I defend this point of view in [Hofstadter 2001]. For more on human categories, see [Sander], [Margolis], [Minsky 1986], [Schank], [Aitchison], [Fauconnier], [Hofstadter 1997], and [Gentner *et al.*].

Page 85   *memories of episodes can be triggered...*   See [Kanerva], [Schank], and [Sander].

Page 86   *That deep and tangled self-model is what "I"-ness is all about...*   See [Dennett 1991], [Metzinger], [Horney 1942], [Horney 1945], [Wheelis], [Nørretranders], and [Kent].

Page 89   *Abstraction piled on abstraction...*   Should anyone care to get a taste of this, try reading [Ash and Gross] all the way to the end. It's a bit like ordering "Indian hot" in an authentic Indian restaurant — you'll wonder why you ever did.

Page 91   *radicals, such as Évariste Galois...*   The great Galois was indeed a young radical, which led to his absurdly tragic death in a duel on his twenty-first birthday, but the phrase "solution by radicals" really refers to the taking of nth roots, called "radicals". For a shallow, a medium, and a deep dip into Galois' immortal, radical insights into hidden mathematical structures, see [Livio], [Bewersdorff], and [Stewart], respectively.

Page 95   *there is a special type of abstract structure or pattern...*   "Real Patterns" in [Dennett 1998] argues powerfully for the reality of abstract patterns, based on John Conway's cellular automaton known as the "Game of Life". The Game of Life itself is presented ideally in [Gardner], and its relevance to biological life is spelled out in [Poundstone].

Page 102   *I am sorry to say, now hackneyed...*   I have long loved Escher's art, but as time has passed, I have found myself drawn ever more to his early non-paradoxical landscapes, in which I see hints everywhere of his sense of the magic residing in ordinary scenes. See [Hofstadter 2002], an article written for a celebration of Escher's 100th birthday.

Page 103   *Is there, then, any genuine strange loop — a paradoxical structure that...*   Three excellent books on paradoxes are [Falletta], [Hughes and Brecht], and [Casati and Varzi 2006].

Page 104   *an Oxford librarian named G. G. Berry...*   Only two individuals are thanked by the (nearly) self-sufficient authors of *Principia Mathematica*, and G. G. Berry is one of them.

Page 108   *Chaitin and others went on...*   See [Chaitin], packed with stunning, strange results.

Page 113   *written in PM notation as...*   I have here borrowed Gödel's simplified version of *PM* notation instead of taking the symbols directly from the horses' mouths, for those would have been too hard to digest. (Look at page 123 and you'll see what I mean.)

Page 114   *the sum of two squares...*   See [Hardy and Wright] and [Niven and Zuckerman].

Page 114   *the sum of two primes...*   See [Wells 2005], an exquisite garden of delights.

Page 116   *The passionate quest after order in an apparent disorder is what lights their fires...*   See [Ulam], [Ash and Gross], [Wells 2005], [Gardner], [Bewersdorff], and [Livio].

Page 117   *Nothing happens "by accident" in the world of mathematics...*   See [Davies].

Page 118   *Paul Erdös once made the droll remark...*   Erdös, a devout matheist, often spoke of proofs from "The Book", an imagined tome containing God's perfect proofs of all great truths. For my own vision of "matheism", see Chapter 1 of [Hofstadter and FARG].

Page 119   *Variations on a Theme by Euclid...*   See [Chaitin].

Page 120   *God does not play dice...*   See [Hoffmann], one of the best books I have ever read.

Page 121   *many textbooks of number theory prove this theorem...*   See [Hardy and Wright] and [Niven and Zuckerman].

Page 122   *About a decade into the twentieth century...*   The history of the push to formalize mathematics and logic is well recounted in [DeLong], [Kneebone], and [Wilder].

Page 122   *a young boy was growing up in the town of Brünn...*   See [Goldstein] and [Yourgrau].

Page 125   *Fibonacci...explored what are now known as the "Fibonacci numbers"...*   See [Huntley].

Page 125   *This almost-but-not-quite-circular fashion...*   See [Péter] and [Hennie].

Page 126   *a vast team of mathematicians...*   A recent book that purports to convey the crux of the elusive ideas of this team is [Ash and Gross]. I admire their chutzpah in trying to communicate these ideas to a wide public, but I suspect it is an impossible task.

Page 126   *a trio of mathematicians...*   These are Yann Bugeaud, Maurice Mignotte, and Samir Siksek. It turns out that to prove that 144 is the only *square* in the Fibonacci sequence (other than 1) does not require highly abstract ideas, although it is still quite subtle. This was accomplished in 1964 by John H. E. Cohn.

Page 128   *Gödel's analogy was very tight...*   The essence and the meaning of Gödel's work are well presented in many books, including [Nagel and Newman], [DeLong], [Smullyan 1961], [Jeffrey], [Boolos and Jeffrey], [Goodstein], [Goldstein], [Smullyan 1978], [Smullyan 1992], [Wilder], [Kneebone], [Wolf], [Shanker], and [Hofstadter 1979].

Page 129   *developed piecemeal over many centuries...*   See [Nagel and Newman], [Wilder], [Kneebone], [Wolf], [DeLong], [Goodstein], [Jeffrey], and [Boolos and Jeffrey].

Page 135   *Anything you can do, I can do better!...*   My dear friend Dan Dennett once wrote (in a lovely book review of [Hofstadter and FARG], reprinted in [Dennett 1998]) the following sentence: "'Anything you can do I can do meta' is one of Doug's mottoes, and of course he applies it, recursively, to everything he does."

Well, Dan's droll sentence gives the impression that Doug himself came up with this "motto" and actually went around saying it (for why else would Dan have put it in quote marks?). In fact, I had never said any such thing nor thought any such thought, and Dan was just "going me one meta", in his own inimitable way. To my surprise, though, this "motto" started making the rounds and people quoted it back to me as if I really had thought it up and really believed it. I soon got tired of this because, although Dan's motto is clever and funny, it does not match my self-image. In any case, this note is just my little attempt to squelch the rumor that the above-displayed motto is a genuine Hofstadter sentence, although I suspect my attempt will not have much effect.

Page 137   *suppose you wanted to know if statement X is true or false...*   The dream of a mechanical method for reliably placing statements into two bins — 'true' and 'false' — is known as the quest for a *decision procedure*. The absolute nonexistence of a decision procedure for truth (or for provability) is discussed in [DeLong], [Boolos and Jeffrey], [Jeffrey], [Hennie], [Davis 1965], [Wolf], and [Hofstadter 1979].

Page 139   *No formula can literally contain...*   [Nagel and Newman] presents this idea very clearly, as does [Smullyan 1961]. See also [Hofstadter 1982].

Page 139   *an elegant linguistic analogy...*   See [Quine] for the original idea (which is actually a variation of Gödel's idea (which is itself a variation of an idea of Jules Richard (which is a variation of an idea of Georg Cantor (which is a variation of an idea of Euclid (with help from Epimenides))))), and [Hofstadter 1979] for a variation on Quine's theme.

Page 147   *"...and Related Systems (I)"...*   Gödel put a roman numeral at the end of the title of his article because he feared he had not spelled out sufficiently clearly some of his ideas, and expected he would have to produce a sequel. However, his paper quickly received high praise from John von Neumann and other respected figures, catapulting the unknown Gödel to a position of great fame in a short time, even though it took most of the mathematical community decades to absorb the meaning of his results.

Page 150   *respect for...the most mundane of analogies...*   See [Hofstadter 2001] and [Sander], as well as Chapter 24 in [Hofstadter 1985] and [Hofstadter and FARG].

Page 159   *X's play is so mega-inconsistent...*   This should be heard as "X's play is omega-inconsistent", which makes a phonetic hat-tip to the metamathematical concepts of *omega-inconsistency* and *omega-incompleteness*, discussed in many books in the Bibliography, such as [DeLong], [Nagel and Newman], [Hofstadter 1979], [Smullyan 1992], [Boolos and Jeffrey], and others. For our more modest purposes here, however, it suffices to know that this "o"-containing quip, plus the one two lines below it, is a play on words.

Page 160   *Indeed, some years after Gödel, such self-affirming formulas were concocted...*   See [Smullyan 1992], [Boolos and Jeffrey], and [Wolf].

Page 164   *Why would logicians...give such good odds...*   See [Kneebone], [Wilder], and [Nagel and Newman], for reasons to believe strongly in the consistency of *PM*-like systems.

Page 165   *not only although...but worse, because...*   For another treatment of the perverse theme of "although" turning into "because", see Chapter 13 of [Hofstadter 1985].

Page 166   *the same Gödelian trap would succeed in catching it...*   For an amusing interpretation of the infinite repeatability of Gödel's construction as demonstrating the impossibility of artificial intelligence, see the chapter by J. R. Lucas in [Anderson], which is carefully analyzed (and hopefully refuted) in [DeLong], [Webb], and [Hofstadter 1979].

Page 167   *called "the Hilbert Program"...*   See [DeLong], [Wolf], [Kneebone], and [Wilder].

Page 170   *In that most delightful though most unlikely of scenarios...*   [DeLong], [Goodstein], and [Chaitin] discuss non-Gödelian formulas that are undecidable for Gödelian reasons.

Page 172   *No reliable prim/saucy distinguisher can exist...*   See [DeLong], [Boolos and Jeffrey], [Jeffrey], [Goodstein], [Hennie], [Wolf], and [Hofstadter 1979] for discussions of many limitative results such as this one (which is Church's theorem).

Page 172   *It was logician Alfred Tarski who put one of the last nails...*   See [Smullyan 1992] and [Hofstadter 1979] for discussions of Tarski's deep result. In the latter, there is a novel approach to the classical liar paradox ("This sentence is not true") using Tarski's ideas, with the substrate taken to be the human brain instead of an axiomatic system.

Page 172   *what appears to be a kind of upside-down causality...*   See [Andersen] for a detailed technical discussion of downward causality. Less technical discussions are found in [Pattee] and [Simon]. See also Chapters 11 and 20 in [Hofstadter and Dennett], and especially the Reflections. [Laughlin] gives fascinating arguments for the thesis that in physics, the macroscopic arena is more fundamental or "deeper" than the microscopic.

Page 174   *leaving just a high-level picture of information-manipulating processes...*   See [Monod], [Berg and Singer], [Judson], and Chapter 27 of [Hofstadter 1985].

Page 177   *symbols in our respective brains...*   See [Hofstadter 1979], especially the dialogue "Prelude... Ant Fugue" and Chapters 11 and 12, for a careful discussion of this notion.

Page 178   *the forbidding and inaccessible level of quarks and gluons...*   See [Weinberg 1992] and [Pais 1986] for attempts at explanations of these incredibly abstruse notions.

Page 178   *the only slightly more accessible level of genes...*   See [Monod], [Berg and Singer], [Judson], and Chapter 27 ("**T**he **G**enetic **C**ode: **A**rbitrary?") in [Hofstadter 1985].

Page 179   *we...best understand our own actions as...*   See [Dennett 1987] and [Dennett 1998].

Page 181   *embellished by a fantastic folio of alternative versions...*   [Steiner 1975] has a rich and provocative discussion of "alternity", and the dialogue "Contrafactus" in [Hofstadter 1979] features an amusing scenario involving "subjunctive instant replays". See also [Kahneman and Miller] and Chapter 12 of [Hofstadter 1985] for further musings on the incessantly flickering presence of counterfactuals in the subconscious human mind. [Hofstadter and FARG] describes a family of computational models of human thought processes in which making constant forays into alternity is a key architectural feature.

Page 182   *housing a loop of self-representation...*   See [Morden], [Kent], and [Metzinger].

Page 186   *as the years pass, the "I" converges and stabilizes itself...*   See [Dennett 1992].

Page 188   *we cannot help attributing reality to our "I" and to those of other people...*   See [Kent], [Dennett 1992], [Brinck], [Metzinger], [Perry], and [Hofstadter and Dennett].

Page 189   *I was most impressed when I read about "Stanley", a robot vehicle...*   See [Davis 2006].

Page 193   *just a big spongy bulb of inanimate molecules...*   I suppose almost any book on the brain will convince one of this, but [Penfield and Roberts] did it to me as a teen-ager.

Page 194   *pioneering roboticist and provocative writer Hans Moravec...*   For some of Moravec's more provocative speculations about the near-term future of humanity, see [Moravec].

Page 194   *from the organic chemistry of carbon...*   See Chapter 22 in [Hofstadter and Dennett], in which John Searle talks about "the right stuff", which underwrites what he terms "the semantic causal powers of the brain", a rather nice-sounding but murky term by which Searle means that when a human brain, such as his own or, say, that of poet Dylan Thomas, makes its owner come out with words, those words don't just *seem* to stand for something, they really *do* stand for something. Unfortunately, in the case of poet Thomas, most of his output, though it sounds rather nice, is so full of murk that one has to wonder what sort of "stuff" could possibly make up the brain behind it.

Page 199   *its symbol-count might well exceed "Graham's constant"...*   See [Wells 1986].

Page 208   *For those who enjoy the taboo thrills of non-wellfounded sets...*   See [Barwise and Moss].

Page 209   *the deeper and richer an organism's categorization equipment is...*   See [Hofstadter 2001].

Page 233   *a devilishly clever bon mot by David Moser...*   One evening not long after we were married, Carol and I invited some friends over for an Indian dinner at our house in Ann Arbor. Melanie Mitchell and David Moser, well aware of Carol's terrific Indian cooking, were delighted to come. It turned out, however, that at the last minute, our oldest guests, in their eighties, called up to tell us that they couldn't handle very spicy foods, which unfortunately torpedoed Carol's cooking plans. Somehow, though, she turned around on a dime and prepared a completely different yet truly delicious repast. A couple of hours after dinner was over, after a very lively discussion, most of our guests took off, leaving just David, Melanie, Carol, and me. We chatted on for a while, and finally, as they were about to hit the road, Carol casually reminded them of what she had originally intended to fix and told them why she hadn't been able to follow through on her promise. Quick as a wink, David, feigning great indignation, burst out, "Why, you Indian-dinner givers, you!"

Page 233   *her personal gemma (to borrow Stanislaw Lem's term...)...*   See "Non Serviam" in [Hofstadter and Dennett], which is a virtuosic philosophical fantasy masquerading as a book review (of a book that, needless to say, is merely a figment of Lem's imagination).

Page 239   *someone trying to grapple with quantum-mechanical reality...*   [Pais 1986], [Pais 1992], and [Pullman] portray the transition period between the Bohr atom and quantum mechanics, while [Jauch] and [Greenstein and Zajonc] chart remaining mysteries.

Page 239   *it might be tempting for some readers to conclude that in the wake of Carol's death...*   See Chapter 15 of [Hofstadter 1997], another place where I discuss many of these ideas.

Page 242   *meaning of the term "universal machine"...*   See [Hennie] and [Boolos and Jeffrey].

Page 248   *concepts are active symbols in a brain...*   See Chapter 11 of [Hofstadter 1979].

Page 252   *a marvelous pen-and-ink "parquet deformation" drawn in 1964...*   For a dozen-plus examples of this subtle Escher-inspired art form, see Chapter 10 of [Hofstadter 1985].

Page 260   *It is not easy to find a strong, vivid metaphor to put up against the caged-bird metaphor...*   The idea of a soul distributed over many brains brought to my mind an image from solid-state physics, the field in which I did my doctoral work. A solid is a crystal, meaning a periodic lattice of atoms in space, like the trees in an orchard but in three dimensions instead of two. In some solids (those that do not conduct electricity), the electrons "hovering" around each atomic nucleus are so tightly bound that they never stray far from that nucleus. They are like butterflies that hover around just one tree in the orchard, never daring to venture as far as the next tree. In metals, by contrast, which are excellent conductors, the electrons are not timid stay-at-homes stuck to one tree, but boldly float around the entire lattice. This is why metals conduct so well.

Actually, the proper image of an electron in a metal is not that of a butterfly fickly fluttering from one tree to another, never caring where it winds up, but of an intensity pattern distributed over the entire crystal at once — in some places more intense, in other places less so, and changing over time. One electron might better be likened to an entire swarm of orange butterflies, another electron to a swarm of red butterflies, another to a swarm of blue butterflies, and so on, with each swarm spread about the whole orchard, intermingling with all the others. Electrons in metals, in short, are anything but tightly bound dots; they are floating patterns without any home at all.

But let's not lose track of the purpose of all this imagery, which is to suggest helpful ways of imagining what a human soul's essence is. If we map each tree (or nucleus) in the crystal lattice onto a particular human brain, then in the tight-binding model (which corresponds to the caged-bird metaphor), each brain would possess a unique soul, represented by the cloud of timid butterflies that hover around it and it alone. By contrast, if we think of a metal, then the cloud is spread out across the whole lattice — which is to say, shared equally among all the trees (or nuclei). No tree is privileged. In this image, then (which is close to Daniel Kolak's view in *I Am You*), each human soul floats among all human brains, and its identity is determined not by its location but by the undulating global pattern it forms.

These are extremes, but nothing keeps us from imagining a halfway situation, with many localized swarms of butterflies, each swarm floating near a single tree but not limited to it. Thus a red swarm might be centered on tree A but blur out to the nearest dozen trees, and a blue swarm might be blurrily centered on tree B, a yellow swarm around tree C, etc. Each tree would be the center of just one swarm, and each swarm would have just one principal tree, but the swarms would interpenetrate so intimately that it would be hard to tell which swarm "belonged" to which tree, or vice versa.

This peculiar and surreal tale, launched in solid-state physics but winding up with imagery of interpenetrating swarms of colored butterflies fluttering in an orchard, gives as clear a picture as I can paint of how a human soul is spread among brains.

Page 264   *Many of these ideas were explored…in his philosophical fantasy "Where Am I?"…*   This classic piece can be found in [Dennett 1978] and in [Hofstadter and Dennett].

Page 267   *internal conflict between several "rival selves"…*   Chapter 13 of [Dennett 1991] gives a careful discussion of multiple personality disorder. See also [Thigpen and Cleckley], from which a famous movie was made. See also [Minsky 1986] and Chapter 33 of [Hofstadter 1985] for views of a normal self as containing many competing subselves.

Page 267   *in such cases Newtonian physics goes awry…*   See [Hoffmann] for a discussion of the subtle relationship between relativistic and Newtonian physics.

Page 271   *every entity…is conscious…*   See [Rucker] for a positive view of panpsychism.

Page 276   *because now they want the symbols themselves to be perceived…*   See the careful debunking in [Dennett 1991] of what its author terms the "Cartesian Theater".

Page 277   *to trigger just one familiar pre-existing symbol…*   This sentence is especially applicable to the nightmare of preparing an index. Only if one has slaved away for weeks on a careful index can one have an understanding of how grueling (and absurd) the task is.

Page 278   *when its crust is discarded and its core is distilled…*   See [Sander], [Kahneman and Miller], [Kanerva], [Schank], [Boden], and [Gentner *et al.*] for discussions of the analogy-based mechanisms of memory retrieval, which underlie all human cognition.

Page 279   *to simplify while not letting essence slip away…*   See [Hofstadter 2001], [Sander], and [Hofstadter and FARG]. To figure out how to give a computer the rudiments of this ability has been the Holy Grail of my research group for three decades now.

Page 279   *There is not some special "consciousness locus"…*   See [Dennett 1991].

Page 282   *but we are getting ever closer…*   See [Monod], [Cordeschi], and [Dupuy 2000] for clear discussions of the emergence of goal-orientedness (*i.e.*, teleology) from feedback.

Page 283   *a physical vortex, like a hurricane or a whirlpool…*   See Chapter 22 of [Hofstadter 1985] for a discussion of the abstract essence of hurricanes.

Page 283   *every integer is the sum of at most four squares…*   See [Hardy and Wright] and [Niven and Zuckerman] for this classic theorem, the simplest case of Waring's theorem.

Page 285   *to see that brilliant purple color of the flower…*   See [Chalmers] for a spirited defense of the notion of qualia, and see [Dennett 1991], [Dennett 1998], [Dennett 2005], and [Hofstadter and Dennett], which do their best to throw a wet blanket on the idea.

Page 287   *There is no meaning to the letter "b"…*   See the dialogue "Prelude… Ant Fugue" (found in both [Hofstadter 1979] and [Hofstadter and Dennett]) for a discussion of how meanings at a high level can emerge from meaningless symbols at a low level.

Page 293   *the notion that consciousness is a novel kind of quantum phenomenon…*   See [Penrose], which views consciousness as an intrinsically quantum-mechanical phenomenon, and [Rucker], which views consciousness as uniformly pervading everything in the universe.

Page 295   *Taoism and Zen long ago sensed this paradoxical state…*   Far and away the best book I have read on these spiritual approaches to life is [Smullyan 1977], but [Smullyan 1978] and [Smullyan 1983] also contain excellent pieces on the topic. These ideas are also discussed in Chapter 9 of [Hofstadter 1979], but from a skeptical point of view.

Page 296   *the story of an "I" is a tale about a central essence…*   See [Dennett 1992] and [Kent].

Page 298   *The…self-pointing loop that the pronoun "I" involves …*   See [Brinck] and [Kent].

Page 299   *This is what John von Neumann unwittingly revealed…*   See [von Neumann] for a very difficult and [Poundstone] for a very lucid discussion of self-replicating automata.

See Chapters 2 and 3 of [Hofstadter 1985] for a simpler discussion of the same ideas. Chapter 16 of [Hofstadter 1979] carefully spells out the mapping between Gödel's self-referential construction and the self-replicating mechanisms at the core of life.

Page 300   *too marbelous for words...*   Borrowing a few words from a love song by Johnny Mercer and Richard Whiting, sung in an unsurpassable fashion by Frank Sinatra.

Page 300   *with alacrity, celerity, assiduity, vim, vigor, vitality...*   My father's friend Bob Herman (a top-notch physicist who famously co-predicted the cosmic background radiation fifteen years before it was observed) loved to recite this riddle, putting on a strong Yiddish accent: "A tramp in the woods happened upon a hornets' nest. When they stung him with alacrity, celerity, assiduity, vim, vigor, vitality, savoir-faire, and undue velocity, 'Oh!', he mused, counting his bumps, 'If I had as many bumps on the left side of my right adenoid as six and three-quarters times seven-eighths of those between the heel of Achilles and the circumference of Adam's apple, how long would it take a boy rolling a hoop up a moving stairway going down to count the splinters on a boardwalk if a horse had six legs?'" And so I thought I'd give a little posthumous hat-tip to Bob.

Page 305   *Dan calls such carefully crafted fables 'intuition pumps'...*   Dennett introduced his term "intuition pump", I believe, in the Reflections that he wrote on John Searle's "Chinese room" thought experiment in Chapter 22 of [Hofstadter and Dennett].

Page 308   *The term Parfit prefers is "psychological continuity"...*   See [Nozick] for a lengthy treatment of the closely related concept of "closest continuer".

Page 309   *what Einstein accomplished in creating special relativity...*   See [Hoffmann].

Page 309   *what a whole generation of brilliant physicists, with Einstein at their core...*   See [Pais 1986], [Pais 1991], and [Pullman].

Page 315   *just tendencies and inclinations and habits, including verbal ones...*   See the Prologue for my first inklings of this viewpoint. See also my Achilles–Tortoise dialogue entitled "A Conversation with Einstein's Brain", which is Chapter 26 in [Hofstadter and Dennett], for more evolved ideas on it.

Page 320   *Dave Chalmers explores these issues...*   See [Chalmers]. I always find it ironic that Dave's highly articulate and subtle ideas on consciousness, so wildly opposed to my own, took shape right under my nose some fifteen or so years ago, in my very own Center for Research on Concepts and Cognition, at Indiana University (although the old oaken table in Room 641 is a bit of a tall tale...). Dave added enormous verve to our research group, and he was a good friend to both Carol and me. Despite our disagreements on qualia, zombies, and consciousness, we remain good friends.

Page 321   *with a nine-planet solar system...*   I'm not about to enter into the raging debate over poor Pluto's possible planethood (is Disney's Pluto a dog?), although I think the question is a fascinating one from the point of view of cognitive science, since it opens up deep questions about the nature of categories and analogies in the human mind.

Page 322   *Z-people...laugh exactly the same as...Q-people...*   See "Planet without Laughter" in [Smullyan 1980], a wonderful tale about vacuously laughing zombies.

Page 324   *Dan Dennett's criticism of such philosophers hits the nail on the head...*   See especially "The Unimagined Preposterousness of Zombies" in [Dennett 1998] and "The Zombic Hunch" in [Dennett 2005] for marvelous Dennettian arguments.

Page 325   *you can quote me on that...*   Actually, the image is Bill Frucht's, so you can quote Bill on that. I had originally written something about a Flash Gordon–style hood ornament, and Bill, probably correctly seeing this 1950's image as too passé, perhaps even camp, pulled me single-handedly into the twenty-first century.

Page 326   *What is this nutty Capitalized Essence all about?*   I concocted the phrase "Capitalized Essences" when I wrote the dialogue "Three-Part Invention" in [Hofstadter 1979].

Page 333   *for all you know, what I am experiencing as redness…*   The most penetrating discussion of the inverted-spectrum riddle that I have read is that in [Dennett 1991].

Page 333   *Bleu Blanc Rouge…*   The colors of the French flag are red, white, and blue, but the French always recite them in the order "blue, white, red". This makes for a tongue-in-cheek suggestion that their color experiences are "just like ours, but flipped".

Page 339   *the so-called problem of "free will"…*   There had to be some arena in which Dan Dennett and I do not quite see eye to eye, and at this late point in my book we have finally hit it. It is the question of free will. I agree with most of Dan's arguments in [Dennett 1984], and yet I can't go along with him that we have free will, of any sort. One day, Dan and I will thrash this out between ourselves.

Page 340   *the analogy to our electoral process is such a blatant elephant…*   This idea of "votes" in the brain is discussed in Chapter 33 of [Hofstadter 1985], as well as in the Careenium dialogue, which is Chapter 25 of the same book.

Page 345   *gentle people such as…César Chávez…*   In the late 1960's and early 1970's, deeply depressed by the assassinations of Martin Luther King and Robert Kennedy, I worked intensely for the United Farm Workers Organizing Committee (later known as the "United Farm Workers of America") for a couple of years, first as a frequent volunteer and then for several months as a boycott organizer (first for grapes, then for lettuce). In this capacity I had the chance to meet with César Chávez a few times, although to my great regret I never truly got to know him as a person.

Page 346   *As far as I can peer back…*   The translation is my own.

Page 347   *This was an abhorrent proposal…*   The translation is my own.

Page 347   *a book entitled 'Le Cerveau et la conscience'…*   This was [Chauchard].

Page 351   *Many performers have been performing…*   The translation is my own.

Page 361   *Riposte: A Soft Poem…*   There is a method to my madness in this section. In particular, both paragraphs were written to an ancient kind of meter called "paeonic". What this means is that three syllables go by without a stress, but on the fourth a stress is placed, without its seeming (so I hope) to have been forced: "And yet to *you*, my *faith*ful *read*er who has *plowed* all through this *book* up to its *near*ly final *page…*" One last constraint upon both paragraphs is simply on their length in terms of "feet" (which means stressed syllables). The number of these "paeons" must be forty, and the reason is, I'm mimicking two paragraphs of forty paeons each on page 5a of *Le Ton beau*.

Page 376   *There is a method to my madness…*   There is a method to my madness in this footnote. In particular, the footnote both describes and represents an ancient meter called "paeonic". What this means is that three syllables go by without a stress, but on the fourth a stress is placed, without its seeming (so I hope) to have been forced. I now will offer one small sample for your pleasure, and respectfully suggest that you try reading it aloud: "There is a *meth*od to my *mad*ness in this *foot*note…" In particular, I've got to use exactly forty feet because I'm mimicking two paragraphs of forty paeons each on page three hundred six-and-seventy of *I Am a Strange Loop.*

BIBLIOGRAPHY

☙ ☙ ☙

Aitchison, Jean. *Words in the Mind: An Introduction to the Mental Lexicon* (second edition). Cambridge, Mass.: Blackwell, 1994.

Andersen, Peter B. *et al.* (eds.). *Downward Causation: Minds, Bodies, and Matter.* Aarhus: Aarhus University Press, 2000.

Anderson, Alan Ross. *Minds and Machines.* Englewood Cliffs, N.J.: Prentice-Hall, 1964.

Applewhite, Philip B. *Molecular Gods: How Molecules Determine Our Behavior.* Englewood Cliffs, New Jersey: Prentice-Hall, 1981.

Ash, Avner and Robert Gross. *Fearless Symmetry: Exposing the Hidden Patterns of Numbers.* Princeton: Princeton University Press, 2006.

Atlan, Henri. *Entre le cristal et la fumée: Essai sur l'organisation du vivant.* Paris: Éditions du Seuil, 1979.

Barwise, K. Jon and Lawrence S. Moss. *Vicious Circles: On the Mathematics of Non-wellfounded Phenomena.* Cambridge, U.K.: Cambridge University Press, 1996.

Berg, Paul and Maxine Singer. *Dealing with Genes: The Language of Heredity.* Mill Valley, Calif.: University Science Books, 1992.

Bewersdorff, Jörg. *Galois Theory for Beginners.* Providence: Am. Mathematical Society, 2006.

Bierce, Ambrose. "An Occurrence at Owl Creek Bridge". In *The Collected Writings of Ambrose Bierce.* New York: Citadel Press, 1946.

Blackmore, Susan. *The Meme Machine.* New York: Oxford University Press, 1999.

Boden, Margaret A. *The Creative Mind: Myths and Mechanisms.* New York: Basic Books, 1990.

Boolos, George S. and Richard C. Jeffrey. *Computability and Logic.* New York: Cambridge University Press, 1974.

Borges, Jorge Luis. *Ficciones.* New York: Grove Press, 1962.

Bougnoux, Daniel. *Vices et vertus des cercles: L'autoréférence en poétique et pragmatique.* Paris: Éditions La Découverte, 1989.

Braitenberg, Valentino. *Vehicles: Experiments in Synthetic Psychology.* Cambridge, Mass.: MIT Press, 1984.

Brinck, Ingar. *The Indexical "I": The First Person in Thought and Language.* Dordrecht: Kluwer, 1997.

Brown, James Robert. *Philosophy of Mathematics.* New York: Routledge, 1999.

Carnap, Rudolf. *The Logical Syntax of Language.* Paterson, N.J.: Littlefield, Adams, 1959.

Casati, Roberto and Achille Varzi. *Holes and Other Superficialities.* Cambridge, Mass.: MIT Press, 1994.

———. *Unsurmountable Simplicities: Thirty-nine Philosophical Conundrums.* New York: Columbia University Press, 2006.

Chaitin, Gregory J. *Information, Randomness, and Incompleteness: Papers on Algorithmic Information Theory.* Singapore: World Scientific, 1987.

Chalmers, David J. *The Conscious Mind: In Search of a Fundamental Theory.* New York: Oxford University Press, 1996.

Chauchard, Paul. *Le Cerveau et la conscience.* Paris: Éditions du Seuil, 1960.

Chandrasekhar, B. S. *Why Things Are the Way They Are.* New York: Cambridge University Press, 1998.

Churchland, Patricia. *Neurophilosophy: Toward a Unified Science of the Mind/Brain.* Cambridge, Mass.: MIT Press, 1986.

Cope, David. *Virtual Music: Computer Synthesis of Musical Style.* Cambridge, Mass.: MIT Press, 2001.

Cordeschi, Roberto. *The Discovery of the Artificial: Behavior, Mind, and Machines Before and Beyond Cybernetics.* Dordrecht: Kluwer, 2002.

Dahl, Roald. *Kiss Kiss.* New York: Alfred A. Knopf, 1959.

Damasio, Antonio. *The Feeling of What Happens: Body and Emotion in the Making of Consciousness.* New York: Harcourt Brace, 1999.

Davies, Philip J. "Are there coincidences in mathematics?" *American Mathematical Monthly* **88** (1981), pp. 311–320.

Davis, Joshua. "Say Hello to Stanley". *Wired* **14** (January 2006).

Davis, Martin (ed.). *The Undecidable: Basic Papers on Undecidable Propositions, Unsolvable Problems, and Computable Functions.* Hewlett, N.Y.: Raven, 1965.

Dawkins, Richard. *The Selfish Gene.* New York: Oxford University Press, 1976.

DeLong, Howard. *A Profile of Mathematical Logic.* Reading, Mass.: Addison-Wesley, 1970. (Reissued by Dover Press, 2004.)

Dennett, Daniel C. *Brainstorms: Philosophical Essays on Mind and Psychology.* Cambridge, Mass.: MIT Press, 1978.

————. *Elbow Room: The Varieties of Free Will Worth Wanting.* Cambridge: MIT Press, 1984.

————. *The Intentional Stance.* Cambridge, Mass.: MIT Press, 1987.

————. *Consciousness Explained.* Boston: Little, Brown, 1991.

————. "The Self as a Center of Narrative Gravity", in F. Kessel, P. Cole, and D. Johnson (eds.), *Self and Consciousness.* Hillsdale, N.J.: Lawrence Erlbaum, 1992.

————. *Kinds of Minds: Toward an Understanding of Consciousness.* New York, Basic, 1996.

————. *Brainchildren: Essays on Designing Minds.* Cambridge, Mass.: MIT Press, 1998.

————. *Sweet Dreams: Philosophical Obstacles to a Science of Consciousness.* Cambridge, Mass.: MIT Press, 2005.

Donald, Merlin. *A Mind So Rare: The Evolution of Human Consciousness.* New York: W. W. Norton, 2001.

Dupuy, Jean-Pierre. *Ordres et Désordres.* Paris: Éditions du Seuil, 1982.

————. *The Mechanization of the Mind: On the Origins of Cognitive Science.* Princeton: Princeton University Press, 2000.

Edson, Russell. *The Clam Theater.* Middletown, Conn.: Wesleyan University Press, 1973.

Enrustle, Y. Ted. *Prince Hyppia: Math Dramatica,* Volumes I–III. Luna City: Unlimited Books, Ltd., 1910–1913.

Falletta, Nicholas. *The Paradoxicon.* New York: John Wiley & Sons, 1983.

Fauconnier, Gilles. *Mental Spaces.* Cambridge, Mass.: MIT Press, 1985.

Flanagan, Owen. *The Science of the Mind.* Cambridge, Mass.: MIT Press, 1984.

Gamow, George. *One Two Three... Infinity.* New York: Mentor, 1953.

Gardner, Martin. *Wheels, Life, and Other Mathematical Amusements.* New York: W. H. Freeman, 1983.

Gebstadter, Egbert B. *U Are an Odd Ball.* Perth: Acidic Books, 2007.

Gentner, Dedre, Keith J. Holyoak, and Boicho N. Kokinov (eds.). *The Analogical Mind: Perspectives from Cognitive Science.* Cambridge, Mass.: MIT Press, 2001.

Gödel, Kurt. *On Formally Undecidable Propositions of Principia Mathematica and Related Systems.* New York: Basic Books, 1962. (Reissued by Dover, 1992.)

Goldstein, Rebecca. *Incompleteness: The Proof and Paradox of Kurt Gödel.* New York: W. W. Norton, 2005.

Goodstein, R. L. *Development of Mathematical Logic.* New York: Springer, 1971.

Greenstein, George and Arthur G. Zajonc. *The Quantum Challenge.* Sudbury, Mass.: Jones and Bartlett, 1997.

Griffin, Donald R. *The Question of Animal Awareness.* New York: Rockefeller U. Press, 1976.

Hardy, G. H. and E. M. Wright. *An Introduction to the Theory of Numbers.* New York: Oxford University Press, 1960.

Hart, Leslie A. *How the Brain Works.* New York: Basic Books, 1975.

Harth, Erich. *Windows on the Mind: Reflections on the Physical Basis of Consciousness.* New York: William Morrow, 1982.

Haugeland, John (ed.). *Mind Design: Philosophy, Psychology, Artificial Intelligence.* Montgomery, Vermont: Bradford Books, 1981.

————. *Artificial Intelligence: The Very Idea.* Cambridge, Mass.: MIT Press, 1985.

Hennie, Fred. *Introduction to Computability.* Reading, Mass.: Addison-Wesley, 1977.

Hoffmann, Banesh. *Albert Einstein, Creator and Rebel.* New York: Viking, 1972.

Hofstadter, Douglas R. *Gödel, Escher, Bach: an Eternal Golden Braid.* New York: Basic Books, 1979. (Twentieth-anniversary edition published in 1999.)

————. "Analogies and Metaphors to Explain Gödel's Theorem". *The Two-Year College Mathematics Journal,* Vol. 13, No. 2 (March 1982), pp. 98–114.

————. *Metamagical Themas: Questing for the Essence of Mind and Pattern.* New York: Basic Books, 1985.

————. *Le Ton beau de Marot: In Praise of the Music of Language.* New York: Basic Books, 1997.

————. "Analogy as the Core of Cognition". Epilogue to D. Gentner, K. Holyoak, and B. Kokinov (eds.), *The Analogical Mind.* Cambridge, Mass.: MIT Press, 2001.

————. "Mystery, Classicism, Elegance: an Endless Chase after Magic". In D. Schattschneider and M. Emmer (eds.), *M. C. Escher's Legacy.* New York: Springer, 2002.

Hofstadter, Douglas R. and Daniel C. Dennett (eds.). *The Mind's I: Fantasies and Reflections on Self and Soul.* New York: Basic Books, 1981.

Hofstadter, Douglas R. and David J. Moser. "To Err Is Human; To Study Error-making Is Cognitive Science". *Michigan Quarterly Review* **28**, no. 2 (1989), pp. 185–215.

Hofstadter, Douglas R. and the Fluid Analogies Research Group. *Fluid Concepts and Creative Analogies.* New York: Basic Books, 1995.

Holland, John. *Hidden Order: How Adaptation Builds Complexity.* Redwood City, Calif.: Addison-Wesley, 1995.

————. *Emergence: From Chaos to Order.* Redwood City, Calif.: Addison-Wesley, 1997.

Horney, Karen. *Self-Analysis.* New York: W. W. Norton, 1942.

————. *Our Inner Conflicts: A Constructive Theory of Neurosis.* New York: W. W. Norton, 1945.

Hughes, Patrick and George Brecht. *Vicious Circles and Paradoxes.* New York: Doubleday, 1975.

Huneker, James. *Chopin: The Man and His Music.* New York: Scribner's, 1921. (Reissued by Dover, 1966.)

Huntley, H. E. *The Divine Proportion: A Study in Mathematical Beauty.* New York: Dover, 1970.

Jauch, J. M. *Are Quanta Real? A Galilean Dialogue.* Bloomington: Indiana University Press, 1989.

Jeffrey, Richard C. *Formal Logic: Its Scope and Limits.* New York: McGraw-Hill, 1967.

Judson, Horace Freeland. *The Eighth Day of Creation.* New York: Simon & Schuster, 1979.

Kahneman, Daniel and Dale Miller. "Norm Theory: Comparing Reality to Its Alternatives". *Psychological Review* **80** (1986), pp. 136–153.

Kanerva, Pentti. *Sparse Distributed Memory.* Cambridge, Mass.: MIT Press, 1988.

Kent, Jack. *Mr. Meebles.* New York: Parents' Magazine Press, 1970.

Klagsbrun, Francine. *Married People: Staying Together in the Age of Divorce.* New York: Bantam, 1985.

Kneebone, G. T. *Mathematical Logic and the Foundations of Mathematics.* New York: Van Nostrand, 1963.

Kolak, Daniel. *I Am You: The Metaphysical Foundations for Global Ethics.* Norwell, Mass.: Springer, 2004.

Kriegel, Uriah and Kenneth Williford (eds.). *Self-Representational Approaches to Consciousness.* Cambridge, Mass.: MIT Press, 2006.

Kuffler, Stephen W. and John G. Nicholls. *From Neuron to Brain.* Sunderland, Mass.: Sinauer Associates, 1976.

Külot, Gerd. "On Formerly Unpennable Proclamations in *Prince Hyppia: Math Dramatica* and Related Stageplays (I)". *Bologna Literary Review of Bologna* **641** (1931).

Laughlin, Robert B. *A Different Universe: Reinventing Physics from the Bottom Down.* New York: Basic Books, 2005.

Le Lionnais, François. *Les Nombres remarquables.* Paris: Hermann, 1983.

Lem, Stanislaw. *The Cyberiad: Fables for the Cybernetic Age* (translated by Michael Kandel). San Diego: Harcourt Brace, 1985.

Livio, Mario. *The Equation that Couldn't Be Solved.* New York: Simon and Schuster, 2005.

Margolis, Howard. *Patterns, Thinking, and Cognition.* Chicago: University of Chicago, 1987.

Martin, Richard M. *Truth and Denotation: A Study in Semantical Theory.* Chicago: University of Chicago Press, 1958.

McCorduck, Pamela. *Machines Who Think.* San Francisco: W. H. Freeman, 1979.

Mettrie, Julien Offray de la. *Man a Machine.* La Salle, Illinois: Open Court, 1912.

Metzinger, Thomas. *Being No One: The Self-Model Theory of Subjectivity.* Cambridge, Mass.: MIT Press, 2003.

Miller, Fred D. and Nicholas D. Smith. *Thought Probes: Philosophy through Science Fiction.* Englewood Cliffs: Prentice-Hall, 1981.

Minsky, Marvin. *The Society of Mind.* New York: Simon & Schuster, 1986.

———. *The Emotion Machine.* New York: Simon & Schuster, 2006.

Monod, Jacques. *Chance and Necessity.* New York: Vintage Press, 1972.

Moravec, Hans. *Robot: Mere Machine to Transcendent Mind.* New York: Oxford University Press, 1999.

Morden, Michael. "Free will, self-causation, and strange loops". *Australasian Journal of Philosophy* **68** (1990), pp. 59–73.

Nagel, Ernest and James R. Newman. *Gödel's Proof.* New York: New York University Press, 1958. (Revised edition, edited by Douglas R. Hofstadter, 2001.)

Neumann, John von. *Theory of Self-Reproducing Automata* (edited and completed by Arthur W. Burks). Urbana: University of Illinois Press, 1966.

Niven, Ivan and Herbert S. Zuckerman. *An Introduction to the Theory of Numbers.* New York: John Wiley & Sons, 1960.

Nørretranders, Tor. *The User Illusion.* New York: Viking, 1998.

Nozick, Robert. *Philosophical Explanations.* Cambridge, Mass.: Harvard University Press, 1981.

Pais, Abraham. *Inward Bound: Of Matter and Forces in the Physical World.* New York: Oxford University Press, 1986.

————. *Niels Bohr's Times.* New York: Oxford University Press, 1991.

Parfit, Derek. *Reasons and Persons.* New York: Oxford University Press, 1984.

Pattee, Howard H. *Hierarchy Theory: The Challenge of Complex Systems.* New York: Braziller, 1973.

Peitgen, H.-O. and P. H. Richter. *The Beauty of Fractals.* New York: Springer, 1986.

Penfield, Wilder and Lamar Roberts. *Speech and Brain-Mechanisms.* Princeton: Princeton University Press, 1959.

Penrose, Roger. *The Emperor's New Mind.* New York: Oxford University Press, 1989.

Perry, John (ed.). *Personal Identity.* Berkeley: University of California Press, 1975.

Péter, Rózsa. *Recursive Functions.* New York: Academic Press, 1967.

Pfeiffer, John. *The Human Brain.* New York: Harper Bros., 1961.

Poundstone, William. *The Recursive Universe.* New York: William Morrow, 1984.

Pullman, Bernard. *The Atom in the History of Human Thought.* New York: Oxford University Press, 1998.

Pushkin, Alexander S. *Eugene Onegin: A Novel in Verse* (translated by James Falen). New York: Oxford University Press, 1995.

————. *Eugene Onegin: A Novel Versification* (translated by Douglas Hofstadter). New York: Basic Books, 1999.

Quine, Willard Van Orman. *The Ways of Paradox, and Other Essays.* Cambridge, Mass.: Harvard University Press, 1976.

Ringle, Martin. *Philosophical Perspectives in Artificial Intelligence.* Atlantic Highlands: Humanities Press, 1979.

Rucker, Rudy. *Infinity and the Mind.* Boston: Birkhäuser, 1982.

Sander, Emmanuel. *L'analogie, du Naïf au Créatif: Analogie et Catégorisation.* Paris: Éditions L'Harmattan, 2000.

Schank, Roger C. *Dynamic Memory.* New York: Cambridge University Press, 1982.

Schweitzer, Albert. *Aus Meiner Kindheit und Jugendzeit.* Munich: C. H. Beck, 1924.

Searle, John. "The Myth of the Computer" (review of *The Mind's I*). *The New York Review of Books,* April 29, 1982, pp. 3–6.

Shanker, S. G. (ed.). *Gödel's Theorem in Focus.* New York: Routledge, 1988.

Simon, Herbert A. *The Sciences of the Artificial.* Cambridge, Mass.: MIT Press, 1969.

Singer, Peter and Jim Mason. *The Way We Eat: Why Our Food Choices Matter.* Emmaus, Pennsylvania: Rodale Press, 2006.

Skinner, B. F. *About Behaviorism.* New York: Random House, 1974.

Smullyan, Raymond M. *Theory of Formal Systems.* Princeton: Princeton Univ. Press, 1961.

————. *The Tao Is Silent.* New York: Harper & Row, 1977.

————. *What Is the Name of This Book?* Englewood Cliffs, New Jersey: Prentice-Hall, 1978.

————. *This Book Needs No Title.* Englewood Cliffs, New Jersey: Prentice-Hall, 1980.

————. *5000 B.C. and Other Philosophical Fantasies.* New York: St. Martin's Press, 1983.

————. *Gödel's Incompleteness Theorems.* New York: Oxford University Press, 1992.

Sperry, Roger. "Mind, Brain, and Humanist Values", in John R. Platt (ed.), *New Views on the Nature of Man*. Chicago: University of Chicago Press, 1965.

Steiner, George. *After Babel*. New York: Oxford University Press, 1975.

Stewart, Ian. *Galois Theory* (second edition). New York: Chapman and Hall, 1989.

Suppes, Patrick C. *Introduction to Logic*. New York: Van Nostrand, 1957.

Thigpen, Corbett H. and Hervey M. Cleckley. *The Three Faces of Eve*. New York: McGraw-Hill, 1957.

Treisman, Anne. "Features and Objects: The Fourteenth Bartlett Memorial Lecture". *Cognitive Psychology* **12**, no. 12 (1980), pp. 97–136.

Ulam, Stanislaw. *Adventures of a Mathematician*. New York: Scribner's, 1976.

Unger, Peter. "Why There Are No People". *Midwest Studies in Philosophy*, **4** (1979).

———. "I Do Not Exist". In G. F. MacDonald (ed.), *Perception and Identity*. Ithaca: Cornell University Press, 1979.

Wadhead, Rosalyn. *The Posh Shop Picketeers*. Tananarive: Wowser & Genius, 1931.

Webb, Judson. *Mechanism, Mentalism, and Metamathematics*. Boston: D. Reidel, 1980.

Weinberg, Steven. *Dreams of a Final Theory*. New York: Pantheon, 1992.

———. *Facing Up*. Cambridge, Mass.: Harvard University Press, 2001.

Wells, David G. *The Penguin Dictionary of Curious and Interesting Numbers*. New York: Viking Penguin, 1986.

———. *Prime Numbers*. New York: John Wiley & Sons, 2005.

Wheelis, Allen. *The Quest for Identity*. New York: W. W. Norton, 1958.

Whitehead, Alfred North and Bertrand Russell. *Principia Mathematica*, Volumes I–III. London: Cambridge University Press, 1910–1913.

Wilder, Raymond L. *Introduction to the Foundations of Mathematics*. New York: John Wiley & Sons, 1952.

Wolf, Robert S. *A Tour through Mathematical Logic*. Washington, D.C.: The Mathematical Association of America, 2005.

Wooldridge, Dean. *Mechanical Man: The Physical Basis of Intelligent Life*. New York: McGraw-Hill, 1968.

Wynne, Clive D. L. *Do Animals Think?* Princeton: Princeton University Press, 2004.

Yourgrau, Palle. *A World Without Time: The Forgotten Legacy of Gödel and Einstein*. New York: Basic Books, 2005.

PERMISSIONS AND ACKNOWLEDGMENTS

ॐ ॐ ॐ

GRATEFUL acknowledgement is hereby made to the following individuals, publishers, and companies for permission to use material that they have provided or to quote from sources for which they hold the rights. Every effort has been made to locate the copyright owners of material reproduced in this book. Omissions that are brought to our attention will be corrected in subsequent editions.

Thanks to William Frucht for the cover photograph of video feedback and for all the photographs in the color insert in Chapter 4.

Thanks to Daniel Hofstadter and Monica Hofstadter for photographs of various loopy structures, used as interludes between chapters.

Thanks to Kellie and Richard Gutman for two photographs in Chapter 4.

Thanks to Jeannel King for her poem "Ode to a Box of Envelopes" in Chapter 7.

Thanks to Silvia Sabatini for the photograph of the lap loop in Anterselva di Mezzo, facing Chapter 8.

Thanks to Peter Rimbey for the photograph of Carol and Douglas Hofstadter facing Chapter 16.

Thanks to David Oleson for his parquet deformation "I at the Center" in Chapter 17.

"Three Kangaroos" logo, designed by David Lance Goines © Ravenswood Winery. Reprinted with permission by Joel Peterson, Ravenswood Winery.

"Three Ravens" logo, designed by David Lance Goines © Ravenswood Winery. Reprinted with permission by Joel Peterson, Ravenswood Winery.

"Peanuts" cartoon, dated 08/14/1960: © United Feature Syndicate, Inc. Reprinted with permission by United Media.

M. C. Escher, *Drawing Hands* © 2006 M. C. Escher Company, Holland. All rights reserved. www.mcescher.com. Reprinted with permission.

Whitehead, Alfred North and Bertrand Russell, *Principia Mathematica* (second edition), Volume I (1927), page 629, reprinted in 1973 © Cambridge University Press.

"Nancy" cartoon: "Sluggo dreaming": © United Feature Syndicate, Inc. Reprinted with permission by United Media.

Morton Salt "Umbrella Girl" © Morton International, Inc. Reprinted with permission of Morton International, Inc.

Excerpt from Karen Horney, *Our Inner Conflicts,* © 1945 by W. W. Norton & Co., Inc. Reprinted with permission by W. W. Norton & Company.

Excerpts from Daniel Dennett, *Consciousness Explained* © 1991 by Daniel C. Dennett. Reprinted with permission by Hachette Book Group USA.

Excerpt from Carson McCullers, *The Heart Is a Lonely Hunter.* Copyright © 1940, renewed 1967, by Carson McCullers. Reprinted by permission of Houghton Mifflin Company. All rights reserved.

Excerpts from Derek Parfit, *Reasons and Persons* © 1984 Oxford University Press. Reprinted with permission of Oxford University Press.

Excerpts from Albert Schweitzer, *Aus meiner Kindheit und Jugendzeit.* © C. H. Beck, Munich, 1924. Personal translation for use in this book only, by Douglas Hofstadter. Reprinted with permission.

INDEX

❧ ❧ ❧

— A —

abbreviations piled on abbreviations, 200–201

aboutness, double, of Gödel's formula, 147–148

absorbing someone else's essence, 236

"abstraction ceiling" in author's mind, *xvi*, 89, 92

abstractions: as causes, 38–41; centrality of, 26: formidable tower of, in human minds, 83–84, 89, 92, 201, 369

abstractly swirling patterns, 283

accidents in mathematics: absence of, 117, 127; possibility of, 126

accretion: of self-model, 82; of soul, 250–254

Ackermann's function, 367

active symbols, *see* symbols

affinity of souls, *see* chemistry (interpersonal)

afterglow of a soul, 258, 274, 316–317

Aimable, the village baker, 152–153, 154

"alacrity, celerity, assiduity", etc., 300, 375

Alfbert, the, 196–199, 201; dream of, 198–199

Alf and Bertie's Posh Shop, 154–155, 160

algorithmic information theory, 108

Ali, Muhammad, 160

Alienware machine, emulated by Macintosh, 242

Alighieri, Dante, 251

alignment: of *PM* theorems and code numbers, 130–131; of truths and *PM* theorems, 129–130; of two souls in married couple dedicated to common goals, 223–224, 228

allegoric license, 199

"Alle Menschen Müssen Sterben" (Bach), 352

Alzheimer's disease, 17, 19, 22, 316–317, 329

ambiguity of operations inside computers, 244–245

américain, pronunciation of the word in movie, 250

amino acids, 174

amplification of input in audio feedback, 54–55; saturation of, 55

analogies: central role of, in this book, *xvii–sviii*; as fabric of human thought, *xviii*, 149; having force proportional to precision and visibility, 153, 155, 158; index entries for, *xvii–xviii*; jumping out automatically, 149–152; made by bees, 369; as mediating reference, 147–161, 245, 305; research on, 25, 26; retrieved automatically by new events, 277; rivalry between two similar ones, 218; seen as simmball patterns in the careenium, 51; self-referential, 62; as source of meaning, 147–161, 245; tossed off effortlessly, 25; trivial-seeming examples, 149; by W. V. O. Quine, 139–143

analogies, serious examples of: between Alfbert and Whitehead/Russell, 196–199; between audio and video feedback, 56; between Aurélie and Pomponnette, 152–153, 154, 157, 244; between the author's mind and others' minds, *xi*; between the author's view of "I" and quantum mechanics, 239; between beer cans and neurons, 29–30, 366; between brain and oil refinery, 194; between brains and countries, 272–273; between brains and *PM* as substrates for strange loops, *xii*, 193–194; between brain structures and genes or atoms, 27; between butterfly swarms and souls, 373–374; between careenium and brain, 45–51, 195–196; between careenium and pinball machine, 48; between cars and dogs, 368–369; between Chantal looking at movie and Russell looking at Gödel's formula, 154; between children with muddy boots, 150; between cookies on same plate, 149; between couples, 151–154; between crystal and orchard, 373–374; between death and eclipse, 227, 258, 274, 316–317; between decision-making and political elections, 340–341; between dedicated machines and music boxes, 243; between dog looking at pixels and Russell looking at Gödel's formula, 153–154, 202; between domino chainium and traffic jam, 39; between donning piece of clothing and identifying with someone else, 236; between Doug and Carol, 228; between edibility and provability, 196–199, 201, 202; between electron clouds and human souls, 373–374; between entwined video loops and entwined souls, 210–211, 253–254; between epiphenomena in brain and in mineral, 30; between formula containing own Gödel number and

brain structures, 25–27, 203
brains: compared to hearts, 27–28; complexity of, as relevant to consciousness, 286; controlling bodies directly *vs.* indirectly, 212–213; eerieness of, *ix*; evolution of, 194, 196; as fusion of two half-brains, 219; as inanimate, *xiii*, 193; inhabited by more than one "I", 248, 354; interacting via ideas, 32, 206; main, 259–260, 268; as multi-level systems, 30–32, 180–181, 202–203; not responsible for color qualia, 335–337; perceiving multiple environments simultaneously, 268–269; receiving sensory input directly or indirectly, 212; resembling inert sponges, 193; unlikely substrate for interiority, 193
Braitenberg, Valentino, 81
bread becoming a gun, 109
Brown, Charlie, *xvi*, 251
Brownian motion, 49
Brünn, Austria (birthplace of Kurt Gödel), 122, 125
buck stopping at "I", 95–96, 182
Bugeaud, Yann, 127, 370
bunnies as edible beings, 14, 18, 19
"burstwise advance in evolution" (Sperry), 32, 206
Bushmiller, Ernie, 144
butterflies: not respecting precinct boundaries, 175; in orchard, as metaphor for human soul, 373–374
Buzzaround Betty, 189

— C —

caged-bird metaphor, 259, 308; as analogous to Newtonian physics, 267; hints at wrongness of, 267–268, 270; as ingrained habit, 271; at level of countries and cultures, 272–273; metaphors opposed to, 260, 272–273, 373–374; normally close to correct, 267–268; as reinforced by language, 270–271; temptingness of, 270
Cagey's doubly-hearable line, 155, 160
cake whose pieces all taste bad, as inferrred by analogy, 149
candles, 300, 350
cantata aria, 220–222
Cantor, Georg, 371
capital punishment, 17, 343
Capitalized Essences, 326–329, 357, 376; canceled, 359
careenium, 45–51, 195–196; growing up, 98–99; self-image of, 98–99; two views of, 48–50, 97–99, 180, 195, 295; unsatisfactory to skeptics, 276, 279
Carnap, Rudolf, 110
Carol-and-Doug: as higher-level entity, 223–224, 228; joint mind of, 223; shared dreads and dreams of, 224, 228

Carolness, survival of, 230, 233–234
Carol-symbol in Doug's brain: being *vs.* representing a person, 238; triggerability of, 238, 254–255
cars: as high-level objects, 28, 33, 40; pushed around by desires, 97
Cartesian Eggo, 306
Cartesian Ego, 305–306, 308, 311, 314–315, 360; as commonsensical view, 306–307; fading of, 316
Cartesian Ergo, 276
Cartier-Bresson, Henri, 251
Caspian Gemstones, allegory of, 126–127
casual façade as Searlian ploy, 30
Catcher in the Rye, The (Salinger), 256
categories and symbols, 73, 75–77; *see also* repertoires
categorization mechanisms: converting complexity into simplicity, 277–279; as determining size of self, 209, 283; efficiency of, 297, 362
Caulfield, Holden, 88, 251
causality: bottoming out in "I", 96; buck of, stopping at "I", 95–96, 182; of dogmas in triggering wars, 33, 35, 179; and insight, 41, 179; schism between two types of, 204, 295; stochasticity of in everyday life, 97–98; tradeoffs in, 98; upside-down, 50; *see also* downward causality
causal potency: of ideas in brain, 39–43, 205–206; of meanings of *PM* strings, 51, 206; of patterns, 37–50
"causal powers of the brain", semantic, 372
cell phones as universal machines, 241, 243
Center for Research into Consciousness and Cognetics, 320, 321
Central Consciousness Bank, 329
central loop of cognition, 277–279
cerulean sardine, 333
chain of command in brain, 31–32
chainium (dominos), causality in, 37–39, 41, 51, 176
Chaitin, Greg, 108
Chalmers, David, 319–323, 324, 330–331, 375; zombie twin of, 322–323, 325, 330–331, 361
chameleonic nature: of integers, 160, 165–166, 243–244; of universal machines, 241, 243
Chantal Duplessix, seeing pixel-patterns as events, 154; missing second level of Aimable's remarks, 154
chaos, potential, in number theory, 114, 117, 118
Chaplin twins, (Freda and Greta), 219–220
character structure of an individual, 185
Chávez, César, 345, 376
chemistry: bypassed in explanation of heredity and reproduction, 174; of carbon as supposed key to consciousness, 194; reduced to physics, 33; virtual, inside computers, 244
"chemistry" (interpersonal); enabling people to live inside each other, 250; as function of musical taste

tennis-playing, 213
Tesler, Larry, 250
"the present work", "the present chapter", 367–368
theorems: as bottom lines of formal derivations, 122,
    135; first, second (etc.) generations of, 129; as
    meaningful patterns, 147–148; as meaningless
    patterns, 147; mirrored by prim numbers, 135, 138
theory of computation, 243
theory of sets, 60–61
theory of types, 60–61, 63, 74, 104, 106–107, 138,
    147; self-referentiality of chapter introducing, 367
therapy sessions for bereaved spouses, 227–228
thermodynamics *vs.* statistical mechanics, 33–34, 295
thermostats, 51, 78, 79, 182, 194, 209, 212, 282
thinking: ; with another's brain, 255; essence of, 25,
    277–279; as synonym of consciousness, 4, 203, 276
thinkodynamics *vs.* statistical mentalics, 34–35, 97–99
thirst: as collective pattern of many beer cans, 30; as
    one beer can popping up, 29, 366
"This formula is not provable", 138, 145
"This formula is provable", 159–160
"this sentence", avoidance of indexical phrase, 138
"This sentence is false", 63, 140, 371
"thit sentence", 62
Thomas, Dylan, 372
"thou" addressed to married couple, 221
thought: basic unit of, 5; as dance of simmballs, 51; as
    dance of symbols, 51, 319; as mere set of habits, 6–
    7; as prime mover in brain, 99
thought experiments: parameters tweaked in, 261–
    262, 263; teleported across Atlantic, 304–305
thoughtmill churned by simms, 51
"three three threes", 60, 367
threshold of complexity: for computational
    universality, 241; for representational universality,
    246, 354
throwaway analogies, random examples of: between
    Buzzaround Betty and Hopalong Cassidy, 189;
    between Cagey and Qéé Dzhii, 157; between car
    buyers and heart surgeons, 28; between
    consciousness and a power moonroof, 325, 375;
    between deconstructing the "I" and deconstructing
    Santa Claus, 294; between Doug/Carol and a
    school of fish, 224; between etymology and an X-
    ray, 345; between exploration of video feedback
    and sea voyage, 65–69; between form–content
    interplay and tail wagging dog, *xviii*; between John
    Searle and Dylan Thomas, 372; between lack of
    imagery and lack of oxygen, 89; between people
    and grasshoppers, 352; between *Principia
    Mathematica* and Newton's *Principia*, 113; between
    reading Euclid's proof and tasting chocolate, 118;

between reading "accessible" version of proof of
    Fermat's Last Theorem and ordering "Indian hot",
    369; between reverberant barking and a chain
    reaction, 67; between Roger Sperry and Hopalong
    Cassidy, 187; between Russell and God, 154;
    between strange loop of "I"-ness and pearl necklace,
    180; between this book and a salad, *xvii*; between
    tired muscles and soft recruits, 96; between top and
    bottom of Shell sign, 90; between TV screen and
    leaf pile, 153; between Twinwirld and Twinnwirrld,
    217; between will's constancy and a gyroscope, 341
throwing-away of information, 35
time-lapse photography, 48, 53
Tinkertoys as substrate for thinking, 29
Titanic baby found floating in life raft, 212
titles of sections in *Reasons and Persons* chapter, 309
toilet paper and pebbles as substrate for thinking, 29
toilets, awareness level of, 78, 79, 194
tomatoes as soulless, 10, 18, 182
Tomonaga, Sin-Itiro, 251
"too marbelous for words", 300
tornado cell, opposing caged-bird metaphor, 260
tower of increasingly abstract definitions, 200–201
toy guns, attempted banishment of, 109
traffic jam, global explanation for, 39–40
trains: identity of, 315–316; who *vs.* that, 315–316
transplanting: a novel to another soil, 224; a soul to
    another soil, 255, 257–258
transportability, differential, of layers of a self, 237
triggering of symbols in brain, 76–77, 85, 87, 91–92,
    98, 186, 277–279
Treisman, Anne, 26
Trenet, Charles, 251
Trento, Italy, 227, 232
Trojan horse, Gödel's, 361
true statements: Gödel numbers of, 172; logicians'
    favorite examples, 178
trustability of sources of information, 90
Truth and Denotation (Martin), 110
truth: as inexpressible notion using *PM* notation, 172;
    preservation of, via rules of inference, 128–129;
    presumed to be equivalent to provability in *PM*,
    129, 130; and unprovability perversely entailing
    each other, 165
tu (second-person singular pronoun) addressed to
    married couple, 221
Turing, Alan Mathison, 242, 243
Turing machines, 28–29
turkey as "which", not "who", 17, 330, 331
TV camera: bolted to TV, 74; on long leash, 75, 194;
    meltdown of, 56; on short leash, 75; universally
    worn on nose, 265